SOLID SUPPORTS AND CATALYSTS IN ORGANIC SYNTHESIS

Ellis Horwood and Prentice Hall

are pleased to announce their collaboration in a new imprint whose list will encompass outstanding works by world-class chemists, aimed at professionals in research, industry and academia. It is intended that the list will become a by-word for quality, and the range of disciplines in chemical science to be covered is:

ANALYTICAL CHEMISTRY
ORGANIC CHEMISTRY
INORGANIC CHEMISTRY
PHYSICAL CHEMISTRY
POLYMER SCIENCE & TECHNOLOGY
ENVIRONMENTAL CHEMISTRY
CHEMICAL COMPUTING & INFORMATION SYSTEMS
BIOCHEMISTRY
BIOTECHNOLOGY

Ellis Horwood   PTR Prentice Hall
ORGANIC CHEMISTRY SERIES

Series Editors:
Ellis Horwood MBE
Dr John Mellor, University of Southampton

Current titles in the
Ellis Horwood   PTR Prentice Hall
Organic Chemistry Series

Hudlicky **CHEMISTRY OF ORGANIC FLUORINE COMPOUNDS:**
          A Laboratory Manual with Comprehensive
          Literature Coverage, 2nd (revised) Edition
Smith   **SOLID SUPPORTS AND CATALYSTS IN ORGANIC SYNTHESIS**

SOLID SUPPORTS AND CATALYSTS IN ORGANIC SYNTHESIS

Editor
Professor K. SMITH, M.Sc., Ph.D.
Head of Department of Chemistry
University College of Swansea
Wales

ELLIS HORWOOD   PTR PRENTICE HALL
NEW YORK   LONDON   TORONTO   SYDNEY   TOKYO   SINGAPORE

First published in 1992 by
ELLIS HORWOOD LIMITED
Market Cross House, Cooper Street,
Chichester, West Sussex, PO19 1EB, England

A division of
Simon & Schuster International Group
A Paramount Communications Company

Printed and bound in Great Britain
by Bookcraft, Midsomer Norton

British Library Cataloguing in Publication Data

A catalogue record for this book is available from the British Library

ISBN 0–13–639998–3

Library of Congress Cataloguing-in-Publication Data

Available from the publisher

Contents

List of contributors

Dr J. A. Ballantine,
Department of Chemistry,
University College of Swansea,
Swansea SA2 8PP
UK

Professor G. Bram,
ICMO, Laboratoire des Reactions Sélectives sur Supports,
URA 478 C.N.R.S.,
Batiment 410,
Université Paris-Sud,
91405 Orsay,
France

Dr M. Butters,
Process Research and Development,
Pfizer Central Research,
Sandwich,
Kent CT13 9NJ,
UK

Dr J. S. Davies,
Department of Chemistry,
University College of Swansea,
Swansea SA2 8PP,
UK

Dr P. Diddams,
Technical Service Petroleum Chemicals,
Grace GmbH,
Postfach 1445,
In der Hollerhecke 1,
D-6520 Worms,
Germany

Dr M. E. Fakley,
ICI Katalco,
PO Box 1,
Billingham,
Cleveland TS23 1LB,
UK

Dr F. King,
ICI Katalco,
PO Box 1,
Billingham,
Cleveland TS23 1LB,
UK

Professor P. Laszlo,
Laboratoires de Chimie fine aux interfaces,
École Polytechnique,
91128 Palaiseau,
France.
and Université de Liège,
Sart-Tilman
par 4000 Liège,
Belgium

Dr A. Loupy,
ICMO, Laboratoire des Reactions Sélectives sur Supports,
URA 478 CNRS,
Batiment 410,
Université Paris-Sud,
91405 Orsay,
France

Dr J. M. Maud,
Department of Chemistry,
University College of Swansea,
Swansea SA2 8PP,
UK

Professor D. Villemin,
ISMRa,
École Nationale Supérieure d'Ingénieurs de Caen,
Laboratoire des Nouvelles Méthodologies Synthétiques,
URA 480 CNRS,
Université de Caen,
14032 Caen Cedex,
France

H. A. White,
Department of Biochemistry and Molecular Biology,
University College London,
Gower Street,
London WC1E 6BT,
UK

Dr J. M. Woodley,
Department of Chemical and Biochemical Engineering,
University College London,
Torrington Place,
London WC1E 7JE,
UK

Preface

Solids have long played a role in organic chemistry. They have been used to moderate reactivity as in Nobel's classic use of a clay to produce dynamite, and to catalyse reactions, typified by the many high-temperature industrial processes carried out over metal oxide catalysts and the use of finely divided metals supported on inorganic solids for catalytic hydrogenation. The petrochemical industry has been particularly active in applying solids to useful processes. Clays and more recently zeolites have been used to bring about catalytic cracking and related processes on a large scale. Despite these important applications, however, the use of solids to control more complex organic reactions is not as well appreciated as it might be.

This situation is beginning to change. Chemists involved in fine chemicals' production are increasingly recognizing the need for use of solids as catalysts or controlling agents in the reactions. Solids can provide a number of advantages: they can be easily recovered and reused; they provide a much lower effluent and waste problem than soluble materials; above all, they may provide the opportunity for influencing isomer ratios, rates of reaction, byproduct formation, and so on. There are some well known reactions that involve the use of solid supported reagents.

In order for the use of solids to become standard, in the way that it is standard for people to try sodium borohydride for reducing organic substrates or chromic acid for oxidizing them, practising organic chemists have to become more familiar with the properties of the solids. Most of the expertise surrounding the solids of interest resides with polymer chemists and solid state chemists and people interested in catalysis of macro-scale reactions. The writings of such chemists concern the synthesis of new solids and their structural identification, and their application as catalysts for petroleum cracking or similar processes. Organic chemists do not customarily read such publications and would not easily translate what they would read into their own research areas even if they were more familiar with such publications. The challenge, therefore, is to persuade organic chemists to learn more about the properties and characteristics of solids in order that they might consider using them in their own research. This is the purpose of this book.

In planning the book, I tried to analyse the way we think as organic chemists. I considered that in general we would not know how to synthesize a solid for use as a catalyst or support. An organic chemist working away at this own research problems would not wish to delve into the seeming black art of solid state synthesis. Equally, if he would have to spend a considerable amount of time becoming expert in the characterization of any solid which he had produced, this would create an inertial barrier over which he would be unlikely to climb. I considered that he would need to know something about the structural properties of the appropriate solids if he were to have any chance of utilizing them intelligently in this research. He would need to know what had been done with the solids before, but also he would need to be able to relate the chemistry carried out to the nature of the solid chosen for the process. With these factors in mind, I planned the book to have several sections.

In Part I, the general structures and characteristics of the various solids are outlined. One of the chapters is devoted to inorganic solids and the other to organic solids. Part II surveys standard organic reactions that have been carried out in the presence of solids as catalysts, supports or controlling agents. It is subdivided into chapters, depending upon the nature of the solid used — amorphous inorganic solids, lamellar solids, zeolite solids, or polymeric solids. Part III deals with the important use of solids in biological chemistry/molecular biology. There are chapters devoted to solid phase peptide synthesis, solid phase oligonucleotide synthesis and immobilized biocatalysts. Finally, in Part IV I decided to introduce some individual topics in order to introduce a flavour of the way research progresses in this area. Chapter 10 deals with the process of catalytic hydrogenation, which is a common process carried out in all kinds of organic synthesis. Chapter 11 is a more personal account of the way in which research in the area can develop, whilst the final chapter is an account of the recently introduced development of microwave technology as applied to reactions in the presence of solids.

In choosing the authors for the book, I was conscious of the need to cater for organic chemists. Consequently, most of the chapters are written by practising organic chemists rather than specialists in solids, though there are one or two exceptions to this principle. All of the authors are people with specific expertise in the use of solids for organic synthesis. What we have tried to produce is a book that caters for the needs of the organic chemist but, at the same time, provides an essential insight into the reasons why solids have been used.

In so far that we have been successful in achieving these goals, it is because of the great efforts of the authors in analysing the literature with an expert eye. I thank them for their efforts and for their tolerance of my editorial idiosyncrasies. If any problems with the book remain, they are entirely the fault of the editor. I should like to thank all of the contributing authors for their dedication to the task and also to thank my secretary, Mrs Beryl Irwin, for her forbearance during the time when the manuscript was in preparation.

Keith Smith
July, 1991

Part I

Structures and nature of solids used as supports/catalysts

1

Inorganic supports and catalysts—
an overview

P. Diddams

1.1 GENERAL CONSIDERATIONS

The range of inorganic solid structures available for use as catalysts or supports is huge, and it is impossible to provide a comprehensive list of the characteristics of the whole range in a book of this type. However, a number of solids have found more widespread use than others because of favourable properties, and attention is therefore concentrated on these.

The structures of polymeric inorganic solids are often complex, being composed of regions of regular repeating sub-units separated by dislocations or regions in which the elemental composition is different. They can generally not be characterized as fully as is usually desired by chemists accustomed to handling organic chemicals, and different samples of the supposedly same solid may have different characteristics. This can present difficulties for choosing a solid for a particular purpose, but such difficulties should not be overstated. Many solids are now produced routinely on a commercial scale to well-defined specifications.

Three general types of useful solids can be distinguished: (a) ones in which the solid is amorphous, with random pores, or more or less impervious to reactant molecules, so that any beneficial reaction takes place at the irregular surface, albeit perhaps in pits or pores thereon; (b) ones in which impervious sheets or layers (lamellae) are separated by interlamellar regions where any useful activity resides; and (c) ones in which the three-dimensional structure comprises a regular array of channels into which molecules of appropriate dimensions can fit. Simple silica and alumina fall into the first category, clays and graphite into the second, and zeolites into the third.

Useful solids may serve several purposes and be comprised of more than one component, including catalytically active species, supports and fillers/binders. Catalytically active species may be highly dispersed metals (e.g. platinum), acidic oxides, or many other species. Activity depends upon the concentration of accessible active sites throughout catalyst particles. In the case of metals, dispersion as very small particles (as small as 0.8–1 nm in some cases) leads to high surface area and therefore a large number of accessible surface sites. Promoters may be added to improve catalyst life or enhance activity (e.g. addition of fluoride or chloride to increase the acidity of alumina catalysts).

Supports (e.g. silicas, aluminas, carbons, etc.) are used as substrates for catalytically active species (e.g. very small metal particles). A common problem with highly dispersed catalysts is sintering (i.e. the coalescence of small particles into larger particles), which leads to lowering of surface area and consequently lower catalytic activity. Sintering is thermodynamically driven by the reduction in surface energy that accompanies decreasing surface area. Catalyst–support interactions can inhibit sintering by reducing the mobility of small particles over the supporting surface.

Fillers are used to provide catalysts with requisite physical and mechanical properties (e.g. density, specific heat capacity, etc.) and in many cases to allow catalysts to be formed into particles of optimum size and shape. Binders (e.g. colloidal silica) are often added to 'glue' the various component particles together, giving catalysts strength and attrition resistance.

An important reason for choosing a solid catalyst or support is to gain selectivity. Interaction with active sites at the surface or within the pores of a solid changes the structure or geometry of sorbate molecules such that the activation energies of particular reaction pathways are lowered (e.g. via protonation, dissociative chemisorption, etc.). Also, geometrical and electronic properties of solids (e.g. within the micropores of a zeolite) can impose constraints on diffusion of species to and from active sites and on the geometry of transition state species. Such constraints lead to kinetic control over reaction pathways, resulting in non-thermodynamic product distributions (i.e. catalyst selectivity).

However, products may participate in secondary reactions, depending upon their contact time with the catalyst. For optimum product selectivity it may be preferred to have primary reaction products desorb back into the liquid or gas phase and be transported away from the catalyst to avoid secondary reactions. Contact time between catalyst and reactants can have a major influence on product selectivity, short contact times generally giving higher selectivity (at lower conversion) than long contact times, and vice versa.

Such considerations may influence design of the reaction process. For example, solid catalysts allow the use of continuous flow type reactors, which have a contained inventory of solid catalyst particles through which a fluid reaction medium is passed. Control of reactant composition and throughput allows contact time to be regulated to optimize activity and product selectivity. Catalyst beds in continuous flow type reactors may be static (fixed-bed reactors) or moving (e.g. circulating-bed reactors, etc.). Catalyst physical and mechanical properties (e.g. particle size, shape, hardness, density, etc.) may have to be controlled for optimum performance.

The purpose of this chapter is to outline the most important characteristics of the common solids used as catalysts and supports in order to provide an understanding of the role played by such solids in reactions described elsewhere in this book. It should also help in the design of new reactions employing solid supports or catalysts.

1.2 SILICA

1.2.1 Introduction

The term 'silica' is used to describe polymorphic forms of silicon dioxide (SiO_2), including hydrated and anhydrous crystalline, microcrystalline and amorphous forms. Silica is ubiquitous amongst minerals, being found in pure crystalline forms (e.g. quartz and crystobalite) and as silicates in combination with many other oxides.

Crystalline silicas are high-density, low-porosity forms that provide a natural source of pure silica. Microcrystalline forms (i.e. individual cystallites typically of sub-micron dimensions) occur in flints, cherts and opals. Amorphous silicas may be either high-density vitreous silicas (e.g. glass) or low-density amorphous silicas (e.g. fumed silica, silica gel, silicic acid, etc.). Low-density amorphous silicas are generally most useful as sorbents and catalyst supports because of their high surface area and porosity.

The principal structural unit of silicas and silicate minerals formally consists of Si^{4+} cations tetrahedrally coordinated to O^{2-} anions. Tetrahedral sub-units are corner linked (via siloxane bridges: Si–O–Si) to form chains, rings, sheets (Fig. 1.1) or three-dimensional assemblages (e.g. zeolites). Unshared tetrahedral apices generally form hydroxyl groups (silanol groups: Si–OH).

1.2.2 Amorphous silica

Amorphous silicas have no long-range crystalline order. However, short-range order may be present (e.g. organization into chains, rings and sheets) depending upon the method of preparation. Amorphous silicas consist of primary particles of colloidal dimensions (1–100 nm) which may be discrete (microparticulate), aggregated into larger secondary particles (1–25 μm) or agglomerated to form a continuous three-dimensional network (e.g. silica gel). In each case the silica may be hydrated or anhydrous.

1.2.2.1 Colloidal silica

Colloidal silicas (silica sols) are aqueous dispersions of discrete primary amorphous silica particles (1–100 nm diameter). They typically contain up to ca. 50 wt% SiO_2 and are stabilized with small amounts of alkali (e.g. sodium or ammonium hydroxide at < 0.05 M) to prevent aggregation or gelling. Drying leads to formation of low-density silica gels with high mesoporosity (2–50 nm) and macroporosity (> 50 nm).

1.2.2.2 Silica gel

Silica gels are continuous three-dimensional agglomerates of primary amorphous silica particles. Voids within silica gels form a continuum of mesopores and macro-pores initially filled with the medium from which the gel was prepared (usually water). Gels containing water within their pore system are termed **hydrogels**.

Drying of hydrogels generally leads to irreversible shrinkage (i.e. reduction in porosity due to structural collapse), resulting in the formation of **xerogels** which are relatively dense and have low surface area and pore volume.

Formula	Name	Structure	
SiO_4^{4-}	orthosilicates	△	(monosilicate)
$Si_2O_7^{6-}$	pyrosilicates	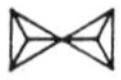 ▷◁	(disilicate)
$(SiO_3)_n^{2n-}$	metasilicates	 △△△	(cyclotrisilicate)
	pyroxenes	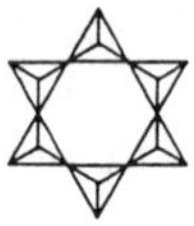 ✡	(cyclohexasilicate)
		 △△△	(single chains)
$(Si_4O_{11})_n^{6n-}$	amphiboles	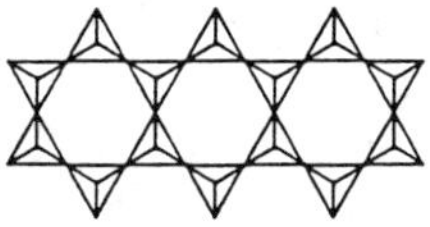	(double sheets)
$(Si_4O_{10})_n^{4n-}$	phyllosilicates	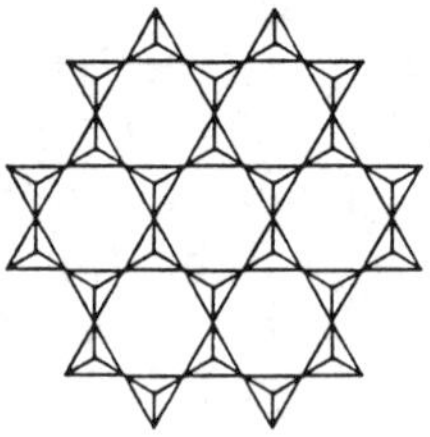	(sheets)

Fig. 1.1 — Classification of silicas (each tetrahedron has Si at the centre and O at each corner).

Hydrogels can be dried under conditions that prevent structural collapse by removing water at temperatures above its critical point (e.g. pyrogenic silica) or by displacing water with a solvent of lower boiling point and removing the solvent at a temperature above its critical point. Drying under conditions that prevent structural collapse results in the formation of low-density **aerogels** in which the hydrogel pore structure is largely retained.

1.2.2.3   Precipitated silica

Silica gels may be produced by precipitation of aqueous silicate solutions (e.g. sodium silicate) or silica sols (e.g. by addition of mineral acid, salt or coagulating agent). Precipitated silicas have a wide range of physical characteristics that can be tailored by adjustment of precipitation and drying conditions. Densities vary between 0.03 and 0.3 $g\,cm^{-3}$. High-density precipitated silicas have high surface areas (up to 800 $m^2\,g^{-1}$), intermediate pore volume (ca. 0.4 $cm^3\,g^{-1}$ and small mean pore diameter (typically < 5 nm). Low-density precipitated silicas have lower surface area (100–$200 \, m^2 \, g^{-1}$), larger pore volumes (up to $2 \, cm^3 \, g^{-1}$) and larger mean pore diameters (up to 25 nm). A range of precipitated silicas is commercially available, with properties intermediate between high- and low-density extremes.

1.2.2.4   Pyrogenic silica
Silica aerosols may be formed in high-temperature (pyrogenic) vapour phase processes (e.g. hydrolysis of silicon tetrachloride). Aerosols formed in such processes are called **pyrogenic** or **fumed** silicas. These are anhydrous, low-density silica gels of high purity (i.e. contain no residues from precipitating agents etc.). Surface area and pore structures of pyrogenic silicas are similar to those of low-density precipitated silicas.

1.2.2.5   Biogenic silica
Biogenic silica occurs in the shells, skeletons and spines of various aquatic organisms (e.g. diatoms, radiolarins and sponges). The most abundant source of biogenic silica is found in sedimentary deposits of diatoms (called **diatomaceous earth, diatomite** or **kieselguhr**) and consists of relatively pure amorphous silica (> 90 wt% SiO_2). In older sedimentary deposits, amorphous silica undergoes transformation to microcrystalline quartz. Diatomite particles (individual shells or skeletons) are elaborately shaped hollow microscopic cylinders and ellipsoids ranging in size from ca. 1 μm to 1 mm, with symmetrical markings and features. Diatomite has a large macropore volume and is particularly useful as a filter medium.

1.2.3   Properties of amorphous silica
Amorphous silicas are sparingly soluble in acidic or neutral aqueous solutions under ambient conditions. However, the solubility of silica markedly increases in alkaline solutions with pH > 9. In aqueous solutions, soluble silica is present as monosilicic acid ($Si(OH)_4$), which is a weak dibasic acid (equations (1.1) and (1.2)).

$$Si(OH)_4 + H_2O \rightarrow SiO(OH)_3^- + H_3O^+ \quad pK_1(20°C) = \ \ 9.8 \qquad (1.1)$$

$$SiO(OH)_3^- + H_2O \rightarrow SiO_2(OH)_2^{2-} + H_3O^+ \quad pK_2(20°C) = 11.8 \qquad (1.2)$$

The solubility of silica decreases with incorporation of impurities within the structure, or when accompanied by a thin coating of organic materials (e.g. on the surface of biogenic silicas). Small concentrations of electrolytes (e.g. sodium chloride) or solvents (e.g. methanol) also lower the solubility of silica in water.

When the solubility of monosilicic acid is exceeded, it undergoes polymerization, particularly when there is a solid phase present to support deposition. Monosolicic acid associates, forming microscopic particles composed of chains, rings, sheets and three-dimensional networks. Under basic conditions, particles increase in size, but the number of particles remains relatively small. However, in acidic conditions, the small particles generally agglomerate to form gels.

The surface of amorphous silicas consists of siloxane (Si–O–Si) and silanol (Si–OH) groups. Silanol groups are hydrophilic, giving silica gels their characteristic water sorption properties. In addition, silanol groups are weakly acidic; hence, amorphous silicas can be used to catalyse facile acid-catalysed reactions. In most cases the silica surface is too weakly acidic to participate chemically in catalysis (i.e. silica does not affect reaction rate constant by lowering activation energy). However, physisorption of reactants from solution or gas phase onto silica surfaces can lead to local increases in concentration, which increases rate of reaction.

Amorphous silicas are most frequently used as supports (or structural promoters) for catalysts (e.g. platinum group metals, etc.). Their high surface areas and large pore volumes enable catalysts to be supported in highly dispersed forms. In addition, catalyst–support interactions can inhibit sintering, thereby improving catalyst life.

Impurities within the structure of amorphous silicas (e.g. aluminium or magnesium) lead to increased surface acidity. Amorphous silica–aluminas (synclysts) are more acidic than either pure silicas or pure aluminas. When Al^{3+} replaces Si^{4+} in a tetrahedral site it results in formation of an acidic hydroxyl group:

$$[(AlO(OH))_x (SiO_2)_{1-x}]$$

Acidic amorphous silica–aluminas were used on a large scale as cracking catalysts before the introduction of zeolite catalysts [1].

Table 1.1 provides information about the structures and properties of some commercially available silicas and other amorphous solids.

1.3   ALUMINA

1.3.1   Introduction

The term 'alumina' is used to describe various hydrated and anhydrous aluminium oxides. Most commercially available aluminas are prepared from the mineral bauxite (a mixture of hydrated aluminas and several non-alumina impurities) by the Bayer process, and are available at $> 99\%$ purity. Thermally stable high-surface-area forms of alumina can be prepared, with acidic or basic surfaces. These may be used as acid or base catalysts or supports for other catalytic materials (e.g. metals, oxides, sulphides, etc.).

1.3.2   Hydrated aluminas

Hydrated aluminas fall into two groups: (a) aluminium trihydrates, $Al(OH)_3$ (equivalent to $Al_2O_3.3H_2O$) and (b) aluminium monohydrate, $AlO(OH)$ (equivalent to $Al_2O_3.H_2O$). Each group has several members with different polymorphic forms as described below.

1.3.2.1   Aluminium trihydrates

Aluminium trihydrates are crystalline aluminium hydroxides with stoichiometry $Al(OH)_3$. The name trihydrate refers to the liberation of three moles of water per Al_2O_3 on dehydration (equation (1.3)):

$$2 \, Al(OH)_3(s) \rightarrow Al_2O_3(s) + 3 \, H_2O(g) \tag{1.3}$$

Table 1.1 — Structure and properties of some amorphous solids

Commercial name	Supplier	Structure type	Pores	Surface area	Acidity
Silicas					
Kieselguhr	Merck	Biogenic silica	Mainly macro	Low	Low
Diatomaceous earth					
Celite	Manville				
Merck-10181	Merck	Silica gel	Macro/meso	675 m^2 g^{-1}	Low
Davisil-633	Grace	(aerogel)		480 m^2 g^{-1}	Low
Davisil-643	Grace			300 m^2 g^{-1}	Low
Cab-o-sil	Cab-o-sil	Pyrogenic	Macro/meso	High	Low
Aerosil	Degussa	(fumed)		High	Low
Aerosil	Degussa	Precipitated	Macro/meso	Varies	Low
Silicic acid	Sigma	Silicic acid	Macro/meso	Varies	Moderate (varies)
Ludox	Du Pont	Colloidal silica	Solution	Not available	Basic
Aluminas					
Alumina	Sigma	Alumina gel (aerogel)	Macro/meso	Typ. 155 m^2 g^{-1}	Varies
Alumina	Merck	Gamma alumina	Macro/meso	High	Mod/high
Corundum	Merck	Alpha alumina	Macro	Low	Low
Silica–alumina					
Synclyst-13	Crosfield	Amorphous fluidized catalytic cracking catalysts	Macro/meso	Mod/high	High
Synclyst-25	Crosfield		Macro/meso	Mod/high	High

There are two main aluminium trihydrate polymorphs: α-alumina trihydrate (gibbsite) and β-alumina trihydrate (bayerite).

Gibbsite is a major constituent of tropical bauxites. However, it is produced commercially in a pure form as a product of the Bayer process. The gibbsite structure consists of Al^{3+} cations octahedrally coordinated to OH^- anions. Octahedra are edge shared to form flat layers that have two thirds of the octahedral sites occupied by Al^{3+} and one third vacant (i.e. **dioctahedral** layer structure, Fig. 1.2). The analogous magnesium mineral, brucite ($Mg_3(OH)_6$), has a **trioctahedral** structure in which all of the layer octahedral sites are occupied by Mg^{2+}. Gibbsite layers are stacked parallel, with surface hydroxyl groups immediately above those of adjacent layers. Gibbsite produced by the Bayer process consists of spherical aggregates of individual gibbsite crystals.

Bayerite is rarely found in natural deposits, but is commercially available as a product of the Bayer process at $> 90\%$ purity, though the world production is small. However, its main use is as a precursor for η-alumina (a transition alumina) which is a constituent of some catalysts.

1.3.2.2 Aluminium monohydrate
Aluminium monohydrate is a crystalline hydrated alumina of stoichiometry AlO(OH). The name 'monohydrate' refers to the liberation of one mole of water per

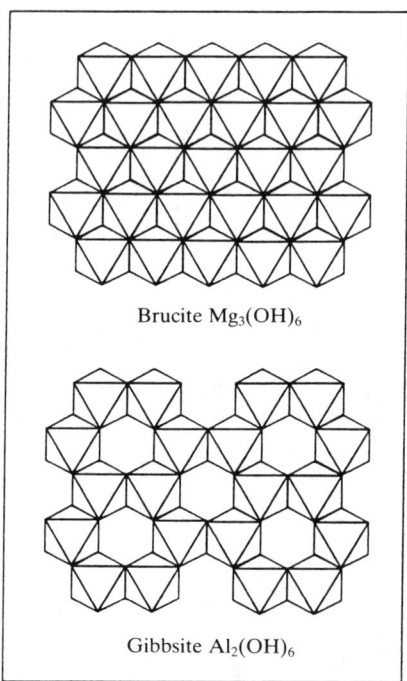

Brucite $Mg_3(OH)_6$

Gibbsite $Al_2(OH)_6$

Fig. 1.2 — Trioctahedral and dioctahedral layer structures of brucite and gibbsite.

Al_2O_3 on dehydration (equation (1.4)):

$$2\,AlO(OH)(s) \rightarrow Al_2O_3(s) + H_2O(g) \tag{1.4}$$

There are two main aluminium monohydrates: α-aluminium monohydrate (boehmite) and β-aluminium monohydrate (diaspore).

Boehmite is a major constituent of Mediterranean bauxites, but is prepared in pure form by the Bayer process for commercial applications. The structure of boehmite consists of Al^{3+} cations octahedrally coordinated to equal numbers of O^{2-} and OH^- anions. Octahedra are edge shared to form 'puckered' sheets (Fig. 1.3). Commercial grades of boehmite generally consist of 30–100 μm aggregates of ca. 1 μm boehmite crystals.

Diaspore is a major constituent of Russian, Greek and Rumanian bauxites, and is also found in some high-alumina clay deposits in the USA. Diaspore is not commercially available from the Bayer process, but has been prepared synthetically on a laboratory scale. When diaspore is heated at $> 450°C$ it undergoes a phase transition to anhydrous α-alumina (corundum). Other hydrated aluminas require heating to ca. 1200°C before α-alumina is formed.

1.3.2.3 *Alumina gels*
Colloidal alumina gels of stoichiometry $AlO(OH)$ can be prepared by hydrolysis of aqueous solutions of aluminium salts or alkoxides. Alumina gels may be amorphous or microcrystalline, depending upon hydrolysis conditions used. Microcrystalline alumina gels are called **pseudoboehmite**.

Pseudoboehmite can be dried under conditions that allow or prevent shrinkage to form alumina xerogels or aerogels respectively. Pseudoboehmite aerogels consist of microcrystalline primary particles of colloidal dimensions. They have high surface areas (up to ca. 500 $m^2\,g^{-1}$) and high mesopore volumes (in the range 2–50 nm). Calcination (heating in air) of pseudoboehmite forms microcrystalline pseudogamma alumina, also with high surface area (up to 500 $m^2\,g^{-1}$) and pore volumes (typically 0.3–0.8 $cm^3\,g^{-1}$). Pseudogamma alumina is thermally stable over a wide temperature range (up to 900°C) and often has a strongly acidic surface. The combination of high surface area, strong surface acidity and good thermal stability makes pseudogamma aluminas useful as catalysts and supports.

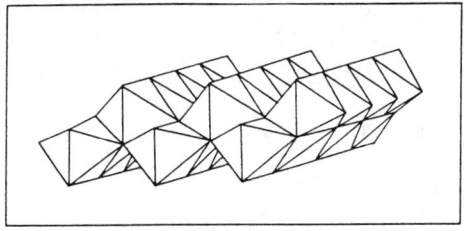

Fig. 1.3 — Boehmite layer structure.

1.3.2.4   Aluminium chlorhydrol

Aluminium chlorhydrol is the chloride salt of an aluminium oxyhydroxide ('Keggin') cation [2]. The structure of the Keggin ion (Fig. 1.4) consists of one tetrahedron, corner shared with 12 mutually edge-sharing octahedra. At the centre of the tetrahedron and octahedra, formally, are Al^{3+} cations. O^{2-} anions occupy apices of the tetrahedron, OH^- anions occupy shared octahedral apices and H_2O molecules occupy unshared octahedral apices.

Aluminium chlorhydrol is water soluble and can be reversibly dried and re-dissolved. However, excessive heating leads to decomposition to anhydrous alumina and evolution of HCl gas. Because of its solubility and thermal decomposition behaviour, aluminium chlorhydrol may be used as a binder in catalysts consisting of aggregates of smaller catalyst particles (e.g. zeolites). Calcination (heating in air) results in formation of an alumina 'glue' that locks the small catalyst particles together.

Aluminium chlorhydrol is commercially available in powdered or solution (up to 50 wt% Al_2O_3) forms or can be prepared in the laboratory by controlled hydrolysis of aqueous aluminium chloride solution with sodium hydroxide [3].

1.3.3   Anhydrous aluminas

1.3.3.1   α-Alumina

The only natural anhydrous alumina is α-alumina (corundum), which has an approximately hexagonal close-packed arrangement of O^{2-} anions with two thirds of the interstitial octahedral sites occupied by Al^{3+} cations. Octahedra are edge shared to form layers (cf. gibbsite), and face shared with octahedra in adjacent layers. The compact three-dimensional structure of α-alumina accounts for its high density, mechanical strength, thermal stability, low surface area and low porosity. α-Alumina is insoluble in most mineral acids, resistant to hydration and largely chemically inactive. The major use of α-alumina is in refractories, ceramics and abrasives. Because α-alumina is chemically relatively inert, it may be used as a low-surface-area catalyst support where reaction conditions are particularly severe (e.g. steam reforming or partial oxidation).

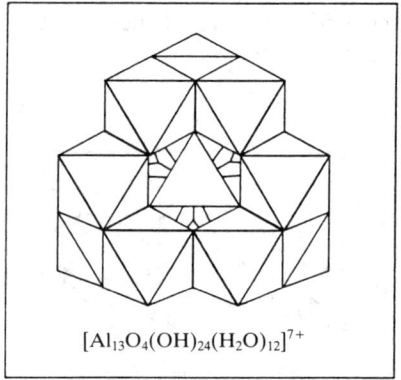

$$[Al_{13}O_4(OH)_{24}(H_2O)_{12}]^{7+}$$

Fig. 1.4 — Aluminium chlorhydrol ('Keggin') cation.

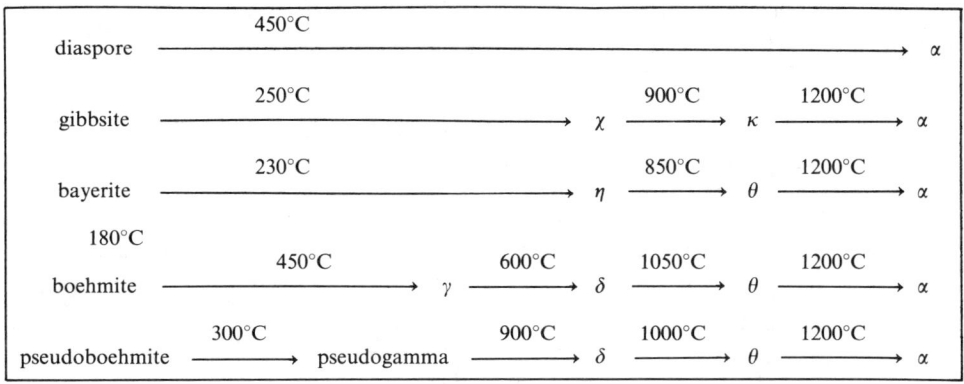

Fig. 1.5 — Decomposition sequence of aluminium hydroxides.

1.3.3.2 Transition aluminas

Calcination of hydrated aluminas leads to dehydration and ultimately formation of α-alumina through a series of meta-stable transition aluminas (γ, δ, θ, κ, χ and/or η, Fig. 1.5). Transition aluminas are distinct crystalline phases, the most important for catalytic applications being γ-alumina, which has a thermally stable high surface area, large mesopore volume and strong surface acidity.

γ-Alumina is used as a bifunctional support for platinum in platforming (platinum reforming) catalysts. Platforming catalysts are used to convert low-octane hydrocarbons (e.g. unbranched alkanes) to higher-octane species (e.g. branched alkanes and aromatics). Catalysts typically contain 0.25–0.5 wt% platinum on high-purity γ-alumina, promoted with 0.5–1.0 wt% Cl^- or F^- to increase surface acidity. Platinum provides sites for hydrogenation/dehydrogenation, and promoted alumina provides acid sites to catalyse isomerization and cyclization reactions.

1.3.3.3 Activated aluminas

Activated aluminas are high-surface-area, high-porosity aluminas which may be amorphous or microcrystalline (often mixtures of transition aluminas) depending upon the method of preparation.

Activation generally involves flash calcination of hydrated aluminas (e.g. bauxite, gibbsite, pseudoboehmite or amorphous aluminas), which results in the explosive release of steam. The steam generates an internal pressure which opens up channels and cavities within the hydrated alumina precursor, giving rise to the high surface area and mesoporosity. Short residence times (1–10 s) at temperatures of 400–800°C are required.

Activated aluminas are amphoteric, containing either acidic or basic sites of varying strength (Fig. 1.6), depending on the method of preparation and post treatment. Surface acidity can be increased by impregnation with promoters (e.g. Cl^- and F^-). The high surface area and porosity facilitate access of reactants and diffusion of products from active sites.

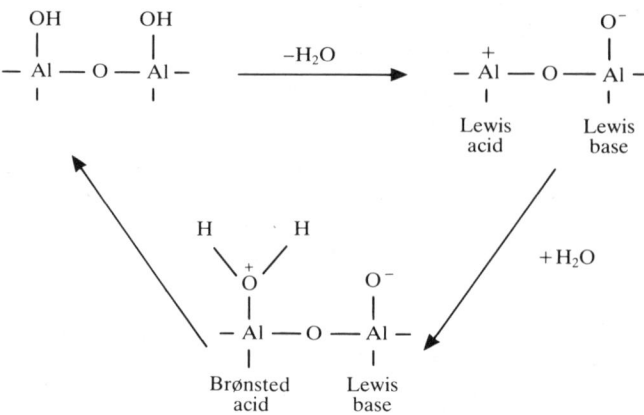

Fig. 1.6 — Acidic and basic sites on aluminas.

Activated aluminas are often used as supports, where they provide a large surface area on which catalysts (e.g. metals) can be highly dispersed. Catalyst–support interactions can inhibit sintering and subsequent loss of catalytic activity.

1.4   CARBON

1.4.1   Introduction
Most carbon is present in nature as compounds (carbonates, organic compounds, etc.), with little being found in its elemental form (e.g. graphite and diamond). Graphite is the thermodynamically most stable form of elemental carbon under ambient conditions. Synthetic carbons (e.g. high-surface-area activated carbons) can be prepared from many carbonaceous compounds (e.g. coals, nut shells, etc.). The most useful forms of carbon for catalytic applications are modified graphites and activated carbons.

1.4.2   Graphite
Most commercially mineable deposits of graphite originate from organic rather than inorganic matter. Graphite is formed by metamorphosis of sedimentary deposits of carbonaceous materials at high temperatures and pressures. Natural graphites contain various impurities (e.g. silica, clay minerals, oxides, etc.) depending upon their source. Crude graphite ores are purified using flotation techniques to give carbon contents typically $> 95\,\mathrm{wt}\%$. Further purification requires various chemical (e.g. halogen gas, sodium carbonate, etc.) and heat treatments.

1.4.2.1   Graphite structure
The structure of graphite consists of stacked layers of sp^2 carbon atoms. Each layer has carbon atoms arranged in a flat hexagonal mesh-like rings (C — C bond length 141.5 pm). Layers are held together by relatively weak van der Waals forces (layer separation 335 pm). Carbon layers are stacked with parallel basal planes in a

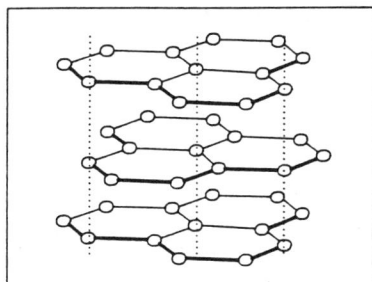

Fig. 1.7 — Graphite layer structure.

staggered arrangement (Fig. 1.7), giving rise to two polymorphs: (a) hexagonal graphite, consisting of a strictly alternating stacking arrangement (ABABAB...) and (b) rhombohedral graphite, consisting of layers stacked in an ABCABC... arrangement. Natural graphites contain a mixture of the two polymorphs (typically up to 30% rhombohedral). Milling and shearing operations increase the proportion of the rhombohedral form.

Interlayer separation (i.e. distance between adjacent layers) depends upon the degree of ordering of stacked layers. **Turbostatic** stacking (i.e. where adjacent parallel layers are rotated by random angles about an axis perpendicular to the layer planes) give rise to weaker van der Waals forces between adjacent layers and consequently increased interlayer separation.

1.4.2.2   Synthetic graphite
Synthetic graphites can be prepared by heating most carbonaceous substances at 2500–3000°C in the absence of oxygen. The structure and properties (e.g. spatial arrangement and size of crystallites, amount and distribution of porosity and stacking disorder, etc.) depend upon the type of carbonaceous precursor used. Carbonaceous materials that pass through a molten mesophase at 450–500°C (graphitizable or soft carbon sources, e.g. lignin) generally form well ordered graphites, whereas materials that remain solid throughout heat treatment (non-graphitizable or hard carbon sources, e.g. cellulose) generally form graphites of lower order (e.g. turbostatic).

1.4.2.3   Properties of graphites
Many of the physical properties of graphites are anisotropic, i.e. depend upon the crystallographic axes. Coefficient of thermal expansion and mohs hardness are greater perpendicular to layer planes. Conversely, thermal conductivity, magnetic susceptibility and elastic modulus are greater within layer planes.

Graphites generally have relatively low surface areas (< 25 m^2 g^{-1}) and low pore volumes compared to activated carbons (Table 1.2) and therefore have limited use as catalysts or supports. However, graphites form intercalation compounds with a wide range of substances, and these are becoming increasingly important as catalysts [4].

Table 1.2 — Typical nitrogen BET surface areas of carbons

Carbon type	Surface area $m^2 g^{-1}$
Graphite (natural)	< 20
Graphite (artificial)	< 25
Graphitized carbon black	20– 100
Carbon black	70–250
Activated carbon:	
(wood charcoal)	300–900
(peat charcoal)	350–1000
(coal)	300–1000
(nutshells)	700–1500
(petroleum needle coke)	1500–3000

1.4.2.4 Graphite intercalation compounds

Graphite intercalation compounds consist of 'guest' species regularly distributed between intact carbon layers of the graphite 'host'. The relatively weak van der Waals forces between graphite layers allow them to be parted so that guest species can enter the graphite interlayers. Intercalation compounds may be formed with a wide range of substances including electron donors (e.g. alkali metals, lanthanides, compounds with polar ligands, etc.) and electron acceptors (e.g. halogens, halides, acids, etc.).

Intercalation leads to increased separation and translation of graphite layers such that the stacking arrangement often becomes eclipsed (AAA...). The greatest mean separation between graphite layers occurs when host (graphite, C) and guest (intercalated species, I) form alternate layers (i.e. CICICI...). This is called **1st-stage intercalation compound**. At lower guest loadings, higher-stage intercalation compounds are formed. An nth-stage intercalation compound consists of n graphite layers per layer of guest (i.e. $C_nIC_nIC_nI...$).

Graphite intercalation compounds can be prepared by contacting powdered graphite with guest species in liquid or gas phases at elevated temperatures and pressures. Intercalation is generally reversible, the stability of the intercalation compound often being dependent upon the stability of the guest species. Intercalation compounds containing electron donors are often readily oxidized in air (and may even be pyrophoric). Intercalates containing electron acceptors are often hygroscopic and of limited use as catalysts.

1.4.3 Activated carbon

1.4.3.1 Structure and properties of activated carbons

Activated carbons are high-surface-area carbons produced using gas or chemical activation processes. The structure and properties of activated carbons vary markedly depending upon the origin of the carbonaceous precursor and processing conditions (e.g. mode of activation, temperature, etc.). The structure of activated carbons generally consists of small turbostatic graphite crystallites (0.5–3 nm) interconnected with an amorphous carbon phase. Activation leads to the formation of numerous cracks and fissures, giving rise to high surface area, mesoporosity (2–50 nm pores) and macroporosity (> 50 nm pores).

Various impurities may be present in activated carbons, depending upon the carbonaceous precursor used. The ash content (wt% residual solids after combustion of carbon to CO/CO_2, e.g. by calcination in air at 500°C) gives an indication of the level of impurity present. Activated carbons prepared from coals generally have high ash contents (up to 20 wt%), whereas activated carbons prepared from phenolic resins can be of the order of parts per million.

In addition to mineral impurities, activated carbons often contain small amounts of chemically bound hydrogen and oxygen. Hydrogen generally arises owing to incomplete pyrolysis of carbonaceous precursors and is present in the form of polycyclic aromatic compounds. The nature of chemically bound oxygen (often called surface oxygen) is not clearly understood. Surface species behaving like carboxyl, phenolic, carbonyl and various other acidic groups have been reported [4]. However, the properties of surface oxygen species depend very much upon the way in which carbons are treated.

Gaseous oxidation (e.g. with NO_2, CO_2, etc.) or reduction (e.g. with H_2) can be used to increase or decrease respectively the concentration of surface oxygen sites. Steam and hydrogen treatments can be performed to remove sulphur, and CO or halogen treatments can be used to remove iron.

Activated carbons are often used as catalyst supports (e.g. for highly dispersed metals, sulphides and halides) because of their high surface area and porosity. Platinum group metals (e.g. palladium and platinum), silver, rhenium and zinc are the metals most commonly supported on activated carbon.

1.4.3.2 *Preparation of activated carbons*

Most carbonaceous materials are suitable precursors for activated carbons (e.g. wood, nut shells, fruit stones, coals, peat, mineral oils, etc.). The degree of activation, pore size distribution and purity depend primarily upon the precursor used. Activation leads to removal of non-graphitic carbon (graphitic carbon removal being minimal) opening up the structure, thereby introducing porosity. Activation is achieved using either a gas or chemical treatments.

Gas activation consists of carbonization of the carbonaceous precursor by heating in an inert atmosphere (e.g. nitrogen), followed by reacting the carbonized product with an oxygen-containing gas (e.g. steam or CO_2) at 800–1000°C. Non-graphitic carbon is oxidized from the outside of carbonized particles inward giving rise to activated carbon with an inhomogeneous pore structure (larger pores towards the exterior of particles). Higher temperatures and longer processing times lead to more extensive carbon removal and consequently more highly activated carbons, but at lower yield.

The proportions of lignin and cellulose in starting materials have a major influence upon the properties of the activated carbon. Lignin forms a molten mesophase, which forms graphitic carbon during carbonization. Cellulose is carbonized via a non-molten state and forms non-graphitic carbon. Hence, the relative proportions of lignin and cellulose in the precursor determine the proportions of graphitic and non-graphitic carbon formed. Nut shells contain a relatively large amount of lignin and form graphitic activated carbons, whereas peats contain a high proportion of cellulosic carbon and form much less graphitic activated carbons. In cases where the concentration of cellulose is high (e.g. > 40 wt%) activated carbons that retain the shape of the carbonized precursor (pseudomorphs) can be prepared. Pre-formed activated carbons in the form of pellets, spheres, mats, fibres, etc. can thus be prepared.

Chemically activated carbons are produced by impregnating the carbonaceous precursor with an activating agent (e.g. zinc chloride or phosphoric acid) and heating at $600-900°C$ in an inert atmosphere. Carbonization and activation occur simultaneously, forming activated carbons with homogeneous porosity. There is some irreversible incorporation of activating agent into the carbon structure. Freshly activated carbons are generally water washed to remove soluble impurities (e.g. activators). Washing with dilute aqueous mineral acids (e.g. hydrochloric or nitric acid) may be performed where very low concentrations of impurities are required.

1.5 CLAY MINERALS

1.5.1 Introduction

Occurrence of clay minerals is ubiquitous and their high surface area, sorptive and ion exchange properties have been exploited for thousands of years. Throughout the twentieth century, clay mineral properties have been turned to catalytic applications. Clay minerals have a wide range of functions in solid catalysts, including: (a) use as catalytically active agents (generally as solid acids); (b) use as bifunctional or 'inert' supports (e.g. for highly dispersed metals, metal complexes, enzymes, etc.) and (c) as fillers to give solid catalysts required physical properties (e.g. attrition resistance, density, specific heat capacity, etc.).

Clay minerals comprise a large family of fine-grained crystalline sheet silicates (phyllosilicates, Fig. 1.1) of which there are four main classes, namely the 1:1, 2:1, 2:1:1 and 2:1 'inverted ribbon' clay minerals (Table 1.3, Fig. 1.8). Each class has a different arrangement of tetrahedral and octahedral layers.

1.5.2 Structure and properties of clay minerals

1.5.2.1 Layer structures

Tetrahedral layers within clay minerals consist of continous sheets of silica tetrahedra linked via three corners to form a hexagonal mesh (cf. phyllosilicate, Fig. 1.1). The fourth corner of each tetrahedron (normal to the plane of the sheet) is corner shared with octahedra in adjacent layers. Each tetrahedron formally contains an Si^{4+} at the centre coordinated to an O^{2-} at each apex. However, replacement of Si^{4+} with Al^{3+} in tetrahedral sites is common.

Table 1.3 — Ideal unit cell composition of clay minerals; $(x + y)$ is the layer charge.

Layer type	Interlayer species	Octahedral cations	Tetrahedral cations	Layer anions
1:1 Layer				
(trioctahedral)	H_2O	Mg_6	Si_4	$O_{10}(OH)_8$
(dioctahedral)	H_2O	Al_4	Si_4	$O_{10}(OH)_8$
2:1 Layer				
(trioctahedral)	(Na, K, Ca, Mg...)	$Mg_{6-y}Li_y$	$Si_{8-x}Al_x$	$O_{20}(OH)_4$
(dioctahedral)	(Na, K, Ca, Mg...)	$Al_{6-y}Mg_y$	$Si_{8-x}Al_x$	$O_{20}(OH)_4$
2:1:1 Layer				
(trioctahedral)	$[Al, Mg (OH)_{12}]$	$Mg_{6-y}Li_y$	$Si_{8-x}Al_x$	$O_{20}(OH)_4$
(dioctahedral)	$[Al, Mg (OH)_{12}]$	$Mg_{6-y}Li_y$	$Si_{8-x}Al_x$	$O_{20}(OH)_4$
2:1 Inverted ribbon				
(trioctahedral)	H_2O	Mg_8	Si_{12}	$O_{30}(OH)_4$
(dioctahedral)	H_2O	Mg_5	Si_8	$O_{20}(OH)_4$

Octahedral layers in clay minerals consist of flat layers of edge-sharing octahedra, each formally containing cations at its centre (usually Mg^{2+} or Al^{3+}) and OH^- or O^{2-} at its apices (O^{2-} occupying apices corner shared with adjacent tetrahedra and OH^- occupying apices that are not shared). Octahedral layers may be trioctahedral or dioctahedral depending upon the degree of occupancy of the octahedral sites. In

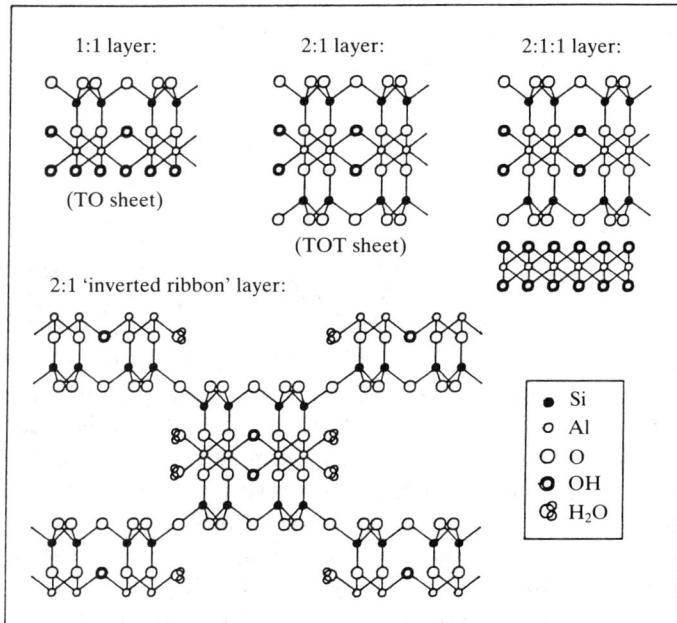

Fig. 1.8 — Layer structures of the clay minerals.

trioctahedral layers, all octahedral sites are occupied, generally by divalent cations (e.g. Mg^{2+}, cf. brucite, Fig. 1.2). Dioctahedral layers have two thirds of their octahedral sites occupied, generally by trivalent cations (e.g. Al^{3+}) and one third of their octahedral sites vacant (cf. gibbsite, Fig. 1.2).

The sheet structures contain combinations of tetrahedral and octahedral layers (Fig. 1.8). 1:1 clay minerals contain one tetrahedral and one octahedral layer in each 1:1 sheet (also called a TO sheet, Fig. 1.8). 2:1 clay minerals contain two tetrahedral layers and one octahedral layer in each 2:1 sheet (also called a TOT sheet, Fig. 1.8). 2:1:1 clay minerals contain alternating 2:1 sheets and gibbsite or brucite layers. 2:1 'inverted ribbon' clay minerals contain 2:1 sheets in which tetrahedra invert at regular intervals to crosss-link octahedral layers, forming a three-dimensional structure (Fig. 1.8).

1.5.2.2   Composition and isomorphous substitution

The majority of tetrahedral sites in clay minerals are occupied by Si^{4+}. However, isomorphous substitution of Si^{4+} is common, resulting in some of the tetrahedral sites being occupied by Al^{3+}.

Trioctahedral clay minerals generally have octahedral sites occupied by Mg^{2+}. Isomorphous substitutions are common in 2:1 structures, leading to some of the octahedral sites being occupied by other divalent cations (e.g. Fe^{2+} or Ni^{2+}) or univalent cations (e.g. Li^{+}).

Octahedral sites in dioctahedral clay minerals are predominantly occupied by Al^{3+}. Isomorphous substitutions of Al^{3+} by other trivalent cations (e.g. Fe^{3+} or Cr^{3+}) or divalent cations (e.g. Mg^{2+} or Fe^{2+}) are common in 2:1 structures.

In addition to cation isomorphous substitutions, there are occasionally anionic isomorphous substitutions. For example, OH^{-} groups may be substituted with F^{-} to form fluorosilicate clay minerals.

1.5.2.3   Layer charge

When cation isomorphous substitutions (tetrahedral or octahedral) result in cations of lower charge replacing the original cations, the sheet gains a net negative layer charge (i.e. becomes macroanionic). In such cases the anionic sheets are charge balanced with cations (e.g. Na^{+}, K^{+}, Ca^{2+}, Mg^{2+}, etc.) that reside between adjacent sheets (called **interlayer** or **interlamellar** cations). In many cases, interlayer cations readily undergo ion-exchange and are referred to as **exchangeable cations**.

In 1:1, 2:1:1 and 2:1 'inverted ribbon' clay minerals the extent of isomorphous substitution is generally small, resulting in low layer charge and low concentrations of interlayer cations. However, in 2:1 structures the extent of isomorphous substitution varies widely, resulting in a series of 2:1 clay minerals, from those with high layer charges (e.g. micas) to those with low layer charges (e.g. talc or pyrophyllite). The properties of 2:1 clay minerals have such a marked dependence upon layer charge that they are subdivided into groups with layer charges in defined ranges (Table 1.4).

Table 1.4 — Layer charge $(x + y)$ per unit
cell of 2:1 clay minerals.

2:1 Clay mineral	Layer charge per unit cell $(x + y)$
Talc/pyrophyllite	ca. 0
Smectites	0.5–1.2
Vermiculites	1.2–1.8
Micas	ca. 2
Brittle micas	ca. 4

1.5.2.4 Swelling

Many clay minerals absorb water (and other molecular or ionic species) between their layers (intercalation). When guest species (e.g. water) enter the interlayer region, the layers part and the clay swells. For swelling to occur, the energy released by cation and/or layer solvation must be sufficient to overcome attractive forces between adjacent layers (e.g. hydrogen bonding). In 1:1 (TO) clay minerals (e.g. kaolinite) water forms strong hydrogen bonds with hydroxyl groups on hydrophilic octahedral layer surfaces, allowing swelling to occur.

There are very few surface hydroxyl groups in 2:1 (TOT) clay minerals, so the ability to swell depends upon solvation of interlayer cations and anionic layer sites. Consequently, the ability to swell is governed by the layer charge and concentration of interlayer cations. Clays with 2:1 structures and low layer charge (e.g. talc and pyrophyllite) have very low concentrations of interlayer cations and therefore do not readily swell. At the opposite extreme, 2:1 clay minerals with very high layer charges (e.g. micas) have strong electrostatic forces holding alternate anionic layers and interlayer cations together, preventing swelling. The ability of 2:1 clays to swell is thus limited to those with an intermediate range of layer charges (e.g. smectites). Swelling properties of smectites depend upon the types of interlayer cations present as well as the layer charge. Smectites with univalent interlayer cations (e.g. Na^+, Li^+, NH_4^+ and H_3O^+) swell most readily, with divalent, trivalent and polyvalent cation forms swelling to increasingly poorer extents. The extent of swelling can be observed by measuring interlayer separations using powder X-ray diffraction.

1.5.2.5 Cation exchange

Interlayer cations present in swelling clay minerals (e.g. smectites) can undergo exchange with cations from external solutions. The concentration of exchangeable cations is called the **cation exchange capacity (cec)**, usually measured in milli-equivalents per 100 g of dried clay. Smectites have the highest concentration of interlayer cations of swelling clays, and consequently have the highest cation exchange capacities (typically 70–120 meq/100 g). Structural defects at layer edges give rise to additional cation exchange capacity and a small amount of anion exchange capacity (typically < 10 meq/100 g).

Cation exchange is generally accomplished by dispersing the clay mineral in a relatively concentrated solution of a salt of the required cation (typically a 1 N solution, i.e. $1/n$ molar solution of M^{n+} cations). The dispersion is mixed for several hours at ambient temperature. Extraneous cations may then be removed by repeatedly washing with distilled water using dialysis or centrifugation. To achieve complete cation exchange it is necessary to repeat the exchange step using fresh salt solution two or three times.

Acidic, proton exchanged, clay minerals can be prepared by ion-exchange using dilute acid solutions. Dioctahedral smectites (e.g. montmorillonite and beidellite) are moderately stable in dilute mineral acids (e.g. 0.5 N) at ambient temperature; however, extensive acid leaching may occur at higher concentrations or temperatures. Trioctahedral smectites (e.g. hectorite and saponite) are much less stable in acidic solutions. Nevertheless, direct proton exchange can be achieved using dilute weak acids (e.g. acetic acid) at low temperature (e.g. 0°C).

Polarizing cations (e.g. Al^{3+}) hydrolyse water giving low pH ion-exchange solutions. In such cases, cation exchange generally results in formation of clay minerals with mixtures of the polarizing cations and protons in their interlayer.

1.5.2.6.   Acidity
Clay minerals with high cation exchange capacity (e.g. smectites) derive most of their acidity from interlayer cations. Some (or all) of the interlayer cations may be protons or polarizing cations (e.g. Al^{3+}) which give rise to strong Brønsted acidity. In additiion, clay minerals have layer surface and edge defects that result in weaker Brønsted and/or Lewis acidity, generally at low concentration. The contribution of surface and edge defects to acidity can be increased by heat treatment or acid leaching, which increase the concentration of such defects. However, such treatments generally result in decreased Brønsted acidity due to interlayer cations. The acid strength of solid acids is usually measured in the Hammett scale [5]. According to this, the acidity of clay minerals can be comparable to that of concentrated sulphuric acid (Table 1.5).

1.5.3   1:1 Clay minerals
The structure of 1:1 clay minerals consists of parallel stacked sheets, each being a composite of one tetrahedral and one octahedral layer (TO layer (Fig. 1.8). The surfaces of each side of the TO layers are different. The tetrahedral side (TO) is silaceous and relatively hydrophobic, whereas the octahedral side (TO) is covered with hydroxyl groups and is very hydrophilic. Adjacent layers are held together by relatively weak hydrogen bonds between T and O faces (i.e. TO.TO.TO...) with a typical interlayer separation of ca. 300 pm.

There are two groups of 1:1 clay minerals, distinguished by their octahedral layers: (a) the serpentine group (trioctahedral) and (b) the kaolin group (dioctahedral), each group having several members with different layer stacking arrangements. The most important 1:1 clay mineral is kaolinite (china clay), the parent member of the kaolin group.

Table 1.5 — Surface acidity of some natural and proton exchanged (hydrogen) clay minerals.

Solid acid	Ho
Natural kaolinite	− 3.0 to − 5.6
Hydrogen kaolinite	− 5.6 to − 8.2
Natural montmorillonite	+ 1.5 to − 3.0
Hydrogen montmorillonite	− 5.6 to − 8.2

1.5.3.1 Kaolinite (china clay)

Kaolinite is an abundant common naturally occurring clay, often found in relatively pure deposits, the largest of which are in Georgia, South Carolina, Cornwall and Brazil. Kaolinites contain a number of mineral impurities (e.g. quartz, feldspars and other clay minerals) at varying levels (ranging from 5 to 50 wt%) depending upon the geological origin. Several deposits of kaolinite contain high concentrations of organic impurities (e.g. kerogen); these are called **ball clays**. Ball clays containing 60–80 wt% kaolinite are particularly common.

Kaolinite has a relatively low surface area, porosity and cation exchange capacity. The weak hydrogen bonding between layers allows swelling in water to form hydrogels. Hydration of hydroxyl groups on octahedral surfaces gives kaolinite gels high plasticity, allowing the gel to be formed into shapes (e.g. granules, pellets, tablets, spheres, etc.) that retain their shape after drying. Calcination leads to dehydroxylation, producing rigid, non-swelling meta-kaolinite. Further calcination gives rise to phase transitions, producing spinels, and ultimately mullite and crystobalite (equations (1.5–1.7)).

$$\underset{\text{(kaolinite)}}{Si_4Al_4O_{10}(OH)_8} \xrightarrow{550°C} \underset{\text{(meta-kaolinite)}}{Si_4Al_4O_{14}} + H_2O \qquad (1.5)$$

$$\underset{\text{(meta-kaolinite)}}{Si_4Al_4O_{14}} \xrightarrow{925-950°C} \underset{\text{(spinel)}}{Si_3Al_4O_{12}} + \underset{\text{(crystobalite)}}{SiO_2} \qquad (1.6)$$

$$\underset{\text{(spinel)}}{Si_3Al_4O_{12}} \xrightarrow{1050°C} \underset{\text{(mullite)}}{Si_2A_6lO_{13}} + \underset{\text{(crystobalite)}}{SiO_2} \qquad (1.7)$$

This makes kaolinite useful as a binder. Catalytically active materials (e.g. zeolites and aluminas) may be added to kaolinite gels, allowing them to be formed into desired-shaped particles (e.g. extrudate, pellets, spheres, etc.). Drying leads to encapsulation of the catalytically active species within a kaolinite matrix. Calcination to meta-kaolinite produces rigid-formed particles with desirable physical and mechanical properties (e.g. strength, density, porosity, specific heat capacity, etc.). The meta-kaolinite itself contributes little to the catalytic activity.

1.5.4 2:1 Clay minerals

The structure of 2:1 clay minerals consists of parallel stacked sheets, each composed of two tetrahedral and one octahedral layers (TOT layer, Fig. 1.8). They can be either trioctahedral or dioctahedral, having octahedral layers based on brucite or gibbsite respectively (Fig. 1.2).

Isomorphous substitutions (in tetrahedral or octahedral sites) give rise to a wide range of chemical compositions, layer charges and interlayer cation concentrations. Layer charge has a major influence over their properties and is used to categorize them into groups with related properties (Table 1.4).

Swelling is an important property for catalytic applications: (a) allowing cation exchange (e.g. with protons to form solid acids) and (b) giving reactants access to active sites between adjacent TOT layers. The members of the smectite group have the greatest ability to swell and are therefore the most important 2:1 clay minerals for catalytic applications.

1.5.4.1 Smectites

Smectites are divided into four sub-classes depending upon (a) the type of octahedral layer (dioctahedral or trioctahedral) and (b) the predominant location of layer charge sites (octahedral or tetrahedral).

Montmorillonite and beidellite groups are dioctahedral smectites with layer charges predominantly in octahedral and tetrahedral sites respectively. Their ideal formulae are:

montmorillonite group: $(M_x^+)^{ex}[Si_8]^{tet}(M(III)_{4-x}M(II)_x)^{oct}O_{20}(OH)_4]^{x-}$

beidellite group: $(M_x^+)^{ex}[(Si_{8-x}Al_x)^{tet}(M(III)_4)^{oct}O_{20}(OH)_4]^{x-}$

where, M^+ is an exchangeable interlayer cation (e.g. Na^+), $M(III)$ and $M(II)$ are non-exchangeable octahedrally coordinated trivalent and divalent cations (e.g. Al^{3+} and Mg^{2+}) respectively, and the layer charge is $0.5 < x < 1.2$.

Hectorite and saponite groups are trioctahedral smectites with layer charges predominantly in octahedral and tetrahedral sites respectively. Their ideal formulae are:

hectorite group: $(M_x^+)^{ex}[Si_8]^{tet}(M(II)_{6-x}M(I)_x)^{oct}O_{20}(OH)_4]^{x-}$

saponite group: $(M_x^+)^{ex}[(Si_{8-x}Al_x)^{tet}(M(II)_6)^{oct}O_{20}(OH)_4]^{x-}$

$M(II)$ and $M(I)$ are non-exchangeable octahedrally coordinated divalent and univalent cations (e.g. Mg^{2+} and Li^+) respectively, and the layer charge is $0.5 < x < 1.2$.

The term **bentonite** is used to describe any clay mineral that predominantly consists of, and has properties governed by, smectites. Montmorillonite is a major constituent of most bentonites (typically 80–90 wt%), the remainder being a mixture of mineral impurities including quartz, crystobalite, feldspars and various other clay minerals, depending upon geological origin. Bentonites containing high levels of montmorillonite are the most abundant and commercially mineable forms of smectites.

1.5.4.2 Montmorillonite

Montmorillonite (the parent member of the montmorillonite group) is the most important smectite used in catalytic applications. Natural montmorillonites contain mixtures of exchangeable cations (most often Na^+, Ca^{2+}, K^+ or Mg^{2+}). In commercially mineable deposits the dominant exchangeable cations are generally either Na^+ or Ca^{2+}, and these forms are called sodium montmorillonite or calcium montmorillonite, respectively.

High cation exchange capacity and good swelling properties allow a wide variety of catalytically active forms of montmorillonite to be prepared (e.g. containing acidic cations, metal complexes, photocatalytically active cations, etc.).

Montmorillonites are most frequently used as Brønsted acid catalysts, where the exchangeable cations are either protons or polarizing cations (e.g. Al^{3+}, Cr^{3+} or Fe^{3+}). Acid site strength depends upon the type of interlayer cations present ($H_3O^+ > Al^{3+} > Ca^{2+} > Na^+$). Higher acid strength generally leads to greater catalytic activity, but poorer product selectivity. Controlling the acid site strength by choice of interlayer cations provides for 'fine-tuning' catalyst selectivity.

Treatment of montmorillonite with mineral acids under harsh conditions (e.g. refluxing) leads to leaching of aluminium (and to a lesser extent silicon) from the TOT layers. This leads to increased surface area and concentration of weak acid sites, but a decrease in concentration of strong acid sites. **Acid leached montmorillonite** is particularly useful for catalytic applications requiring only weak acid sites, where strong acid sites give rise to poor selectivity.

Acidic montmorillonites can be prepared by directly impregnating natural montmorillonite with an acid (e.g. sulphuric acid). This results in partial proton exchange of original interlayer cations, and the formation of an extraneous salt (e.g. equation (1.8)):

$$Ca^{2+}\text{-montmorillonite} + H_2SO_4 \rightarrow H^+\text{-montmorillonite} + CaSO_4 \quad (1.8)$$

Provided that the salt is inert, acidic montmorillonites prepared by impregnation have similar properties to those prepared by cation exchange (e.g. strong Brønsted acidity, swelling, surface area, etc.).

1.5.4.3 Pillared smectites

Large cationic species (e.g. aluminium chlorhydrol 'Keggin' ion, Fig. 1.4) can be inserted between TOT layers of smectites (e.g. montmorillonite) by cation exchange. Such bulky species act like pillars, propping apart the TOT layers (typically by 0.3–1.5 nm). Slit-shaped pores are formed betweeen 'pillars' and TOT layers, which give rise to a uniform two-dimensional micropore system between adjacent TOT layers. Pore entry sizes are governed by the height of the pillars and the distance between them.

A wide range of bulky cationic species have been used to prepare pillared smectites, including organic cations (e.g. quaternary ammonium cations and 1,4-diazabicyclo-[2,2,2]-octane–DABCO [6]) and inorganic cations (e.g. polyoxyhydroxides of aluminium and zirconium [2]). The most widely studied pillared smectites are those with 'alumina' pillars derived from aluminium chlorhydrol (ACH). Aqueous solutions of aluminium chlorhydrol are acidic (equation 1.9).

$$[Al_{13}O_4(OH)_{24}(H_2O)_{12}]^{7+} \rightarrow [Al_{13}O_4(OH)_{24+x}(H_2O)_{12-x}]^{(7-x)+} + x\,H^+ \quad (1.9)$$

Hence, cation exchange generally results in mixed ACH and proton smectites which are acidic.

Aluminium chlorhydrol exchanged 'pillared' smectites have strong Brønsted acidity and retain a limited ability to swell. Heating at low temperatures (e.g. $< 110°C$) leads to dehydration. Calcination at higher temperatures (e.g. 450°C) leads to irreversible conversion of ACH cations to small particles of alumina $(AlO(OH)/Al_2O_3)$ which remain between the smectite's TOT layers, propping them apart by typically 0.8–0.9 nm. Alumina pillared smectites have rigid, non-swelling, interlayer micropore structures (pore volume typically 0.1 cm^3 ml^{-1}) and moderate surface areas (typically 150–300 m^2 g^{-1}), but considerably lower Brønsted acidity than their ACH exchanged precursors. They retain their interlayer micropore structure at high temperatures that would normally cause non-pillared smectites to collapse (i.e. lose sorbate from the interlayer, causing layers to sit directly on top of each other). The ability to retain microporosity at high temperatures makes pillared smectites attractive, as catalysts or supports, for gas phase reactions where retention of surface area is important.

1.5.4.4   *Commercial availability of clay minerals*
A number of clays with well-defined characteristics can be obtained from commercial sources. Table 1.6 lists some of the more useful examples.

Table 1.6 — Structure and properties of some commercial clay minerals

Commercial name	Supplier	Structure type	Properties/comments
Bentonite	Merck	Montmorillonite (2:1 layer)	Calcium or sodium forms; need exchanging with acidic cations for use as acid catalysts; medium–high cation exchange capacity
K10	Sud chemie	Montmorillonite (2:1 layer)	Acid treated; high surface area, acidic clay; low–medium cation exchange capacity
KSF	Sud chemie	Montmorillonite (2:1 layer)	Acid impregnated, high acidity clay; medium–high cation exchange capacity
Sepiolite	Tolsa SA	2:1 Inv. ribbon	Low acidity; low cation exchange capacity
Kaolin/ china clay	Merck	1:1 Layer	Low acidity; low cation exchange capacity

1.6   ZEOLITES

1.6.1   Introduction

Zeolites are crystalline, microporous aluminosilicates with molecular-sized intracrystalline channels and cages. Intracrystalline pores generally contain cations and water molecules, which have access to the zeolite's external environment, allowing hydration, dehydration and cation exchange. Guest molecules with molecular diameters smaller than zeolite pore apertures can enter the interior of zeolite crystals (intercalation) giving rise to shape and size selective sorption (i.e. molecular sieving).

Modification of zeolites by cation exchange (e.g. to introduce protons, forming solid acids), impregnation (e.g. with platinum group metals), etc., gives rise to a diverse range of physical and chemical properties, allowing zeolites to be adapted for numerous applications. The properties of individual zeolites depend upon: (a) the framework structure (size, shape, tortuosity, interconnectivity of channels and cages); (b) framework composition (concentration and distribution of framework isomorphous substitutions); (c) intracrystalline cations (type, charge, concentration, location); and (d) additional phases (e.g. non-framework oxides, impregnated metals, intercalated organic compounds, etc.).

There are various families of zeolites based on different framework compositions (e.g. aluminosilicates, aluminophosphates, etc.). This section will mainly refer to aluminosilicate zeolites, which are currently the most important group for catalytic applications. Zeolites of the A type are commonly sold as molecular sieves, type A; zeolite X is sold as molecular sieve, type X; and zeolite Y is commonly sold under that name. Other zeolites are available from the manufacturers.

1.6.2 Zeolite framework structures

The primary building blocks for aluminosilicate zeolites are the $[SiO_2]$ and $[AlO_2]^-$ tetrahedra (i.e. formally Si^{4+} or Al^{3+} at the centre of a tetrahedron with O^{2-} at each apex). Tetrahedra are corner shared to form ordered three-dimensional macrostructures. Zeolite framework structures are usually represented by joining the tetrahedral centres (T) with straight lines, ignoring the O atom between the centres.

It is useful to think of zeolite structures as collections of tetrahedra, arranged into secondary building units (SBUs, Fig. 1.9). SBUs may then be assembled in different ways to construct polyhedral cages (Table 1.7, Fig. 1.10) and channels. Polyhedra may be linked via faces (acting like windows) and/or interconnecting channels to give uniform intracrystalline pore networks. Additional cavities and channels may be formed in voids between linked polyhedra. Intracrystalline voidage can be as high as 50 vol% (e.g. zeolite Y). Consequently, the bulk density of zeolites is generally low.

1.6.2.1 Zeolites X, Y (faujasite)

Zeolite Y (synthetic form of the rare mineral faujasite) is one of the most important zeolites manufactured. Its major use is as an additive in powdered detergents where it prevents powder agglomeration and functions as a builder. Zeolite Y is also one of the most important zeolite catalysts produced (e.g. used widely in catalytic cracking). Zeolite X has the same crystalline form but a different Si/Al ratio (see later).

The structure of zeolites X, Y (Fig. 1.11) consists of **sodalite cages** (also called β-cages or 14-hedron(I), Fig. 1.10) interlinked through face-shared hexagonal prisms (8-hedron). A three-dimensional pore system is formed within sodalite cages and hexagonal prisms with pore apertures formed by rings of six tetrahedra (6-T rings).

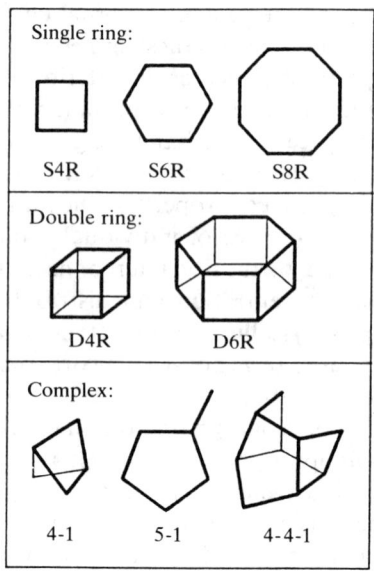

Fig. 1.9 — Zeolite secondary building units.

Table 1.7 — Polyhedra found in zeolite structures.

Polyhedron	n-Hedra	Number of T-ring apertures				Examples
		4	6	8	12	
Cube	6	6	—	—	—	Zeolite A
Hexagonal prism	8	6	2	—	—	Zeolites Y, X
Delta-cage	10	8	—	2	—	Paulingite
Eta-cage	11	6	5	—	—	Offretite
Beta-cage	14(I)	6	8	—	—	Zeolites A, Y, X
14-Hedron(II)	14(II)	9	2	3	—	Offretite
17-Hedron(I)	17(I)	9	5	3	—	Levynite
17-Hedron(II)	17(II)	6	11	—	—	Losod
Gamma-cage	18	12	—	6	—	Paulingite
20-Hedron	20	12	2	6	—	Chabazite
23-Hedron	23	12	5	6	—	Erionite
Alpha-cage	26(I)	12	8	6	—	Zeolite A
Supercage	26(II)	18	4	—	4	Zeolites Y, X

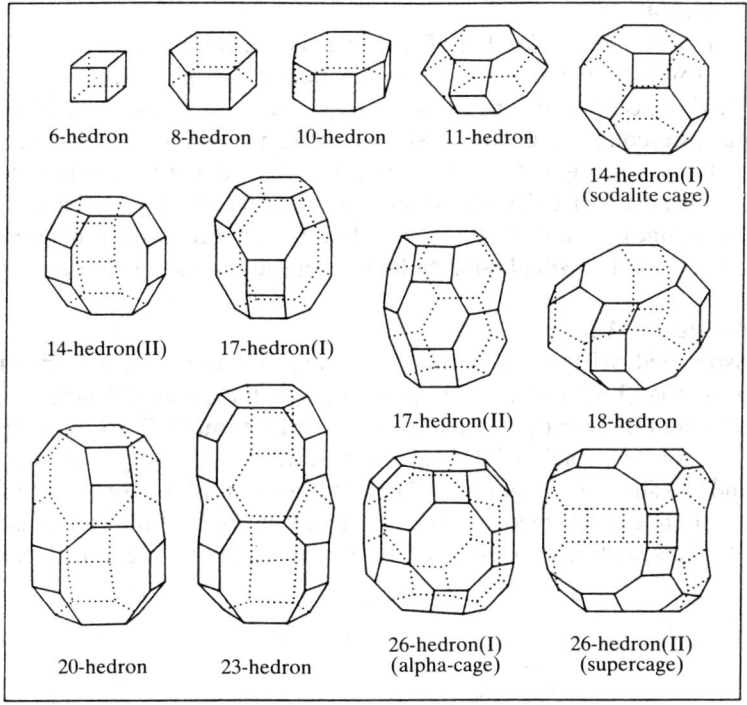

Fig. 1.10 — Zeolite polyhedral cage structures.

Void space outside the sodalite cages and hexagonal prisms forms a second three-dimensional pore system consisting of **supercages** (26-hedron(II), Fig. 1.10) face shared through 12-T ring apertures. The 12-T ring apertures (diameter 740 pm) allow molecules with kinetic diameters (i.e. minimum cross-sectional diameters based on van der Waals surfaces) smaller than 740 pm to enter the supercage pore system.

The 6-T ring apertures (diameter 220 pm) of the sodalite cages are large enough to permit passage of unsolvated cations, but too small to allow molecules to pass through. Hence, molecular species are confined to the supercage pore system and do not enter sodalite cages.

12-T ring apertures are the largest pore entries known for aluminosilicate zeolites, and the supercage of zeolite Y has the largest inscribed diameter (1.18 nm). Zeolites with 12-T ring apertures are termed **large pore zeolites**. There are examples of aluminophosphate zeolites with larger pore aperture (e.g. 18-T [7,8]), but their aluminosilicate analogues have not yet been synthesized successfully. Another commercial zeolite with 12-T ring apertures is large part mordenite. In this case the large voids are in one dimension only, with no large intersections. Thus, the internal spaces are somewhat smaller than in zeolites X, Y and diffusion in also more restricted.

1.6.2.2 *Zeolite A*

The structure of zeolite A, like that of zeolite Y, is based on interlinked sodalite cages (Fig. 1.11). However, the sodalite cages are linked through face-sharing cubes (6-hedron). Void space outside the sodalite cage and cube framework forms a three-dimensional interconnecting pore system consisting of α-cages (26-hedron(I), Fig. 1.10) face shared through 8-T ring apertures. The 8-T ring apertures (diameter 410 pm) permit only molecules with small kinetic diameters (e.g. unbranched alkanes) to enter the α-cages (inscribed diameter 1.14 nm). Zeolites with a maximum of 8-T ring apertures are called **small pore zeolites**, even though cage volumes may be large.

1.6.2.3 **Zeolite ZSM-5**

ZSM-5 (also called MFI) has a framework structure based on cross-linked chains of pentasil sub-units (Fig. 1.11). Void space outside the pentasil framework forms a three-dimensional intersecting system of 10-T ring channels. Two types of 10-T ring channels are formed, each with oval cross-section: (a) straight channels (530 × 560 pm) and (b) sinusoidal channels (510 × 550 pm) in planes normal to the axis of the straight channels. Polyhedral cavities formed at channel intersections are asymmetrical. Zeolites with a maximum of 10-T ring apertures are called **medium pore zeolites**.

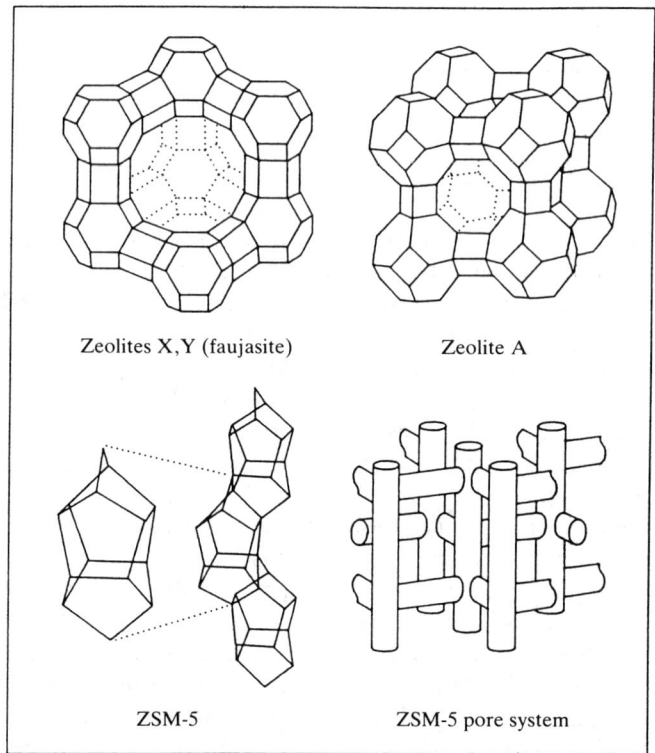

Zeolites X,Y (faujasite)   Zeolite A

ZSM-5   ZSM-5 pore system

Fig. 1.11 — Zeolites X, Y; zeolite A and ZSM-5 structures and ZSM-5 pore system.

ZSM-5 is a commercially important zeolite because the diameter of its channels gives rise to molecular sieving between isomers of commercially important organic molecules (e.g. xylenes, branched alkanes, etc.).

1.6.3 Zeolite composition

The composition of zeolites can vary to a great extent owing to: (a) isomorphous framework substitutions; (b) cation exchange; (c) impregnation and (d) intercalation. However, the general formula for 'silica' zeolites is:

$$C_x[(T_ySi_{1-y})O_2]X_zwM$$

where: x, z and w are the numbers of non-framework cations (C), anions (X) and molecules (M, e.g. water) respectively; $(T_ySi_{1-y})O_2$ is the framework composition and y is the number of tetrahedrally coordinated framework elements other than silicon (T). Natural zeolites usually contain a mixture of cations (e.g. Na^+, K^+, Mg^{2+} and Ca^{2+}). However, these may be ion-exchanged for others. Most of the discussion in this section is restricted to aluminosilicate zeolites, where T is Al, M is water and the concentration of anions (X) is negligible.

The extent of isomorphous substitution in aluminosilicate zeolites is expressed as a molar framework Si/Al ratio, which is generally between 1 and 5. However, high silica zeolites (e.g. ZSM-5) can have much higher Si/Al ratios (e.g. 10–100). The minimum framework Si/Al ratio for an aluminosilicate zeolite is 1 (i.e. strictly alternating Si and Al tetrahedra throughout the zeolite framework). No two adjacent tetrahedral sites are occupied by Al^{3+} (i.e. there are no framework Al–O–Al linkages in zeolites (Lowenstein's rule [9]). In the case of faujasite zeolites, the low silica forms (Si/Al = 1–1.5) are termed zeolite X and higher silica forms (Si/Al > 1.5) are termed Y.

1.6.4 Exchangeable cations

A zeolite framework consisting purely of $[SiO_2]$ tetrahedra would be electrostatically neutral and have no exchangeable cations. However, isomorphous substitution of some of the Si^{4+} with Al^{3+} results in the formation of anionic framework sites $[AlO_2]^-$, which are balanced by non-framework cations to maintain electroneutrality. Charge-balancing cations reside within intracrystalline channels and polyhedral cages. Cations that occupy sites accessible through 8-T ring (or larger) apertures are usually hydrated, while those occupying sites only accessible through 6-T ring (or smaller) apertures are generally unsolvated.

Solvated cations are mobile, and able to undergo rapid solvent-mediated migration between accessible anionic framework sites at ambient temperature. Unsolvated cations (e.g. trapped in polyhedral cages) undergo much slower migration between anionic framework sites until they enter solvent-accessible channels and cages.

Solvated cations from external solutions (M_b) may diffuse into zeolite pore systems with 8-T ring (or larger) apertures and readily undergo ion-exchange with the zeolite's original cations (M_a) (equation (1.10)):

$$M_a\text{-zeolite(s)} + M_b^+(aq) \rightarrow M_b\text{-zeolite(s)} + M_a^+(aq) \tag{1.10}$$

Cations present in zeolite sites accessible through 6-T ring (or less) apertures are less readily exchanged, often requiring heating under reflux.

Cation exchange is a reversible process, the extent of exchange depending largely on the numbers of cations in exchange solution and zeolite. Repeated treatments using fresh solution are required to achieve complete exchange. It is readily carried out in simple apparatus and can therefore be tackled by researchers with little or no prior experience of zeolites.

1.6.5   Acidity

In most catalytic applications, zeolites are used as solid acids or bifunctional supports. Acidity is usually introduced by cation exchange with ammonium or polyvalent cations, followed by calcination (equations (1.11)–(1.14)):

$$Na^+\text{-zeolite(s)} + NH_4^+(aq) \rightarrow NH_4^+\text{-zeolite(s)} + Na^+(aq) \tag{1.11}$$

$$NH_4^+\text{-zeolite(s)} \xrightarrow{>450°C} H^+\text{-zeolite(s)} + NH_3(g) \tag{1.12}$$

$$Na^+\text{-zeolite(s)} + M(H_2O)^{n+}(aq) \rightarrow M(H_2O)^{n+}\text{-zeolite(s)} + Na^+(aq) \tag{1.13}$$

$$M(H_2O)^{n+}\text{-zeolite(s)} \xrightarrow{>450°C} M(OH)^{(n-1)+}, H^+\text{-zeolite(s)} \tag{1.14}$$

Direct cation exchange with mineral acids is generally avoided because of acid leaching of zeolite framework alumina (dealumination). However, dealumination is less facile with high-silica zeolites, where direct proton exchange using acids is possible at ambient temperature.

Protons within zeolites are generally associated with anionic framework sites, giving rise to readily ionizable hydroxyl groups with strong Brønsted acidity. Calcination leads to formation of Lewis acid sites via dehydroxylation of two Brønsted acid sites (Fig. 1.12). Lewis acid sites are initially oxygen-deficient sites, which are unstable and undergo dealumination to form non-framework alumina (NFA) with Lewis acidity.

Acid site strength is influenced by the distribution of aluminium throughout the zeolite framework. Each $[AlO_2]^-$ site has four $[SiO_2]$ neighbours. However, the next nearest neighbours (NNN) can be either $[AlO_2]^-$ or $[SiO_2]$. $[AlO_2]^-$ sites with fewer $[AlO_2]^-$ next nearest neighbours are more strongly acidic (i.e. 0-NNN > 1-NNN > 2-NNN, etc.). Consequently, the fewer total $[AlO_2]^-$ sites (i.e. the higher the framework Si/Al ratio), the stronger the acid site strength. Thus, Hzeolite Y is more acidic than Hzeolite X. The number and strength of acid sites are important factors in catalytic applications. The framework Si/Al ratio can be controlled during synthesis or subsequent modifications (e.g. dealumination, silicon reinsertion, etc.) of zeolites.

1.6.6   Dealumination

The process of removing aluminium from zeolite framework sites is called dealumination. Dealumination decreases the framework Si/Al ratio, thereby: (a) lowering cation exchange capacity; (b) decreasing total number of acid sites; (c) increasing mean acid

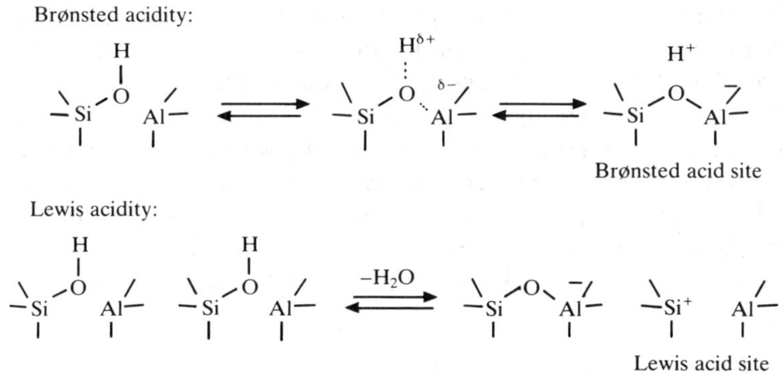

Brønsted acidity:

Brønsted acid site

Lewis acidity:

Lewis acid site

Fig. 1.12 — Brønsted and Lewis acidity in zeolites.

site strength and (d) increasing hydrothermal stability. Dealumination may be achieved using various techniques including steaming, acid leaching, and various chemical treatments (e.g. silicon tetrachloride and fluoro compounds). These techniques would not normally be carried out by organic chemists, but samples of zeolites so treated can be obtained from zeolite producers.

1.6.6.1 Steaming

Steaming zeolites (e.g. at 400–800°C in 50–100% steam for 1–24 h) leads to removal of aluminium from framework sites and the production of non-framework alumina. Vacant framework sites pick up protons from water to form weakly acidic silanol groups. Steaming conditions vary widely, depending upon the extent of dealumination required and the properties of the starting zeolite.

Under steaming conditions, framework silicon migrates to fill tetrahedral vacancies left after dealumination (silicon re-dispersion). Migrating silicon originates from regions within zeolite crystals (e.g. neighbouring sodalite cages in zeolite Y), which collapse to form supermicropore (1–2 nm) and mesopore (2–50 nm) defects.

Non-framework alumina takes on a variety of forms, depending upon the types of exchangeable cations present, subsequent heat treatments, etc. The initial alumina species formed is likely to be $Al(OH)_3$, which reacts with steam to produce a mixture of hydroxides and hydrated Al^{3+}. Non-framework alumina is initially located within zeolite channels and cages. Extensive polymerization leads to some pore occlusion and migration of some non-framework alumina to external crystal surfaces. The properties of non-framework alumina can have a significant effect on the catalytic properties of zeolites, for example by providing non-selective acid sites outside the confines of the zeolite pore structure. Consequently, it is often desirable to remove non-framework alumina, for example by washing with chelating agents (e.g. EDTA).

1.6.6.2 Silicon reinsertion

Treatment of zeolite crystals with silicon tetrachloride ($SiCl_4$) vapour in an inert carrier gas (e.g. nitrogen) at temperatures > 450°C leads to consecutive dealumination and silicon resinsertion into the framework vacancies produced. Aluminium released from the framework forms volatile aluminium trichloride ($AlCl_3$), which is removed by the carrier gas rather than forming non-framework alumina. However, $AlCl_3$ can react with sodium cations to form non-volatile $NaAlCl_4$, which causes partial pore occlusion, limiting further access of $SiCl_4$ to the zeolite's interior. $NaAlCl_4$ is soluble in water and may be removed by washing. High-silica zeolites can be formed from sodium exchanged zeolites by repeated $SiCl_4$ and water wash treatments. However, $SiCl_4$ treatment of ammonium exchanged zeolites allows more extensive dealumination and silicon reinsertion in a single step, owing to the formation of volatile product (NH_3, HCl, $AlCl_3$) rather than $NaAlCl_4$.

1.6.6.3   *Fluorine treatments*

Dealumination and silicon reinsertion can be achieved by treatment of zeolite with fluorine compounds (e.g. aqueous ammonium hexafluorosilicate, $(NH_4)_2SiF_6$ solution). Treatment with aqueous $(NH_4)_2SiF_6$ allows orderly dealumination and silicon reinsertion to be achieved under mild conditions, without damaging the zeolite crystals (equation (1.15)):

$$NH_4^+[(AlO_2)(SiO_2)_x](s) + (NH_4)_2SiF_6(aq) \rightarrow [(SiO_2)_{1+x}](s) +$$
$$+ (NH_4)_2AlF_5(aq) + NH_4F(aq) \qquad (1.15)$$

Solution dealuminatiion techniques are not as effective for treating small pore zeolites, where reagents have limited access through zeolite pore apertures.

1.6.7   **Sorption and molecular sieving**

Under ambient conditions, zeolites contain water (typically up to 10 wt%) within their channels and cages. Water can be removed by heating in a dry gas (e.g. nitrogen) or under reduced pressure. After dehydration the zeolite structure remains intact, leaving the intracrystalline channels and cages available for the sorption of gases, vapours or solutions. Zeolite pore apertures impose steric limits on the size and shape of molecules that may be sorbed; thus zeolites behave as molecular sieves. Pore aperture dimensions depend primarily upon framework topology, and typical pore diameters are shown in Table 1.8.

Intracrystalline cations also influence the effective pore aperture size of zeolites. This is most pronounced with small pore zeolites (e.g. zeolite A) where cations markedly decrease the mean pore aperture diameter. Commercially available Linde 3A, 4A and 5A molecular sieves are based on K^+, Na^+ and Ca^{2+} exchanged zeolite A respectively, the mean free dimension of their pore apertures increasing with decreasing cation radius and/or decreasing cation concentration. Calcium exchanged zeolite A (Linde 5A molecular sieve) is used to sorb unbranched alkanes selectively from hydrocarbon mixtures. Potassium exchanged zeolite A (Linde 3A molecular sieve) is used to sorb water selectively from hydrocarbon mixtures.

Exchangeable cations and framework substitutions give rise to electronic field gradients within zeolite channels and cages. The electronic field gradients are 'felt' by all sorbate species because of the short distances between zeolite walls in channels and

Table 1.8 — Maximum free diameters of zeolite framework T-ring apertures.

No. of tetrahedra in ring (T-ring)	Maximum free aperture diameter (pm)	
	(Breck)[a]	(Barrer)[a]
4	160	115
5	150	196
6	280	280
8	430	450
10	630	630
12	800	800
18	1500	—

[a]See bibliography on zeolites.

cages. Consequently, sorbate species have properties, for example degree of dissociation, conductivity, etc., which differ from the species in their normal liquid state. Polar molecules (e.g. water and ammonia) interact more strongly with intracrystalline electronic field gradients than non-polar molecules (e.g. hydrocarbons). Zeolites with extensive framework substitution (i.e. low-framework Si/Al ratio) have stronger electronic fields and therefore preferentially sorb polar molecules. Conversely, non-polar molecules are more easily sorbed by high-silica zeolites (i.e. with high-framework Si/Al ratio).

1.6.8 Catalytic properties of zeolites
The combination of high internal surface area, strong acid sites, selective sorption and molecular sieving properties makes zeolites among the most useful and versatile heterogeneous catalysts. High internal surface area and acidity give rise to high activity, while selective sorption and molecular sieving result in high reaction selectivity. Reaction selectivity may be diffusion controlled (reactant or product selective) or may be geometrically controlled (transition state selective) (Fig. 1.13) [10]. Reactant selectivity occurs when the zeolite preferentially sorbs some reactants from a mixture, discriminating by size, shape or polarity. Product selectivity applies when only products small enough to leave through zeolite pores are found in reaction product mixtures (even though larger product molecules may be formed as intermediates within the zeolite cages). Transition state selectivity occurs when zeolite cavities impose geometrical constraints on the size or shape of transition states.

Zeolites may be used as bifunctional supports for other catalytically active species (e.g. platinum group metals), where the zeolite provides acidic sites and the supported metal provides hydrogenation activity. The activity of supported metals is related to their surface area (i.e. number of active sites). Highly dispersed metals (e.g. platinum) can be loaded into zeolites (e.g. zeolite Y) by cation exchange (e.g. as aqueous

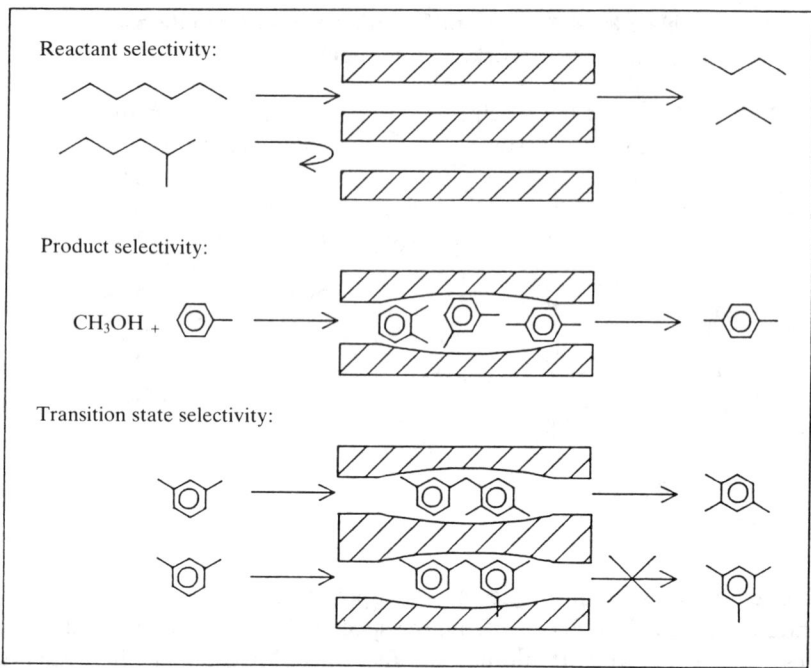

Fig. 1.13 — Examples of reactant, product and transition state selectivity in zeolite-catalysed reactions (based on information in ref. [10]).

$Pt(NH_3)_4^{2+}$), followed by reduction in hydrogen (equations (1.16) and (1.17)):

$$Na - Y(s) + Pt(NH_3)_4^{2+}(aq) \rightarrow Pt(NH_3)_4^{2+} - Y(s) + 2\,Na^+(aq) \qquad (1.16)$$

$$Pt(NH_3)_4^{2+} - Y(s) + H_2(g) \rightarrow Pt(0),2H^+ - Y(s) + 4\,NH_3(g) \qquad (1.17)$$

Cation exchange ensures that platinum ions are dispersed throughout the zeolite structure, occupying cation exchange sites within zeolite Y supercages. Hydrogen reduction leads to the formation of small platinum clusters within zeolite Y super-cages (maximum diameter 1.18 nm) and generation of protons as exchangeable cations (providing acid sites).

1.7   CONCLUSION

Although giving only an outline of the structure and properties of inorganic solids useful as supports or catalysts, this chapter should provide the basis upon which the organic chemist can understand the role played by such solids and help in the design of new reactions. For further detail, the publications cited in the general bibliography provide excellent sources of information.

Organic chemists would not generally attempt the synthesis of most of the solids described herein, but would purchase them from commercial sources. However, simple treatments such as on-exchange, calcination or deposition of a reagent onto a solid require no special expertise and can be undertaken with confidence. In many cases the small step into unfamiliar territory will be handsomely rewarded through better reactions.

BIBLIOGRAPHY

There are numerous reviews covering the types of inorganic solids used in heterogeneous catalysts. The following list is far from exhaustive, but provides a starting place for further reading.

Heterogeneous catalysis and catalyst design
D. L. Trimm, *Design of Industrial Catalysts, Chem. Eng. Monogr. 11*, Elsevier, Amsterdam, 1980.

T. Seiyama and K. Tanabe, *New Horizons in Catalysis*, Studies in Surface Science and Catalysis 7B, Elsevier, Amsterdam, 1980.

Kirk Othmer Encyclopaedia of Chemical Technology, 3rd edn, Wiley-Interscience, New York, 1980.

M. S. Whittingham (ed.), *Intercalation Chemistry*, Academic Press, London, 1982.

B. E. Leach (ed.), *Applied Industrial Chemistry*, Vol. 2, Academic Press, London, 1983.

G. Poncelet and P. Grange (eds), *Preparation of Catalysts*, Vol. 3, Elsevier, Amsterdam, 1985.

R. Setton (ed.), *Chemical Reactions in Organic and Inorganic Constrained Systems, NATO ASI Series C*, Vol. 165, Reidel, Dordrecht, 1986.

A. B. Stiles, *Catalyst Supports and Supported Catalysts, Theoretical and Applied Concepts*, Butterworth, Boston, 1987.

J. F. Le Page *et al.*, *Applied Heterogeneous Catalysis: Design, Manufacture, Use of Solid Catalysts*, Editions Technip., Paris, 1987.

M. J. Phillips and M. Ternan (eds), *Catalysis: Theory to Practice, Proc. 9th Int. Congr. Catal.*, Vols 1–4, The Chemical Institute of Canada, Ontario, 1988.

J. T. Richardson, *Principles of Catalyst Development, Fundamentals and Applied Catalysis Series*, Plenum, New York, 1989.

E. G. Christoffel (ed.), *Laboratory Studies of Heterogeneous Catalytic Processes, Studies in Surface Science and Catalysis*, **42**, Elsevier, Amsterdam, 1989.

T. Inui (ed.), *Successful Design of Catalysts, Studies in Surface Science and Catalysis*, **44**, Elsevier, Amsterdam, 1989.

K. Tanabe *et al.* (eds), *New Solid Acids and Bases, Their Catalytic Properties, Studies in Surface Science and Catalysis*, **51**, Elsevier, Amsterdam, 1989.

M. Misono, Y. Moro-oka and S. Kimura (eds), *Future Opportunities in Catalytic and Separation Technology, Studies in Surface Science and Catalysis*, **54**, Elsevier, Amsterdam, 1989.

Silicas
W. Eital, *The Physics and Chemistry of Silicates*, University of Chicago, Chicago, 1954.

R. K. Iler, *The Chemistry of Silica*, John Wiley & Sons, New York, 1979.

K. K. Unger, *Porous Silica*, Elsevier, Amsterdam, 1979.

Aluminas

W. H. Gitzen, *Alumina as a Ceramic Material*, The American Ceramic Soc. *Special Publication No. 4*, Columbus, Ohio, 1970.

E. Dorre and H. Hubner, *Alumina. Processing and Properties*, Springer-Verlag, Weinheim, 1984.

Condea Chemie GmbH, Technical Bulletin: *Pural and Dispural High Purity Catalytic Aluminas*.

Degussa, Technical Bulletin No. 72: *Aerosil, Aluminium Oxide C and Titanium Dioxide P25 for Catalysts*.

Kaiser Chemicals, Technical Bulletin: *Alumina Products and Technology*.

Carbons

R. L. Bond, *Porous Carbon Solids*, Academic Press, London, 1967.

M. Smisek and S. Cerny, *Active Carbon. Manufacture, Properties and Applications, Topics in Inorganic and General Chemistry, Monogr 12*, Elsevier, Amsterdam, 1970.

R. C. Bansal, J. B. Donnet and F. Stoeckli, *Active Carbon*, Marcel Dekker, New York, 1988.

Sutcliffe Speakman Carbons Ltd, Technical Data.

Clay minerals

R. E. Grim, *Applied Clay Mineralogy*, McGraw-Hill, London, 1962.

W. A. Deer, R. A. Howie and J. Zussman, *Rock Forming Minerals: Sheet Silicates*, Vol. 3, John Wiley & Sons, New York, 1971.

B. K. G. Theng, *The Chemistry of Clay-Organic Reactions*, John Wiley & Sons, New York, 1974.

H. van Olphen and J. J. Fripiat, *Data Handbook for Clay Minerals and other Non-Metallic Minerals*, Pergamon, Oxford, 1979.

G. W. Brindley and G. Brown, *Crystal Structures of Clay Minerals and their X-Ray Identification*, Miner. Soc. Monogr. No. 5., London, 1980.

T. J. Pinnavaia and R. A. Schoonheydt (eds), *Metal Complex Catalysts in Intracrystalline Environment*, J. Molec. Catal., 1982, **27**.

A. C. D. Newman (ed.), *Chemistry of Clays and Clay Minerals*, Miner. Soc. Monogr. No. 6., London, 1987.

I. V. Mitchell (ed.), *Pillared Layered Structures: Current Trends and Applications*, Elsevier, London, 1990.

Zeolites

D. W. Breck, *Zeolite Molecular Sieves*, John Wiley & Sons, New York, 1974.

J. A. Rabo, *Zeolite Chemistry and Catalysis*, ACS Monograph 171, ACS, Washington, 1976.

R. M. Barrer, *Zeolites and Clay Minerals as Sorbents and Molecular Sieves*, Academic Press, London, 1978.

R. P. Townsend (ed.), *Properties and Applications of Zeolites, Chemical Society Special Publication No. 33*, London, 1980.

R. M. Barrer, *Hydrothermal Chemistry of Zeolites*, Academic Press, London, 1982.

G. D. Stucky and F. G. Dwyer (eds), *Intrazeolite Chemistry, ACS Symposium Series 218*, Washington, 1983.

J. Dwyer and A. Dyer, *Zeolites for Industry, Chemistry and Industry*, 1984, pp. 237–269.

W. M. Meier and D. H. Olson (eds), *Atlas of Zeolite Structure Types*, 2nd Edition, Butterworths, London, 1987.

A. Dyer, *An Introduction to Zeolite Molecular Sieves*, John Wiley & Sons, New York, 1988.

S. Kaliaguine (ed.), *Keynotes in Energy Related Catalysis, Studies in Surface Science and Catalysis*, **35**, Elsevier, Amsterdam, 1988.

H. G. Karge and J. Weitkamp (ed.), *Zeolites as Catalysts, Sorbents and Detergent Builders, Applications and Innovations, Studies in Surface Science and Catalysis*, **46**, Elsevier, Amsterdam, 1989.

P. A. Jacobs and R. A. van Santen (eds), *Zeolites: Facts, Figures, Future, Studies in Surface Science and Catalysis*, **49**, Elsevier, Amsterdam, 1989.

J. Klinowski and P. J. Barrie (eds), *Recent Advances in Zeolite Science, Studies in Surface Science and Catalysis*, **52**, Elsevier, Amsterdam, 1989.

J. Scherzer, *Octane Enhancing Zeolite FCC Catalysts: Scientific and Technical Aspects, Chemical Industries Series*, **42**, Marcel Dekker, New York, 1990.

REFERENCES

[1] P. B. Venuto and E. T. Habib, *Fluid Catalytic Cracking with Zeolite Catalysts*, Marcel Dekker, New York, 1979.

[2] G. Johansson, *Acta Chem. Scand.*, 1960, **14**, 771.

[3] R. Burch (ed.), *Pillared Clays, Catalysis Today*, Vol. 2, 1988.

[4] D. L. Trimm, *Royal Soc. Chem. Special. Periodical Reports*, Catalysis Vol. 4, 1980, pp. 210–235.

[5] K. Tanabe, *Solid Acids and Bases, Their Catalytic Properties*, Academic Press, New York, 1970.

[6] J. Shabtai, N. Frydman and R. Lazar, *Proc. 6th Int. Congr. Catal.*, Paper B5, pp. 1-7, *The Chemical Soc.*, London, 1976.

[7] M. E. Davies, C. Saldarriaga, C. Montes, J. Garces and C. Crowder, *Zeolites*, 1988, **8**, 362. For a recent review of large and extra-large pore molecular sieves, see M. E. Davies, *Chem. 2nd (London)*, 1992, 137.

[9] W. Lowenstein, *Amer. Mineral.*, 1954, **39**, 92.

[10] S. M. Csicsery, *Zeolites*, 1984, **4**, 202.

2

Organic supports — an overview

J. M. Maud

2.1 INTRODUCTION

The use of organic polymers for the support of substrates, reagents and catalysts, together with their subsequent application in synthetic organic chemistry, is illustrated in subsequent chapters of this monograph. This chapter describes the polymeric organic supports that are suitable for such application. Although in general, organic chemists will not wish to prepare the basic polymeric supports, some knowledge of the processes of formation of the commercial polymers is essential if the properties of the polymers are to be understood. In addition, it is recognized that the combination of the support and the supported reagent or substrate must be viewed as a whole rather than as if the support were a simple immobilizing agent or one which would merely serve to confer insolubility on the desired substrate or reagent. (Nevertheless, for convenience, functionalized polymers may still be drawn in terms of their most significant functional group attached to an encircled upper case letter 'P'.) Accordingly, where possible, some of the aspects of the chemical and physicochemical nature of the polymeric supports that influence substrate or reagent reactivity, and that ultimately determine, limit or enhance their applicability to synthetic organic chemistry, are discussed. Within this framework, the account aims to be representative rather than exhaustive. A number of reviews and monographs [1–5] provide more or less detailed accounts of the subject.

2.2 THE PREPARATION OF ORGANIC POLYMERIC SUPPORTS

The history of polymer supports in synthetic organic chemistry is dominated by functionalized cross-linked polystyrene in its various forms. This is presumably because of: (a) its wide availability, both in its virgin form and when functionalized, in the form of commercial ion-exchange resins which permit the support of a wide range of ionic substrates and reagents; and (b) its notably successful use by the early workers in the field of solid-phase peptide synthesis. Accordingly, this account concentrates on styrene or styrene-like polymers. However, many of the general principles that are discussed are applicable, immediately or with little modification, to other polymeric supports, particularly those of the polyvinyl type. Important examples of polymeric supports not based on polystyrene are discussed, if only briefly, towards the end of the chapter.

2.2.1 Polystyrene resins

Addition polymerization of vinyl monomers (1), with use of cationic, free radical, anionic or Ziegler–Natta initiators, can give linear high molecular weight polymers (2) (equation (2.1)). Thus, vinylbenzene (3), or styrene, gives polystyrene (4). Polymerization of styrene in the presence of a small amount of a divinyl compound such as 1,4-divinylbenzene (DVB) (5) gives an insoluble styrene polymer in which DVB serves to cross-link different chains at different points.

$$\hspace{10cm} (2.1)$$

2.2.1.1 Suspension polymerization

Free radical initiated polymerization may be carried out under conditions of so-called *suspension polymerization* [5–8], when the insoluble cross-linked polymer is obtained in the form of small spherical particles often described as *beads*. These beads represent an extremely useful physical form of the cross-linked polymer and permit ease of handling in any subsequent chemical modification of the polymeric support and in its actual use in synthetic organic chemistry. It should be noted, however, that the actual performance of suspension polymerization is a far from trivial process. The procedure involves, in the case of the preparation of polystyrene, the vigorous stirring of the lipophilic monomer styrene in an aqueous solution, and the course of the preparation may depend critically upon actual reactor and stirrer design. The control of particle or bead size in particular may require careful attention to these details. The polymerization medium may also contain dispersing agents or surfactants in order to lower interfacial tension, and these agents also have a great influence on particle size. For polymer beads of a size (100–500 μm) which is normally employed for polymer-supported reactions, the simplest way to guarantee a well-defined particle size is to use a sieving process [9].

2.2.1.2   Gel-type resins

The structure and morphology of the cross-linked polymer depend critically upon the relative amount of cross-linking agent incorporated into the polymer network. If the polymer is suspended in a compatible solvent then it can expand, or swell, forming a so-called *gel*. Within the gel, the polymeric network limits the extent to which the material can *flow* while the solvent provides molecular mobility, including that for solutes. If the degree of cross-linking is small (< 2% of the backbone residues involved in cross-linking, say) then the gel network largely comprises solvent with only a small part of the total mass being made up of polymer. Such gels are truly porous. However, as the degree of cross-linking is increased, true porosity becomes limited. With incompatible solvents, no expansion or swelling occurs, even for lightly cross-linked polymers, and there is no true porosity. Gel-type resins of the polystyrene –DVB type date back some 40 years or more, and they have found wide commercial use since that time, as ion-exchange resins for example.

2.2.1.3   Macroporous resins

If the copolymerization of the vinyl monomer with the divinyl cross-linker is carried out in the presence of an inert diluent, called a porogen in this context, such as toluene in the case of a styrene–DVB polymerization, then a so-called *macroporous* resin may result. Solvation of the polymer during its growth permits the formation of a much more extended network, and even when all of the solvent has been removed, significant porosity remains, even in the dry state. The porosity depends critically upon a number of factors, particularly the degree of cross-linking and the actual nature of the inert solvent or diluent used in the polymerization. A macroporous polymer remains a gel, and may swell in the presence of compatible solvents, although the degree of swelling may be limited by high degrees of cross-linking. However, if the polymerization is carried out to give a high degree of cross-linking then a particularly rigid polymer may result, retaining its shape completely when solvent is removed. Such a polymer is often described as a *macroreticular* resin. Such resins have large discreet pores and it has been suggested [10,11] that this type of support offers advantage in polymer-supported reactions. If the polymerization is carried out in the presence of a non-solvating porogen, or in the presence of an inert polymeric porogen, then it is possible to obtain macroporous resins that contain only relatively large pores. Much more extensive discussions [12,13] of the factors that affect polymer structure and morphology are available.

2.2.1.4   Polymer homogeneity

The use of divinylbenzene as a cross-linking agent in the preparation of styrene-based polymers inevitably leads to inhomogeneity in the density of the cross-links. The problem arises because the first vinyl group in divinylbenzene reacts some seven times faster than that in styrene, to leave the second group attached to a polymer chain and with a reactivity comparable to that of styrene itself [14]. This leads at the beginning of the polymerization to regions containing a high density of cross-links, with subsequent polymerization producing regions that are less highly cross-linked. To some extent the problem may be alleviated by use of divinyl cross-linking agents, such as ethylene glycol dimethacrylate, in which the reactivities of the two double bonds are identical and independent.

The situation is further complicated in the commercial production of styrene –DVB polymers. Commercial divinylbenzene comprises a mixture of para- and meta-divinylbenzenes together with some ethylvinylbenzenes, vinyltoluene and traces of other impurities. The first double bond to react in meta-divinylbenzene does so at a rate which is approximately four times as fast as that in styrene. Thus, laboratory-prepared styrene–DVB polymers, using pure para-divinylbenzene as the cross-linking agent, cannot necessarily be expected to reproduce apparently similar commercial materials exactly.

2.2.2   Functionalized polystyrene resins

A desired functionality may be introduced into a polymeric resin during the actual formation of the resin, by polymerization or copolymerization of a functionalized monomer, or it may be introduced by chemical modification of preformed polymer. Both methods have found wide application.

2.2.2.1   *Polymerization of functionalized monomers*

For many polymers, such as poly-4-vinylpyridine, polymerization of a functionalized monomer represents the only possible route. For others, synthesis by this route may represent a method of choice. Homopolymerization of a monomer **(6)** can produce a polymer **(7)** in which every repeat unit, apart from those involved in cross-linking, carries the desired functional group (equation (2.2)).

$$(2.2)$$

                    (6)   (7)

Alternatively, copolymerization with an inert monomer, such as styrene, permits the production of a polymer **(8)** with a known and well defined stoichiometry which depends upon the molar ratio of the comonomers used (equation (2.3)).

$$(2.3)$$

                              (8)

This type of approach has been utilized in the production of polystyrene polymers modified with phosphines [15,16], pyridine [17,18] units and ring halogens [15,19]. In the latter case the position of halogenation is unequivocal, being determined by the monomer constitution, in contrast to that occurring on direct halogenation of an unfunctionalized polymer.

The copolymerization of monomer mixtures is not without some problems. Some functionalized monomers may not be commercially available and may require unacceptably expensive synthetic routes for their preparation. The use of comonomers with different reactivities creates particular problems. Attempts to polymerize two monomers of widely different reactivities result in sequences of repeat units that are far from being random. Although this problem can be surmounted at an industrial level by careful control of monomer concentrations, reproduction in the laboratory may prove to be technically difficult. Fortunately, in many cases, such as the copolymerization of styrene and 4-vinylpyridine [20,21], reactivities are sufficiently similar for this not to be a problem.

2.2.2.2   *Chemical modification of cross-linked polymers: general considerations*

In many ways the chemical modification of preformed polymers represents a more attractive route to functionalized supports than the copolymerization of monomers, particularly for those not versed in the methods of polymer preparation. Polystyrene beads are readily available and, in principle at least, a wide variety of synthetic methods exists for their modification. However, a number of problems may occur. Not all reagents that work successfully in solution are necessarily compatible with a polymer matrix. Aqueous reagents, for example, might not penetrate a lipophilic polystyrene network, so that any modification might be confined to the support surface. A wide variety of reactions involve solid reagents, and clearly these cannot be expected to be effective. Also, many reactions are accompanied by side reactions; when small molecules are the substrates, byproducts are easily removed, but following chemical modification of a polymer, any unwanted functionality remains bound to the polymer network. Thus, electrophilic reaction at polystyrene is expected to give a mixture of para- and meta-substitution when, ideally, only the former might be desired. Finally, it may not be possible accurately to control the degree of chemical modification or, in other words, the loading of the resin. In some cases this may be critical for successful and reproducible application of the functionalized resin.

Chemical modification of a polymeric resin affects structure and morphology and it may be difficult to control or even characterize these changes. Certainly some reduction in pore volume can be expected, and when modification involves the introduction of a highly polar group, such as quaternary ammonium or sulphonate, much greater changes are likely to occur. Notwithstanding these reservations, the chemical modification of preformed polymers has proved to be successful in an enormous number of cases, and it can readily be tackled by organic chemists not expert in polymer synthesis.

The two most frequently used methods of initial functionalization of virgin polystyrene resins are chloromethylation and lithiation. They are complementary in that the first provides an electrophilic centre within the polymer (9), thus permitting

subsequent reaction with a nucleophile to give polymer **(10)** (equation (2.4)), whereas
the second introduces a nucleophilic site, in polymer **(11)**, at which an electrophile can
subsequently react, to give **(12)** (equation (2.5))

$$(2.4)$$

                    (9)   (10)

$$(2.5)$$

                    (11)   (12)

In principle, at least, the chloromethyl group in **(9)** might serve both purposes, with
nucleophilic reactivity being provided after conversion to an organolithium or
Grignard reagent. Until recently, all attempts to effect such a conversion appear
to have been unsuccessful. However, a recent report [2] suggests that conversion to
the Grignard reagent may be carried out using the magnesium–anthracene–THF
complex, with subsequent conversion of the Grignard reagent **(13)** to the carboxylic
acid functionalized polymer **(14)** (equation (2.6)).

$$(2.6)$$

                    (13)   (14)

2.2.2.3   *Chloromethylation and subsequent modification*

Chloromethylation of polystyrene may be carried out using chloromethyl methyl ether and a Lewis acid catalyst such as zinc chloride [22], boron trifluoride etherate [23], stannic chloride [24], etc. For low degrees of functionalization, the first two of these catalysts have been particularly recommended. At a practical level, the extreme hazard of chloromethyl methyl ether, a potent carcinogen, should be recognized, together with uncertainty in the polymer structure, both in terms of (a) the position of the newly introduced group, i.e. whether *meta* or *para*, and (b) the additional degree of cross-linked polymer **(15)** which can occur via alkylation of a phenyl group on one chain by a chloromethyl group on another. Problems associated with this cross-linking or methylene bridging may be avoided via use of an alternative route to chloromethylpolystyrene that involves hypochlorite oxidation [25] of poly-*p*-methyl-styrene **(16)**, or one of its copolymers (equation (2.7)).

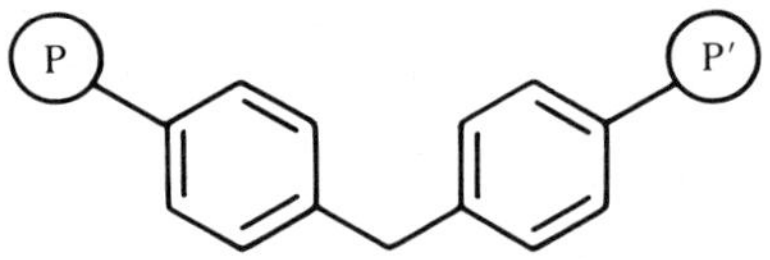

(15)

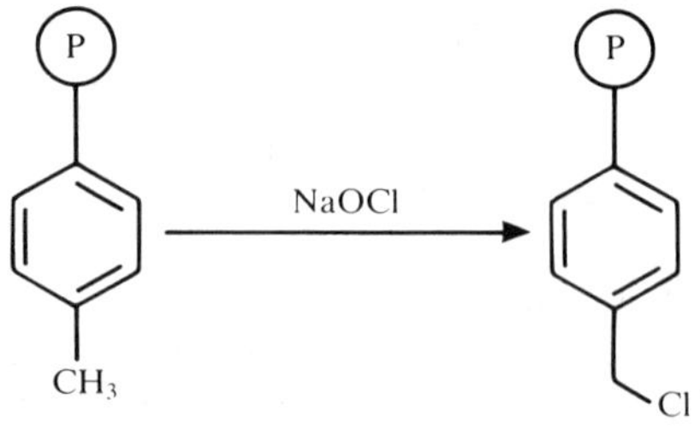

(2.7)

(16)

The chloromethyl group readily lends itself to a wide variety of subsequent modification processes. Reactions with amines [26] are, of course, the principal route to quaternary ammonium functionalized anion-exchange resins **(17)** (equation (2.8)).

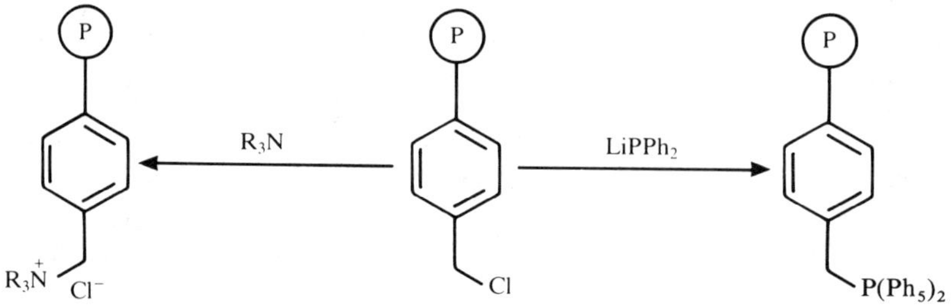

(2.8)

(17)   **(18)**

Any one of a wide variety of other nucleophiles may be caused to react, an important example being the diphenylphosphine anion [27,28] which gives the polymer-supported phosphine **(18)** (equation (2.8)). Moffat oxidation [29] permits conversion to formyl-substituted polymer **(19)**, which can subsequently be oxidized to the carboxylic acid **(20)** (equation (2.9)).

$$(2.9)$$

2.2.2.4 *Lithiation and subsequent modification*
Direct lithiation of cross-linked polystyrene may be achieved using *n*-butyllithium-tetramethylethylenediamine (TMEDA) in heptane [27,30]. Subsequent reactions with carbon dioxide [30] or diphenylchlorophosphine [27] afford the important carboxylic acid **(20)** and triaylphosphine **(21)** functionalized polymers respectively (equation (2.10)).

$$(2.10)$$

Highly functionalized polymers are not usually accessible by this route, loadings being limited to 1–2 mmol g^{-1}. Better results may be obtained indirectly, via lithium halogen exchange. Although iodinated resins [31] have been utilized, most work has been done in connection with brominated polymers. Reaction between cross-linked polystyrene and bromine in the presence of thallium triacetate [27,32,33] permits the incorporation of halogen at approximately 4 mmol g^{-1} resin, representing nearly 100% ring monosubstitution. The brominated polymers undergo relatively efficient lithium bromine exchange, although more than one treatment with *n*-butyllithium may be required.

The attractiveness of the lithiated polymer for the functionalization of polystyrene lies in the very wide number of electrophiles with which it may subsequently be reacted. Nevertheless, there are difficulties, as indicated above, and poly(2-vinylthiophene) [34] **(22)** has been suggested as an alternative support to polystyrene. Butyllithium effects lithiation smoothly at the alpha position to give **(23)** and subsequent reaction with an electrophile gives the functionalized polymer **(24)** (equation (2.11)). In spite of the attractiveness of this approach, the method is likely to be limited by the enhanced sensitivity to oxidation of the thiophene residue.

$$(2.11)$$

(22)   (23)   (24)

2.2.2.5   *Other important methods of polystyrene functionalization*
Many other routes to functionalized polystyrenes have been suggested or described, and the subject has been extensively reviewed [35,36]. Some important examples are described below.

Treatment [31] of cross-linked polystyrene with sulphuric acid gives the sulphonated polymers **(25)**, which are commercially important as acidic ion-exchange resins.

(25)

Acetylation followed by oxidation provides [37] an alternative method for the production of carboxylic acid functionalized resins **(20)** (equation (2.12). The acid is readily converted into the acid chloride **(26)**, which can be reacted with a wide variety

$$(2.12)$$

(20)   (26)

of nucleophiles. Reaction [38–40] between bromine functionalized polymer (27) and lithium diphenylphosphide provides another, and important, route to triarylphosphine functionalized polymeric resins (21) (equation (2.13)).

$$(2.13)$$

(27)   (21)

2.2.3   Other polymeric supports

The requirement for chemical inertness under the conditions of a polymer-supported reaction severely limits the number and types of polymer that can be used for the support. Even polystyrene cannot be used in the presence of powerful electrophiles or under vigorously oxidizing conditions.

Polymers that have been suggested as alternatives to polystyrene include those based on methacrylate esters (28) and amides (29) [41,42]. Monomers are readily available, or easily prepared, and polymerization proceeds easily. Again, some limitation to the type of reaction that can be supported must apply, in this case because of the possibility of hydrolysis. This problem also militates against the widespread use of polyesters, polyamides, polycarbonates and polyurethanes, although Nylons have found some application, as their N-chloro derivatives (30) [43–45], as oxidizing agents.

(28)   (29)

(30)

Polyaromatic ethers (31), etheretherketones (32) and etherethersulphones [46] (33) would appear to be attractive candidates for use as polymeric supports in view of their high thermal stability. The polyether (34) can be brominated to give the bromomethyl polymer (**35**) (equation (2.14)) [47], which has been used as a soluble support although not, apparently, for polymer-supported reactions.

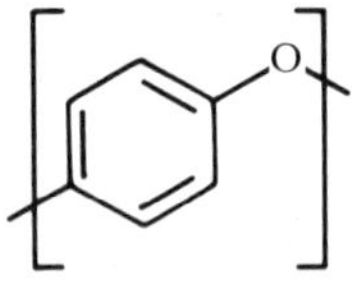

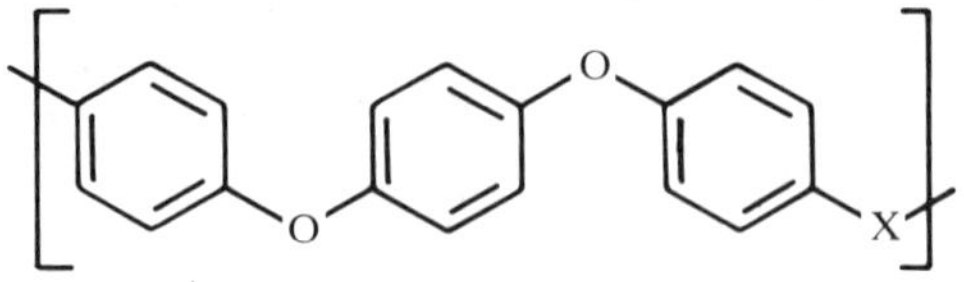

(31)

$$X = \overset{O}{\underset{}{\bigwedge}} \quad (32)$$

$$X = -SO_2- \quad (33)$$

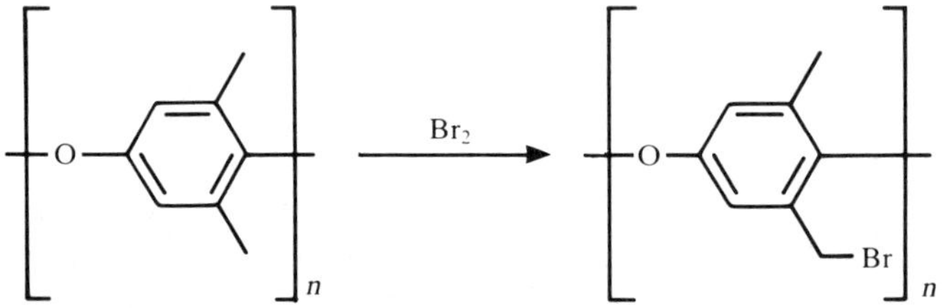

(2.14)

(34)   (35)

It should be emphasized that many of the reactions described for the functionalization of polymeric supports are not necessarily reproduced as easily as the corresponding reactions in solution, with either small molecules or soluble linear polymers as substrates. The exact course of a reaction may depend very heavily upon the degree of polymer cross-linking and upon the nature of the solvent, which will affect the degree of swelling. The role of any catalyst may also prove to be more critical.

2.3   THE 'NATURE' OF THE POLYMERIC SUPPORT

As indicated earlier, a supported substrate or reagent may interact intimately with the support itself in such a way that its reactivity, when compared with that in solution, is modified. In some cases this may be advantageous and may represent the principal reason for actually performing a polymer-supported reaction. In others, the interaction may be deleterious and actually militate against the use of the supported reaction, or at least place it at a disadvantage relative to the analogous solution phase reaction. This chapter attempts to identify and illustrate some of the factors involved in this area. As always, more comprehensive accounts [12,13] are available.

2.3.1   Solvent compatibility
If a reagent in solution is actually to react with a supported substrate, or vice versa, then it must 'find' it first, and it is necessary that the solvent should provide good solvation of the one and compatibility with the other. The same requirements apply to supported catalysts.

One of the first asymmetric hydrogenation catalysts, the DIOP–rhodium system, proved to be remarkably effective in the homogeneous hydrogenation [48] of N-acetyldehydrophenylalanine **(36)**, giving N-acetylphenylalanine **(37)** in quantitative yield and with an optical yield of 86% (equation (2.15)).

When the system was first supported on a polymer, via use of chloromethylpoly-styrene, no reaction occurred. However, by changing the nature of the polymer-supported ligand, to the styrene–hydroxmethyl methacrylate copolymer **(38)**, excellent yields, both material and optical, were once again obtained [49,50]. This set of observations vividly illustrates the requirement for good compatibility between the polymeric support and the liquid medium, which includes the substrate, surrounding the polymer.

$$(2.15)$$

$$(36) \qquad\qquad (37)$$

$$(38)$$

A particularly important aspect is the nature of the solvent itself. The use of a polymer-supported reagent or substrate may permit the use of a solvent not normally applicable to the corresponding homogeneous reaction, and often with some advantage. Thus, phenoxides supported on Amberlite IRA-900, a macroreticular anion exchange resin, show enhanced reactivity to alkyl halides when the solvent is changed from ethanol to benzene [51]. Similar trends are observed in the alkylation of resin-supported β-diketonates and in the reaction between resin-supported nitrite and α-halocarboxylate esters. The enhanced reactivity is attributed to the lack of specific solvation of the anions.

In other cases, specific solvation may be important. Thus, polymer-supported periodate oxidation of dibenzyl sulphide gives a 99% yield of sulphoxide when methanol is used as solvent, but only a 33% yield of product in the presence of chloroform under similar conditions [52].

2.3.2 Polymer microenvironment

The reactivity of a particular functional group supported on a polymeric matrix will depend upon the actual chemical nature of the polymeric support and most particularly on the nature of those parts of the support closest in space to it. Although of fundamental importance, the nature of this polymer microenvironment is difficult to investigate, and relatively few attempts have been made. In an important example of such a study, the polarities of the microenvironments of several polymers, including polystyrene, poly(4-vinylpyridine) and poly(methyl methacrylate), were studied using the shifts in solvatochromic charge transfer bands of reporter monomers embedded in the polymer chain [53]. Reporter residues included the methacrylate pyridinium iodide **(39)** and the merocyanine **(40)**. For each of a very wide range of solvents, and for all the polymers studied, a red shift in the solvatochromic band of the reporter was observed when compared to the reporter spectrum in solution. This indicates that the polymer microenvironment polarity is less than that of the bulk solution. Solvents were confined to those normally regarded as quite polar and included alcohols, ketones, amides, nitriles and pyridines. The smallest differences were observed for poly(methyl acrylamide) and poly(2-vinylpyridine) amongst others. Larger differences were seen in the case of poly(4-vinylpyridine), poly(methyl methacrylate) and polystyrene. In a similar vein, photochromic spiropyranmerocyanines [54] have been suggested as probes for studying steric and polar effects in polymers.

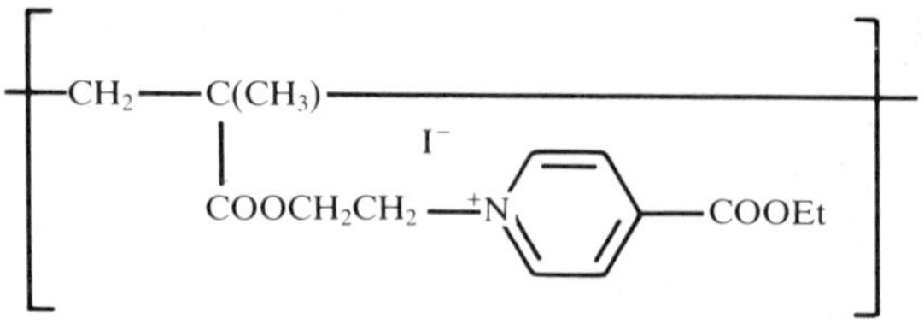

(39)

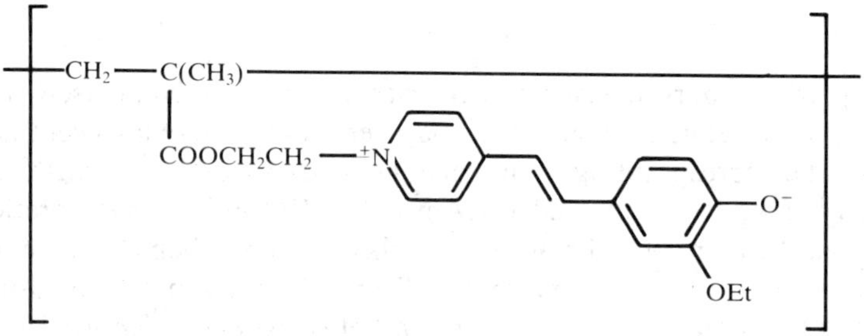

(40)

The suggested polarity differences described above are in contrast to, but not inconsistent with, those suggested to account for differences in product distributions between particular reactions in solution and the analogous reactions involving polymer-supported reagents. Thus benzylic bromination of cumene **(41)** with *N*-bromosuccinimide in carbon tetrachloride, a solvent of low polarity, gives the expected product **(42)** in 80% yield [55].

However, use of the polymeric maleimide-derived brominating agent **(43)** gives **(44)** as the major product (48% yield) together with smaller amounts of **(45)** and **(46)**. These results have been interpreted in terms of dehydrobromination of the product **(42)** to give the methylstyrene **(47)** which then reacts further. It is argued that the elimination of HBr from **(42)** is favoured by the increased polarity of the polymer microenvironment [55].

A rather different example involves the phosphine–carbon tetrachloride mediated conversion [56] of octanol into octyl chloride where relative rates of reaction for monomer phosphine **(48)** and linear and cross-linked phosphine polymers **(49)** are in the approximate ratios 1:10:25. It is argued that as the reaction proceeds, the phosphonium centres that are generated during its course must remain close to unreacted phosphines in the polymers, with a consequent increase in microenvironment polarity and effect on reaction rate [56]. Conformational changes which might reduce the effect in the linear polymer are not accessible to the cross-linked polymer, where the enhancement of reaction rate is at a maximum.

2.3.3 Location of functionality

In principle, the reactivity or effectiveness of a particular functional group might depend upon whether it is located at or near a cross-linking point or somewhere on the chain well removed from such a point. This idea has been investigated in the case of the phosphine-mediated reaction between alcohols and carbon tetrachloride to give alkyl halides [57]. Two triphenylphosphine-modified polystyrenes were prepared. In the case of the first polymer **(49)**, pendant phosphine groups were introduced via copolymerization of *p*-styryldiphenylphosphine with styrene and DVB. In the case of the second **(50)**, phosphines were located at the cross-linking points via use of bis(*p*-styryl)phenylphosphine as cross-linking agent. In fact, no clear difference emerged between the two types of polymer [57].

(48)   (49)   (50)

2.3.4   Accessibility of functional groups; grafted polymers

When a polystyrene, or other, resin is prepared via copolymerization of a monomer, some fraction of the functional groups may be buried within the polymer matrix, in regions of high cross-linking perhaps, such that they are inaccessible to, or unavailable for, subsequent reaction. To some extent this problem can be overcome by delayed introduction of the functionalized comonomer to the polymerization medium during synthesis of the resin. Thus, a series of chloromethylpolystyrenes was prepared [13] by the copolymerization of chloromethylvinylbenzene with styrene and divinylbenzene. When the chloromethylbenzene was introduced at the beginning of the polymerization process, the accessibility of the polymer-supported chloromethyl groups to nucleophilic attack by lithium diphenylphosphide was only 40%. However, when the chloromethylvinylbenzene was introduced near the end of the reaction, when nearly all the divinylbenzene had been consumed, the accessibility of the chloromethyl groups on the polymer rose to ca. 70%.

An obvious extension of this idea involves the 'grafting' of functionalized polymers via the reaction of functionalized monomers with a preformed cross-linked resin. The accessibility of the functional groups should then become very good because the density of cross-linking in the regions of functionalization will be very small. The premise has been proven by the reaction of trimethylamine with chloromethyl-modified polystyrene prepared by post-copolymerization of chloromethylvinylbenzene with a preformed macroporous styrene–DVB polymer [13]. Quaternization was essentially complete after 80 min at 35°C. Reaction involving a conventional resin [58] required 8 h at 50°C.

When a functionalized polymer is grafted onto an inert core, there results a so-called 'pellicular' resin. One of the first examples [59] involved the reaction between non-porous silica particles and the difunctionalized benzene, $Cl_3SiCH_2C_6H_4NO_2$. Reaction between nucleophilic sites on the silica surface and the trichlorosilyl group served to attach the organic functionality to the inert inorganic core. Reduction followed by diazotization provided an initiator for the radical copolymerization of styrene and vinylpyridine. The resulting polymer served as a support for copper salts. The functionalized methacrylate, $H_2C = C(CH_3)CO_2(CH_2)_3Si(OCH_3)_3$, a commercially available material, was, in a similar vein, grafted onto macroporous silica, and the supported vinyl group used in a copolymerization with other vinyl monomers [60]. Although the resulting pellicular resin was used for chromatographic separation, the possibilities for polymer-supported reactions are clear.

Pellicular resins are not confined to those based on inorganic cores. For example, polypropylene particles have been functionalized via treatment with ozone in a fluidized bed, and the resulting peroxide-functionalized particles have been used to initiate the radical polymerization of functionalized monomers [13]. Chloromethyl-polystyrene prepared in this way was subsequently converted into a polymer-supported sulphoxide and it was also used as the basis for an ion-exchange resin. More recently, the use of polystyrene grafted onto polyethylene sheets has been described as a support in solid-phase peptide synthesis [61].

2.3.5   Improved accessibility via the use of spacer groups

Intuitively, it might appear that functional groups directly attached to polymer backbones should be less accessible than those attached via some sort of spacer group. This idea would seem to be supported [62] by lineshape analysis of the esr signals of nitroxyl radicals (51): (a) covalently attached to the backbone of a cross-linked polymer (52); and (b) simply dissolved within a similar polymer matrix. Further, it has been shown [63] that, in particular cases, the rates of oxidation of particular soluble ferrocene-derivatized polymers increase with an increasing length of spacer group. It is not surprising, therefore, that there have been various attempts to improve polymer-supported functional group accessibility via the use of spacer groups.

(51)   (52)

An important recent development [64, 65] in this area involves the Friedel Crafts functionalization of lightly (1–4%) cross-linked polystyrene with an ω-bromoalk-1-ene in a reaction catalyzed by triflic acid (equation (2.16)). The resulting bromoalkyl-functionalized polystyrene (53) can be reacted with trialkylamines or phosphines to give polymer-supported 'onium salts having greater activity as phase transfer catalysts than the corresponding polymer-supported salts without spacers. Covalent attachment of the spacer via formation of a carbon–carbon bond is preferable to the use of amide, ester, ketone or ether linkages, which may have limited stability under the desired reaction conditions.

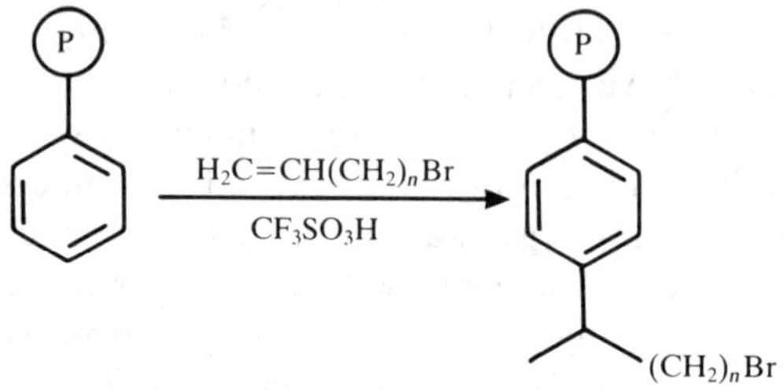

$$H_2C=CH(CH_2)_nBr$$
$$CF_3SO_3H$$

(2.16)

$(CH_2)_nBr$

(53)

As an alternative to functionalization of a preformed polymer, the use of vinyl monomers containing incorporated functional spacer groups has been investigated. Examples include the styrene ethers (54) and (55) [66,67], and the 3-bromopropyl-styrene (56) [68], which has been converted into the latent silyl protecting group (57) and the nucleophilic catalyst precursor (58) (equation 2.17)) [2].

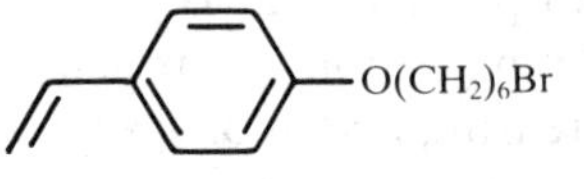

$O(CH_2)_6Br$

(54)

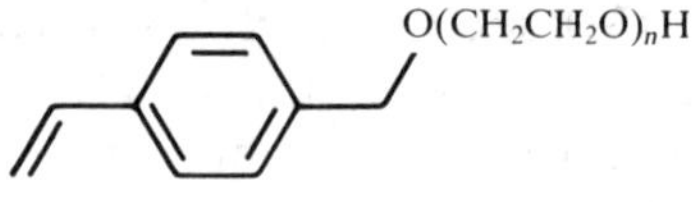

$O(CH_2CH_2O)_nH$

(55)

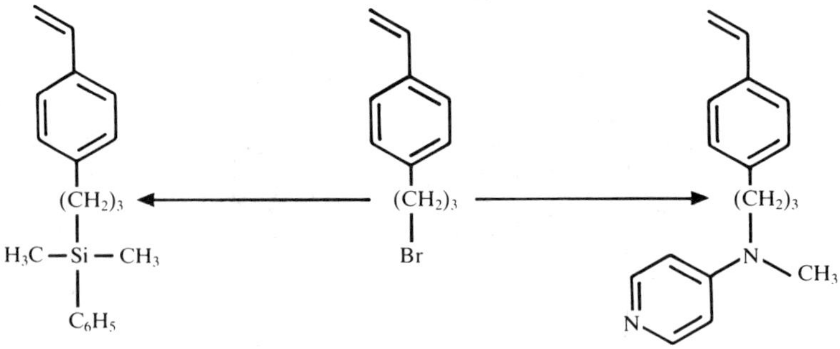

$(CH_2)_3$
$H_3C-Si-CH_3$
C_6H_5

(57)

$(CH_2)_3$
Br

(56)

$(CH_2)_3$
$N-CH_3$

(58)

2.3.6   Site-isolation

Reaction between two different components in solution is often complicated by self-reaction between two molecules of the same component. Crossed-condensations between two different carbonyl compounds are notable examples, and in particular cases the self-condensation product can be the major or even exclusive product. In principle, at least, attachment to a polymeric support might prevent reaction between two identical molecules of the same component, via so-called 'site-isolation', and thereby permit smooth and subsequent reaction with a second component. The actual degree of site-isolation that can be obtained will depend upon a number of factors, and it appears that polymer flexibility may be more important than the actual degree of polymer functionalization.

Early attempts to demonstrate the principle of site-isolation were not unequivocal. For example, reports describe the alkylation and acylation of polymer-supported ester enolates at room temperature, albeit with low product yield, without the self-condensation product which would occur via site–site interaction [69–72]. On the other hand, in a series of very detailed experiments, site–site reactions were demonstrated in attempted Dieckmnann cyclizations using polymer-bound diesters [73, 74]. In both cases, the substrate loadings were limited to not more than 0.1 mmol g^{-1} polymer, and the styrene–DVB polymer was only lightly (2–4%) cross-linked. This degree of cross-linking appears to be insufficient to prevent significant polymer motion, and site–site interaction is therefore possible. The observation of intrapolymer anhydrides [75], organocobalt dimers [76] and disulphides [77] on styrene –DVB polymers of similar constitution supports this view.

The hypothesis has been elegantly demonstrated [78] by Ford's use of highly (10–20%) cross-linked styrene–DVB supports. Supported esters were converted into their enolates at room temperature, and subsequent reactions with acylating and alkylating reagents proceeded with high yields. Substrate loadings as high as 0.67 mmol g^{-1} polymer were used. Furthermore, better results were obtained with a 20% cross-linked macroporous polymer than with a 10% cross-linked gel. A macroporous morphology was used for the 20% cross-linked polymer in order to minimize the time required for reagent diffusion, from the external solution to a reactive site within the polymer matrix.

One of the most widely quoted examples of site-isolation involves polymer-supported titanocene **(59)** [79]. Titanocene itself can act as a hydrogenation catalyst in solution. However, the metallocene is prone to dimerization, and its catalytic lifetime is short. When supported on a highly (20%) cross-linked polystyrene at a moderate (0.14 mmol g^{-1}) loading, the catalyst lifetime is greatly increased and the activity is one to two orders of magnitude greater than that of the soluble analogues.

(59)

2.4 CONCLUSIONS

The number of different currently accessible functionalized polymers, a majority of them based on polystyrene, suitable for polymer-supported reactions, is great. Within the polystyrene series, and notwithstanding the considerable advances in the understanding of the subtleties of the support which have been made in recent years, there remains scope for further advancement. Outside the polystyrene series, that scope is very much greater.

For organic chemists making their first attempts at utilizing polymer supports or catalysts for synthesis, it is probably best to make use of a commercial resin already possessing appropriate functionality and having appropriate physical characteristics. Table 2.1 gives a list of some of the more common resins that are available. However, the modification of a preformed resin is not a difficult operation, given proper

Table 2.1 — Some commercially available functionalized resins

---

Polystyrene cross-linked with (typically) 1–2% divinylbenzene

Available with the following functional groups: aminomethyl ($-CH_2NH_2$), α-aminobenzyl ($-CH(NH_2)C_6H_5$), chloromethyl ($-CH_2Cl$), bromo (Br) and diphenylphosphino ($-P(C_6H_5)_2$)

Amberlyst resins — macroreticular styrene–divinylbenzene ion-exchange resins (with a high divinylbenzene content) designed for use with organic solvents.

(Amberlyst is a Registered Trademark of Rohm and Haas)

Amberlyst A15 — a strongly acidic resin functionalized with $-SO_3H$

Amberlyst A21 — a weakly basic resin functionalized with $-N(CH_3)_2$

Amberlyst A26 — an anion-exchange resin functionalized with $-N(CH_3)_3^+$

Available as the fluoride (F^-), chloride (Cl^-), bromide (Br^-), iodide (I^-), acetate (CH_3COO^-), cyanide (CN^-), sodium carbonate ($NaCO_3^-$), nitrite (NO_2^-), nitrate (NO_3^-), thiocyanate (SCN^-), tribromide (Br_3^-), borohydride (BH_4^-) and hydrogenchromate ($HCrO_4^-$)

Poly-4-vinylpyridine and pyridium resins — cross-linked with (typically) 2% divinylbenzene

---

Pyridinium resins available as the chloride (Cl^-), toluene-4-sulphonate ($CH_3C_6H_4SO_3^-$), tribromide (Br_3^-), chlorochromate ($CrClO_3^-$) and dichromate ($1/2Cr_2O_7^{2-}$)

Poly-4-vinylpyridine–sulphur trioxide complex

Poly-2-vinylpyridine–borane complex

---

attention, and the outlines given in this chapter should allow decisions about the design of an appropriately modified resin to be made.

REFERENCES

[1] D. C. Sherrington and P. Hodge (ed.), *Syntheses and Separations using Functional Polymers*, John Wiley & Sons, New York, 1988.

[2] J. M. J. Fréchet, G. D. Darling, S. Itsuno, P.-Z. Lu, M. V de Meftahi and W. A. Rolls Jr., *Pure Appl. Chem.*, 1988, **60**, 353.

[3] P. Laszlo (ed.), *Preparative Chemistry Using Supported Reagents*, Academic Press, London, 1987.

[4] W. T. Ford (ed.), *Polymeric Reagents and Catalysts*, A.C.S. Symp. Ser., No. 308, American Chemical Society, Washington D.C., 1986.

[5] P. Hodge and D. C. Sherrington (eds), *Polymer-supported Reactions in Organic Synthesis*, John Wiley & Sons, New York, 1980.

[6] H. Jacobelli, M. Bartholin and A. Guyot, *J. Appl. Polym. Sci.*, 1979, **23**, 927.

[7] E. Farber in *Encyclopedia of Polymer Science and Technology* (eds H. Mark, N. G. Gaylord and N. M. Bikales), Wiley-Interscience, New York, London, 1970, Vol. 13, p. 552.

[8] M. Munzer and E. Tromsdorff in *High Polymers* (eds C. E. Schildknecht and I. Skeit), Wiley-Interscience, New York–London, 1977, Vol. 29, p. 106.

[9] M. Tomoi and W. T. Ford, *J. Am. Chem. Soc.*, 1981, **103**, 821.

[10] A. Guyot and M. Bartholin, *Progr. Polym. Sci.*, 1982, **8**, 277.

[11] T. Brunelet, G. Gelbard and A. Guyot, *Polym. Bull.*, 1981, **5**, 145.

[12] A. Guyot, in ref. [1], p. 1.

[13] A. Guyot, *Pure Appl. Chem.*, 1988, **60**, 365.

[14] P. W. Kwant, *J. Polym. Sci., Polym. Chem. Ed.*, 1979, **17**, 1331.

[15] S. V. McKinley and J. W. Rakshys, *J. Chem. Soc., Chem. Commun.*, 1972, 134.

[16] F. Camps, J. Castells, J. Font and F. Vela, *Tetrahedron Lett.*, 1971, 1715.

[17] J. A. Greig and D. C. Sherrington, *Polymer*, 1978, **19**, 163.

[18] J. M. J. Frechet, M. J. Farrall and L. J. Nujens, *J. Macromol. Sci. A*, 1977, **11**, 507.

[19] L. R. Melby and D. R. Strobach, *J. Am. Chem. Soc.*, 1967, **89**, 450.

[20] R. M. Fuoss and G. I. Cathers, *J. Polymer Sci.*, 1949, **4**, 493.

[21] T. Tamikado, *J. Polymer Sci.*, 1960, **43**, 489.

[22] R. S. Feinberg and R. B. Merrifield, *Tetrahedron*, 1974, **30**, 3209.

[23] J. T. Sparrow, *Tetrahedron Lett.*, 1975, 4637.

[24] K. W. Pepper, H. M. Paisley and M. A. Young, *J. Chem. Soc.*, 1953, 4097.

[25] S. Mohanraj and W. T. Ford, *Macromolecules*, 1986, **19**, 2470.

[26] S. L. Regen and D. P. Lee, *J. Am. Chem. Soc.*, 1974, **96**, 294.

[27] M. J. Farall and J. M. J. Frechet, *J. Org. Chem.*, 1976, **41**, 3877.

[28] R. H. Grubbs and L. C. Kroll, *J. Am. Chem. Soc.*, 1971, **73**, 3062.

[29] H. W. Gibson and F. C. Bailey, *J. Polym. Sci., Polym. Chem. Ed.*, 1975, 13, 1951.

[30] R. H. Grubbs and S. C. H. Su, *J. Organometallic Chem.*, 1976, **122**, 151.

[31] W. Heitz and R. Michels, *Makromol. Chem.*, 1971, **148**, 9.

[32] G. A. Crosby, N. M. Weinshenker and H. S. Uh, *J. Am. Chem. Soc.*, 1975, **97**, 223.

[33] N. M. Weinshenker, G. A. Crosby and J. Y. Wong, *J. Org. Chem.*, 1975, **40**, 1966.

[34] A. A. H. Al-Kadhumi, P. Hodge and F. G. Thorpe, *Polymer*, 1985, **26**, 1695.

[35] D. C. Sherrington, in ref. [5], p. 1.

[36] P. Hodge, in ref. [1], p. 43.

[37] R. L. Letsinger, M. J. Kornet, V. Mahadevan and D. M. Jerina, *J. Am. Chem. Soc.*, 1964, **86**, 5163.

[38] S. L. Regen and D. P. Lee, *J. Org. Chem.*, 1975, **40**, 1669.

[39] P. Hodge and G. Richardson, *J. Chem. Soc., Chem. Commun.*, 1975, 622.

[40] H. M. Relles and R. W. Schluenz, *J. Am. Chem. Soc.*, 1974, **96**, 6469.

[41] J. M. J. Fréchet and E. Bald, *Reactive Polym.*, 1982, **1**, 21.

[42] A. Akelah, M. Hassanein, A. Selim and E.-R. Kenawy, *Europ. Polym. J.*, 1986, **22**, 983.

[43] H. Schuttenberg, G. Klump, U. Kaczmar, R. Turner and R. C. Schultz, *J. Macromol. Sci., Chem. A*, 1973, **7**, 1085.

[44] H. Schuttenberg, G. Klump, U. Kaczmar, R. Turner and R. C. Schultz, *Polym. Prepr., Am. Chem. Soc., Div. Polym. Chem.*, 1972, **13**, 866.

[45] H. Schuttenberg and R. C. Schultz, *Angew. Makromol. Chem.*, 1971, **18**, 175.

[46] C. Bailly, D. J. Williams, F. E. Karasz and W. J. MacKnight, *Polymer*, 1987, **28**, 1009.

[47] C. Pugh and V. Percec, *Macromolecules*, 1986, **19**, 65.

[48] W. Dumont, J. C. Poulin, T. P. Dang and H. B. Kagan, *J. Am. Chem. Soc.*, 1973, **95**, 8295.

[49] N. Takashi, E. Imai, C. A. Bertelo and J. K. Stille, *J. Am. Chem. Soc.*, 1976, **98**, 5400.

[50] T. Masuda and J. K. Stille, *J. Am. Chem. Soc.*, 1978, **100**, 268.

[51] G. Gelbard and S. Colonna, *Synthesis*, 1977, 113.

[52] P. Hodge in ref. [5], p. 89.

[53] P. Strop, F. Mikes and J. Kálal, *J. Phys. Chem.*, 1976, **80**, 694; *ibid.* 702.

[54] P. H. Vandewyer and G. Smets, *J. Polym. Sci., Part A-1*, 1970, **8**, 2361.

[55] C. Yaroslavsky, A. Patchornik and E. Katchalski, *Tetrahedron Lett.*, 1970, **11**, 3629.

[56] C. R. Harrison, P. Hodge, B. J. Hunt, E. Khoshdel and G. Richardson, *J. Org. Chem.*, 1983, **48**, 3721.

[57] W. M. McKenzie and D. C. Sherrington, *J. Polym. Sci., Polym. Chem. Ed.*, 1982, **20**, 431.

[58] C. Lucas, I. Poinescu, E. Avram, A. Ioanid, I. Petrariu and A. Carpov, *J. Appl. Polym. Sci.*, 1983, **28**, 3701.

[59] P. J. Verlaan, J. P. C. Bootsma and G. Challa, *J. Mol. Catal.*, 1982, **14**, 211.

[60] G. Wulff, D. Oberkobusch and M. Minarik, *Reactive Polym.*, 1985, **3**, 261.

[61] R. H. Berg, K. Almdal, W. B. Pedersen, A. Holm, J. P. Tam and R. B. Merrifield, *J. Am. Chem. Soc.*, 1989, **111**, 8024.

[62] S. L. Regen, *Macromolecules*, 1975, **8**, 689.

[63] G. J. Oxford and J. M. Maud, unpublished results.

[64] M. Tomoi, N. Kori and H. Kakiuchi, *Reactive Polym.*, 1985, **3**, 341.

[65] M. Tomoi, N. Kori and H. Kakiuchi, *Makromol. Chem.*, 1986, **187**, 2735.

[66] M. Tomoi, S. Shiiki and H. Kakiuchi, *Makromol. Chem.*, 1986, **187**, 357.

[67] T. Hamaide, A. Revillon and A. Guyot, *Europ. Polym. J.*, 1984, **20**, 855.

[68] M. L. Hallensleben, *Angew, Makromol. Chem.*, 1973, **31**, 147.

[69] A. Patchornik and M. A. Kraus, *J. Am. Chem. Soc.*, 1970, **92**, 7587.

[70] M. A. Kraus and A. Patchornik, *Isr. J. Chem.*, 1971, **9**, 269.

[71] M. A. Kraus and A. Patchornik, *J. Polym. Sci., Polym. Symp.*, 1974, No. 47, 11.

[72] A. Patchornik and M. A. Kraus, *Pure Appl. Chem.*, 1975, **43**, 503.

[73] J. I. Crowley and H. Rapoport, *Acc. Chem. Res.*, 1976, **9**, 135.

[74] J. I. Crowley and H. Rapoport, *J. Org. Chem.*, 1980, **45**, 3215.

[75] L. T. Scott, J. Rebek, L. Ovsyanko and C. L. Sims, *J. Am. Chem. Soc.*, 1977, **99**, 625.

[76] S. L. Regen and D. P. Lee, *Macromolecules*, 1977, **10**, 1418.

[77] G. Wulff and I. Schulze, *Angew. Chem., Int. Ed. Engl.*, 1978, **17**, 537.

[78] Y. H. Chang and W. T. Ford, *J. Org. Chem.*, 1981, **46**, 5364.

[79] W. D. Bonds Jr., C. H. Brubaker Jr., E. S. Chandrasekaran, C. Gibbons, R. H. Grubbs and L. C. Kroll, *J. Am. Chem. Soc.*, 1975, **97**, 2128.

Part II

Traditional organic reactions aided by solid supports or catalysts

3

Reactions assisted by amorphous inorganic solids — a survey

M. Butters

3.1.1 General scope of survey

This chapter deals with the utilization of synthetic reagents or reactants adsorbed on insoluble amorphous inorganic supports. Silica gel and alumina are typical examples of amorphous inorganic oxides (others include titania and zirconia). The structural properties of such materials are discussed in Chapter 1. It is the surface structures of these materials that are of special interest, and organic chemists continue to explore new methods to carry out chemical transformations on those surfaces. Many applications of supported reagents have been discovered in recent years, and some of these developments have been catalogued in two key reviews [1, 2], which provide a comprehensive account up to 1978. There is also a recent review [3] on graphite intercalation compounds in organic synthesis. It is intended that this survey will highlight the usefulness of 'amorphous-supported' reagents in situations where conventional solution phase conditions are unable to offer the necessary/desired control.

The utilization of amorphous inorganic solids in organic transformations generally falls into three categories:

(a) as insoluble supports for stoichiometric inorganic reagents (e.g. oxidants), extending the usefulness of water-soluble reagents and allowing reactions to be performed in organic solvents;

(b) as catalysts or supports for catalysts (e.g. KF base) which can promote organic reactions in an heterogeneous phase; such materials offer remarkable simplicity in their practical application;

(c) as unique environments for bimolecular reactions (particularly in the absence of solvents), which can differentiate structures by selective adsorption/binding and thereby influence the regiochemistry of certain reactions (e.g. monoalkylation of diols).

3.1.2   Advantages of amorphous-supported reagents

- Several 'supported reagents' are now commercially available and can be used 'off-the-shelf'.
- These reagents can offer a simple procedure and purification, and can extend the choice of solvent for a given reaction.
- Some of the new methodologies associated with these reagents introduce superior reaction selectivity and control (e.g. chemo- and regioselectivity) in comparison to homogeneous counterparts.
- These 'solid-reagents' are safe and easy to handle; in general they are non-volatile and odourless.

3.1.3   Disadvantages of amorphous-supported reagents

- The most serious drawback is that these materials are ill-defined in terms of their physical and chemical properties; it is difficult to quantify surface properties in familiar solution terms (e.g. pK_a).
- There are many commercial suppliers and manufacturers of amorphous inorganic oxides. Structural composition and properties can vary significantly, resulting in unpredictable results.
- Some procedures for preparation and handling require the use of specialist apparatus (e.g. high-temperature furnace, glove box). The lifetime of some reagents can be limited.
- Some reagents require a low 'loading' of reactants, resulting in poor efficiency for large-scale work (sometimes high dilution; voluminous operations). However, these problems can potentially be avoided by doing reactions in a 'column-mode'.

3.1.4   Specifications of commercial silica and alumina

The surfaces of amorphous silica and alumina have properties that are not duplicated in the solution or gas phase, and because of this, entirely new chemistry may occur. In some circumstances, inorganic oxides can have a profound effect on the chemical reactivity of synthetic reagents. Before performing a synthetic transformation with the aid of a solid catalyst or supported reagent it is important to obtain a solid that offers the necessary physical characteristics. Unfortunately, many of the early publications on supported reagents do not report details of the source or nature of solid employed and frequently refer to them by chemical name only. It can be appreciated (see Chapter 1) that the analysis and characterization of inorganic oxides is complex and beyond the scope of most organic chemistry research facilities. In general, however, the organic chemist need only consider a limited number of physical properties when selecting a solid support.

Several grades of silica gel are available commercially, and E. Merck offer a useful selection which are manufactured to closely defined specifications. Silica gel is a highly active system that is thermodynamically and kinetically unstable and it is therefore important to be aware of its characterization. For its application in different types of reactions, not only the skeletal structure but also its porosity, surface area and acidity become important. The change in physical parameters associated with change in pore size for a selection of silicas is shown in Table 3.1.

Table 3.1 — Characteristics of several commercial silicas

Silica	Mallinckrodt Silicic acid	BDH chromatography grade	Merck type 40	Merck type 60	Merck type 100
Mean pore size diameter (nm)	3.4	1.8	4	6	10
Pore volume (ml g^{-1})	0.5	0.39	0.68	0.82	1.05
Specific surface area (m^2 g^{-1})	585	733	675	550	420
pH of a 10% aqueous slurry	4.5	4.7	5.5	7.0	7.5

From a catalysis viewpoint, the rate of reaction is partly dependent on the surface area of the active phase. For organic reactions at the solid/solution interface, the rate is also limited by diffusion processes (to and from the solid surface). Pore size distribution will clearly influence the rate of reactant diffusion. It is apparent from Table 3.1 that surface acidity is closely related to mean pore size. Grades of silica with micropores (i.e. pores with diameter ≤ 2 nm) have the potential to be strongly acidic. However, as pore size decreases, the surface area and amount of atmospheric moisture that may be adsorbed by the silica gel increases. Presence of physisorbed water has the effect of 'deactivating' the silica surface covering the active acidic sites. Thus the properties of catalytic activity and moisture content are closely related and are susceptible to changes in ambient humidity and temperature. It is noticeable that these properties are dynamic and are liable to change with storage/lifetime of the solid. Manufacturers will usually dry/activate silicas and aluminas before supply. If there is any doubt regarding the integrity of a silica sample it is usually safer to start a reagent preparation (or reaction) with a fresh unopened bottle. In fact for most synthetic applications of inorganic oxide supports, excess water is detrimental (obvious exceptions include hydrolysis reactions). As a general guideline, silica can be reactivated (physical water removed) by heating at 120°C for 10 h. At temperatures in the range 120 to 300°C, the 'chemical water' of the surface silanol groups is removed and this is detrimental to the surface acidity (formation of new siloxane bonds SiOSi). The relative amount of water lost over the temperature range 120 to 300°C reflects the degree of surface acidity. Thus a sample of BDH chromatography grade silica (see Table 3.1) was shown by thermogravimetric analysis to possess ca. 1.3% w/w 'chemical water' compared with 0.5% w/w for silica gel type 60. This difference in 'chemical water' is paralleled by the relative acidity as measured by pH of 10% aqueous slurry (see Table 3.1).

Aluminium oxide is usually prepared from the dehydration of aluminium hydroxide at 900°C in a stream of CO_2. The resultant alumina particles are coated with a thin layer of aluminium oxycarbonate; this corresponds to a 'basic' grade of alumina pH 10% aq. slurry $\simeq 10$. Water content and alkalinity are then adjusted by washing with dilute acids (e.g. HCl) to give different grades. Thus a range of mean pore sizes (60, 90 and 100) are available as basic, neutral or acidic (pH 9.5, 7.0, 6.0 and 4.5 for 10% aqueous slurry) forms. The basic and neutral grades of alumina have wide applications as supports and catalysts in organic synthesis. The deactivation of alumina by physical hydration is an important feature of its surface chemistry. Brockmann and Schodder [4] have defined a series of activities (I to V), for silica and alumina, based on water content (at 0, 20, 40, 60 and 80% relative humidity). The Brockmann activity scale has been adopted in chromatography (and to a limited extent in supported reagent chemistry) and depends on the elution of specified dyes in a predetermined order from the solid.

3.2   SELECTIVE METHODS FOR THE PREPARATION OF ESTERS

3.2.1   Use of alumina as a support catalyst

Posner and Oda have reported a convenient and very mild method for the selective acetylation of primary alcohols [3]. The alcohol is dissolved in an excess of ethyl acetate, and Woelm-200-neutral chromatographic alumina is added to catalyse the transesterification. In this simple procedure, ethyl acetate functions as both solvent and reagent. The conditions are mild enough to permit successful acetylation of 'base sensitive' starting materials. Compounds (1) and (2) have been prepared by this method in 70% and 63% yields respectively.

(1)   (2)

It is difficult to obtain monoesters of dicarboxylic acids by conventional esterification methods, but neutral alumina can be used to promote selective monomethylation of such acids [5]. Dimethyl sulphate and diazomethane are both effective methylating reagents for this purpose. For example, dicarboxylic acids are quantitatively monoesterified using diazomethane in DMF (30°C). To avoid homogeneous reaction it is important to preadsorb the dicarboxylic acid on the alumina (e.g. Scheme 1) before adding the diazomethane because the technique relies upon the different reactivities of the adsorbed and free carboxyl groups. However, the strength of carboxyl adsorption is adversely affected by surface water; thus addition of water (10% w/w) gives a decrease in 'monomethyl esterification selectivity' from 99% to 49%, which is equal to the selectivity found in the homogeneous reaction.

Reactions using dimethyl sulphate show a slight difference in behaviour. Dicarboxylic acids are preadsorbed on alumina from a solution in DMF, and the solid is collected and dried. Good monomethylation selectivity is observed (see Table 3.2) when the solid is slurried in cyclohexane with dimethyl sulphate.

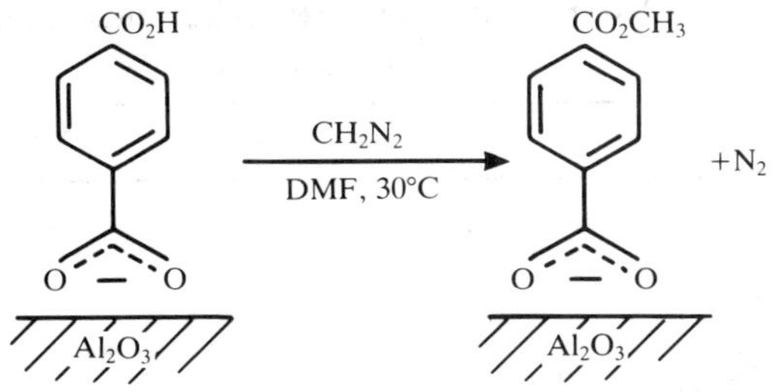

Scheme 1.   Selective monoesterification of a dicarboxylic acid over alumina.

3.2.2   Use of an acidified graphite

Electrolysis of 98% sulphuric acid with a graphite anode gives deep-blue crystals of the electrolytic lamellar compound, $C_{24}^+HSO_4^-.2H_2SO_4$. Kagen and coworkers have shown that this solid is an excellent catalyst for a wide range of esterification reactions [6]. Primary and secondary alcohols are readily esterified at room temperature using cyclohexane as a solvent. Various aliphatic carboxylic acids (see Table 3.3) are tolerated, and aromatic or conjugated acids require higher reaction temperatures (ca. 70°C). The procedure works well even with sparingly soluble acids (such as tartaric acid), which gradually dissolve as esterification proceeds. Moreover, it is not necessary to use excess alcohol. Unfortunately, tertiary alcohols often dehydrate in the presence of this catalyst, leading to complex mixtures. Ethyl esters are also readily obtained in excellent yield by treatment of acids with triethyl orthoformate in the presence of a catalytic amount of 'graphite hydrogen sulphate'.

3.3   BASE CATALYSIS FOR THE ACTIVATION OF 'C–H ACIDS'

For many years, basic alumina has been recognized as an effective catalyst for the condensation of ketones with aldehyde [7]. More recently, Clark *et al.* have introduced 'KF-on-neutral-alumina' as a potent basic catalyst [8]. This material is now commercially available from Fluka.

Table 3.2 — Selective monomethyl esterification of dicarboxylic acids using alumina and dimethyl sulphate

Substrate	(%) Yield of methylated compound	
	mono-	di-
Terephthalic acid	72	12
Isophthalic acid	63	19
Phthalic acid	80	8.0
Adipic acid	93	6.7
Suberic acid	99	0
1,12-Dodecanedioic acid	97	3.2

Table 3.3 — Esterification of carboxylic acids catalysed by
$C_{24}^+ HSO_4^-.2H_2SO_4$

Alcohol	Acid	Yield of ester (%)
	HCO_2H	98
	$PhCOCO_2H$	50
	$AcOH$	94
	Ph⌒⌒CO_2H	74
2 MeOH	HO⧸⧸⧸⌒CO_2H / HO▾⌒CO_2H	99

The formation of a hydrogen bond between a fluoride anion and an organic compound will result in the transfer of electron density from the anion to the organic compound, thus enhancing the nucleophilicity of the organic compound. The base strength of an ionic fluoride is dependent on the solvent in which it is dissolved, the amount of water present, and the nature of the counter cation. In a systematic study of different sources of fluoride, Clarke *et al.* [8] found that KF or CsF supported on alumina (at a loading of 2.5 molecules nm^{-2}) represents the best source of 'naked fluoride' in several types of reactions. For example, in the Michael reaction of nitroethane with 3-buten-2-one, traditional reagents like KF/18-crown-6 gave 5% yield in 8 min, whereas KF/alumina gave 100% yield in 3 min. Table 3.4 presents a series of reactions which have been promoted by KF/Al_2O_3. Reactions can be carried out in 1,2-dimethoxyethane, dichloromethane, or acetonitrile solvents.

Some of the examples [11] in Table 3.4 were performed using a column (5 cm³ polypropylene syringe) packed with KF/Al_2O_3. Pure products were obtained by simple elution of starting materials in dichloromethane at 25°C at a rate of 2 mmol min^{-1}.

In a recent investigation, Bergbreiter and Lalonde [12] have noted some advantages of using 'basic alumina' to prepare the KF/Al_2O_3 reagent. Excellent yields (96 to 100%) were obtained in the Michael addition of various nitroalkanes (4 equivalents) to α,β-unsaturated carbonyl compounds (Table 3.5) using KF/basic-Al_2O_3 in THF. A disadvantage of KF/neutral-Al_2O_3 is that primary

Table 3.4 — Base catalysis using KF/Al_2O_3

C–H acid	Electrophile	Product	Yield (%)	Reference
$EtNO_2$			100	8
			80	9
$H—\equiv—Ph$			54	9
	PhCHO	(100% Z)	97	11
	PhCHO		82	10
	PhCHO		76	11
	$CS_2 + 2MeI$		85	11

Table 3.5 — Michael addition of nitroalkanes to α,β-unsaturated carbonyl compounds using KF/basic-Al_2O_3

Nitroalkane	α,β–Unsaturated carbonyl	Yield (%)
	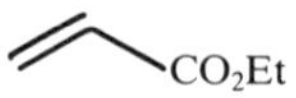	97
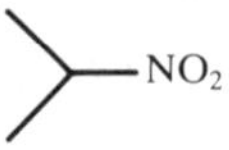	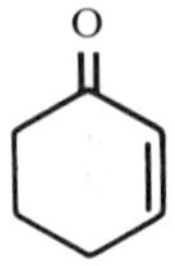	100
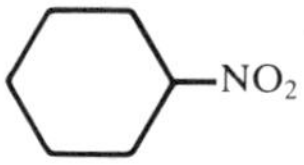		90

nitroalkanes tend to give multiple Michael additions. Fortunately, this is not a problem with KF/basic-Al_2O_3. Bergbreiter has demonstrated that this method is effective on a reasonable scale (20 g KF/Al_2O_3 furnished 25 g product in 750 ml THF) and that the KF/basic-Al_2O_3 catalyst can be recycled. Unfortunately, α,β-unsaturated carbonyl substrates that are hindered at the β-carbon (e.g. 4-methyl-3-penten-2-one) do not react.

Furans and isoxazoline *N*-oxides can be synthesized with the aid of KF/neutral-Al_2O_3 catalysis [13]. Secondary nitroalkanes react with a range of arylidene-ketones, esters, sulphones and nitriles. Thus KF/neutral-Al_2O_3 promotes reaction at 25°C to give the expected Michael adducts (isolated by the addition of water to the reaction mixture). Reaction at 80°C (15 h) over KF/neutral-Al_2O_3 (in acetonitrile) results in conversion of the Michael adduct to a 4,5-dihydrofuran (equations (3.1) and (3.2)) via intramolecular nitrite displacement. A possible mechanism could involve ring closure of the Michael adducts to afford cyclopropanes. Thus, Alonso and Morales have reported that cyclopropyl ketones undergo stereoselective ring enlargement over alumina to give 4,5-dihydrofurans at room temperature (equation (3.3)) [14].

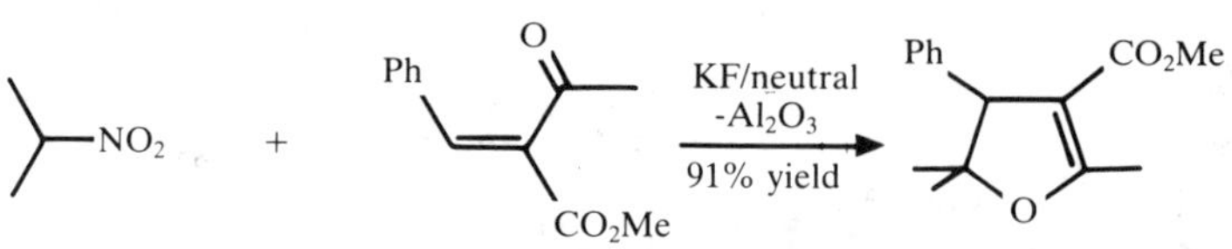

(3.1)

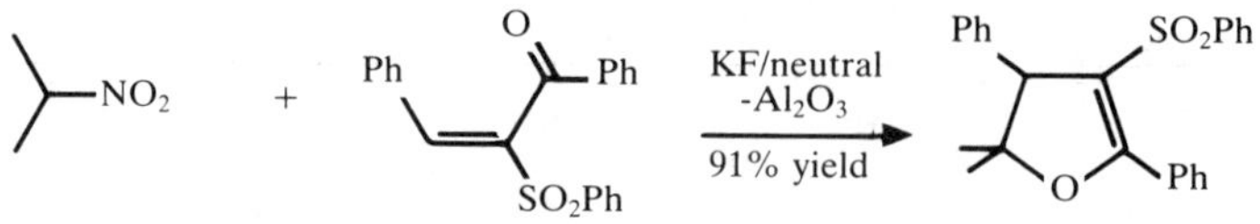

(3.2)

$$(3.3)$$

In the presence of KF/neutral-Al_2O_3, primary nitroalkanes react to give a mixture of dihydrofurans and furans. Formation of furans involves a Nef reaction of the intermediate Michael adduct.

When secondary nitroalkanes are reacted with nitroalkenes (acetonitrile solvent) in the presence of KF/Al_2O_3, isoxazoline N-oxides are formed in moderate yields (equation (3.4)) [13]. The reaction is complicated by base-catalysed polymerization of the nitroalkene.

$$(3.4)$$

The palladium-catalysed alkylation of allylic acetates is a useful synthetic procedure [15]. Intermediate palladium-bound allyl cations are readily attacked by carbon nucleophiles such as dimethyl sodiomalonate or dimethyl malonate in the presence of homogeneous-hindered bases (e.g. 1,5-diazabicyclo[5.4.0]undec-5-ene, DBU). Muzar *et al.* [16] have used KF/neutral-Al_2O_3 as a heterogeneous base for the assistance of palladium-catalysed allylic aklylations (see Table 3.6). Alkylations using carbon acids (pK_a range 5 to 13) are realized in good yield ($> 75\%$) at room temperature without the need for preformation of the corresponding carbanion. Moreover, the basic support can activate even Meldrums acid [3] (reported to be unreactive under homogeneous conditions). The heterogeneous base KF/Al_2O_3 compares favourably with triethylamine and DBU. Unlike with DBU, no decarboxylation occurs in the presence of KF/Al_2O_3.

The Z-kinetic dithioenolization of methyl dithiopropanoate is favourable using lithium diisopropylamide (LDA) in THF (at $-79°C$); alkylation of the so-formed thioanion with benzyl chloride gives a 70/30 Z/E mixture of dithioenol ethers in 70% yield [17]. The lithium cation is an important factor in the stereoselectivity (KO^tBu gives a 57/43 mixture). KF/Al_2O_3 [17] in acetonitrile solvent (20°C) gives even better kinetic selectivity; reaction with benzyl chloride gives an 85/15 Z/E ratio of dithioenol ethers in 96% conversion (equation (3.5)). Thus KF/Al_2O_3 is now the base of choice for kinetic enolate formation of dithioesters.

$$(3.5)$$

(85% this isomer)

Table 3.6 — Palladium-catalysed allylic alkylations promoted by $KF/neutral-Al_2O_3$

C–H acid	Product from allyl acetate	Yield (%)
NO₂ structure	allyl product with NO₂, CO₂Me	95[a]
CO₂Et structure	allyl product with CO₂Et, CO₂Et	75[a]
CN structure	allyl product with CN, SO₂Ar	75[a]
dioxane structure (3)	diallyl dioxane product	82[b]

[a] Catalyst = Pd (dppe)₂. [b] Catalyst = Pd (dba)₂ /PPh₃
[dppe is 1,2-bis(diphenylphosphino)ethane; dba is dibenzylideneacetone].

KF/Al_2O_3 has also been used to catalyse diazo group transfer under mild conditions
[18]. A mixture of ethyl acetoacetate, tosyl azide and KF/Al_2O_3 (base catalyst) in
THF at 0°C gave a 94% yield of the corresponding α-diazocarbonyl compound
(equation (3.6)). The method is efficient and convenient; the relatively polar byproduct (tosyl amide) is easily removed as an alumina sorbate.

(3.6)

3.4 SILICA GEL AS A MEDIUM FOR THE FORMATION AND SELECTIVE CLEAVAGE OF ACETALS AND KETALS

3.4.1 Aluminium chloride on silica gel for acetalization of carbohydrates
Unfortunately, most of the reported methods for catalysis of acetal formation are
rather forcing and of limited use for more sensitive organic compounds. By contrast

AlCl$_3$ on silica gel is a mild and convenient catalyst for acetalization of carbohydrates [19]. Thus, 4-deoxy-DL-*threo*-pentopyranose dissolved in acetone is readily converted to 4-deoxy-1,2-*O*-isopropylidene-β-DL-*threo*-pentopyranose (equation (3.7)).

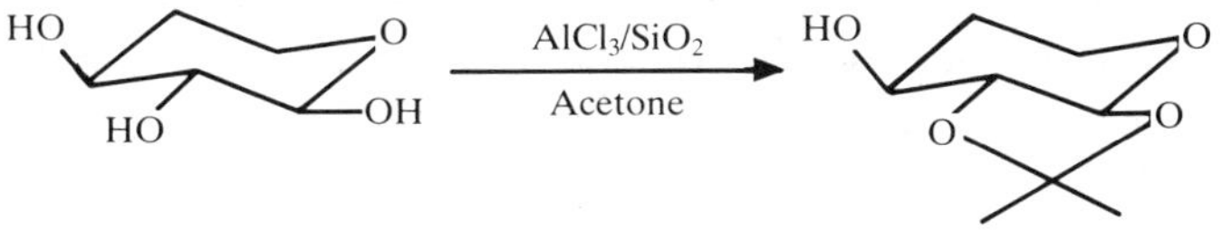

$$(3.7)$$

3.4.2 Silica gel–thionyl chloride for thioacetalization of aldehydes

Silica gel treated with thionyl chloride produces a 'surface-chlorinated' silica (having Si $-$ Cl bonds). This material is an excellent reagent for the chemoselective thioacetalization of aldehydes in the presence of ketones (much slower to react) [20]. Reactions are carried out in refluxing toluene with one mole equivalent of dithiol and one gram of SOCl$_2$–SiO$_2$ per mmole starting material. The striking chemoselectivity for aldehydes over ketones (insensitive to amount of SOCl$_2$–SiO$_2$ and temperature) was rationalized on the basis of a hindered silica environment. Thus surface Si–Cl groups might be easily accessed by formyl functionalities to give carbocationic centres (Scheme 2).

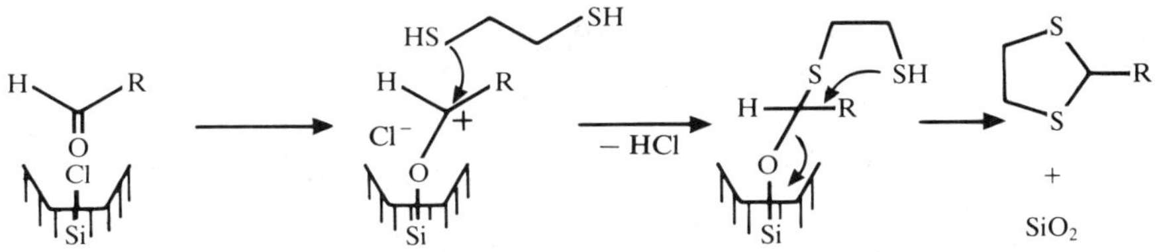

Scheme 2.   Thioacetalization of aldehydes over SiO$_2$–SOCl$_2$.

At the end of the reaction, the pure product is obtained by filtration, aq NaOH washing and evaporation of solvent. High yields are obtained from a range of aliphatic and aromatic ketoaldehydes.

3.4.3 Silica gel–sulphuryl chloride for dethioacetalization

Treatment of thioacetals with sulphuryl chloride in the presence of wet silica gel (at 0 to 25°C) results in smooth removal of the protecting group and generation of the carbonyl compounds in excellent yield. These reactions most probably proceed via initial oxidation of one of the sulphur atoms to the corresponding sulphoxide. For this procedure, the most effective grade of silica gel is Mallinckrodt silicic acid (100 mesh). This silica is quite acidic (pH aq. slurry ca. 4.5) and presumably this aids the polarization of SO$_2$Cl$_2$ into an electrophilic source of chlorine. It is suggested that water hydrolyses the sulphoxide intermediate [21].

3.4.4 Wet silica gel for deacetalization

Conia and coworkers have shown that wet silica gel is an excellent medium for deacetalisation [22]. Relatively unstable enones can be obtained from parent acetals in excellent yields and the reactions proceed without decomposition or double bond migration. A similar reagent made from oxalic acid and wet silica gel has likewise been applied to the chemoselective hydrolysis of α-epoxyketone acetals to α-epoxyketones [23]. Stapleford has shown that the acidity of these reagents can be modified to effect the selective hydrolysis of mixed enol ether/acetals (equations (3.8) and (3.9)) [24].

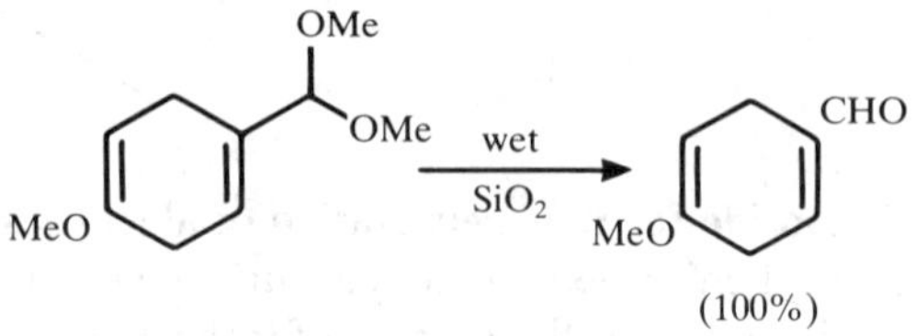

(3.8)

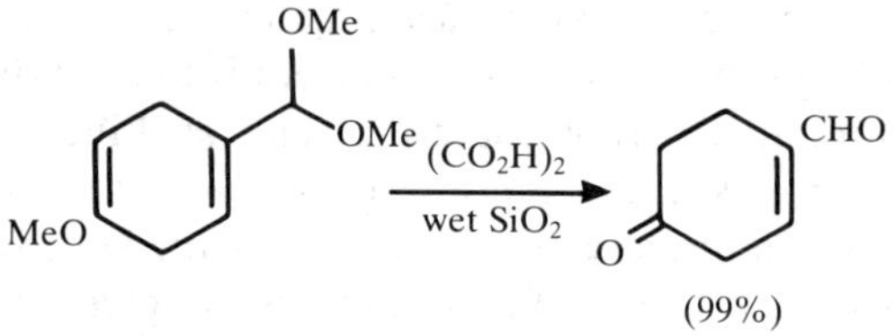

(100%)

(3.9)

(99%)

The reagent oxalic acid–wet silica gel has also recently been applied in the mild hydrolysis of oxazolidines (equation (3.10)) [81].

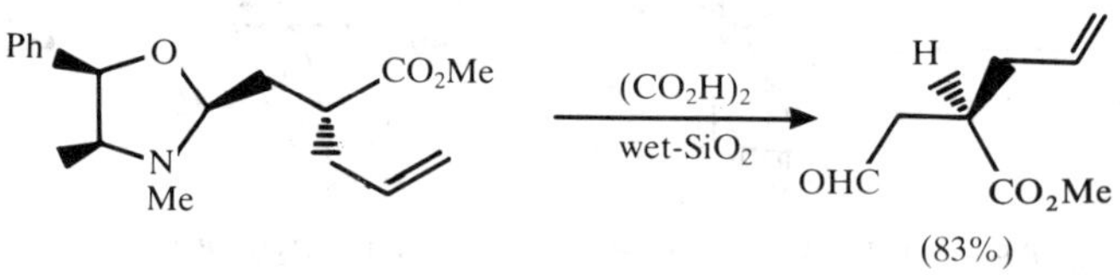

(3.10)

(83%)

3.5   REAGENTS AND CATALYSTS FOR ELECTROPHILIC SUBSTITUTION OF AROMATIC AND HETEROCYCLIC COMPOUNDS

3.5.1   Friedel–Crafts alkylations

It is well known that conventional Friedel–Crafts-type alkylations do not always proceed successfully on sensitive heterocyclic aromatic compounds. Furan and thiophene are troublesome because of their instability towards strong Lewis acids or mineral acids. In the presence of acidic silica gel (Mallinckrodt), however, thiophene, furan, benzothiophene and 1-methylindole are all cleanly alkylated in good yields [25]. Using ^{t}BuBr, alkylation occurs at the 2- and 5-positions of thiophene and furan, and at the 3-positions of benzofuran, benzothiophene and 1-methylindole.

The alkylation of benzenes with alkenes is catalysed under mild conditions using metal oxide–tantalum halide-oxide catalysts [26]. Johnson has reported that a catalyst prepared by deposition of $TaCl_5$ on silica promotes the 1-pentene alkylation of benzene at 25°C in 30 min (100% yield of pentylbenzene). The catalyst is also effective for a wide range of monosubstituted benzenes including anisole and phenol.

Similarly, $BF_3 . Et_2O$ on alumina is an effective reagent for the allylic alkylation of reactive phenols [27]. Reaction of the allylic alcohol (+)-*p*-methadiene-1-ol with olivetol in the presence of $BF_3 . Et_2O$ on alumina gives cannabidiol in 55% yield (equation (3.11)). This represents a remarkable improvement over the traditional synthetic routes [82].

$$(3.11)$$

(55%)

3.5.2 Electrophilic chlorination

Electrophilic chlorination of aromatic compounds in the presence of inorganic oxides has been studied extensively (Table 3.7). Tertiary-butyl hypochlorite in the presence of silica gel (BDH chromatography, pH aqueous slurry ca. 4.7) generates a potent source of electrophilic chlorine which is capable of aromatic substitution reactions under very mild conditions [28]. The only side product of the reaction is tertiary-butanol, and unlike with Cl_2 or SO_2Cl_2, HCl gas is not generated. The reactions are clean and high yielding, and workup is simple. This method is particularly useful for the chlorination of aromatic substrates of moderate reactivity, providing outstanding selectivity for monochlorinated products.

N-Chlorodialkylamines in the presence of silica gel are useful reagents for the controlled chlorination of phenols [29]. Indeed, this reagent appears to be chemoselective for phenolic compounds; anisoles and acetanilides are not affected (although anilines are oxidized). *N*-Chlorobis(2-chloroethyl)amine over silica gel (BDH chromatography) is particularly useful for controlling the mono : dichlorination selectivity, as well as providing high ortho/para ratios. It has been suggested [29] that the origin of the regioselectivity stems from binding of the phenolic OH group onto the silica surface in close proximity to the site of *N*-chloroamine protonation. Neutral alumina-supported $CuCl_2$ is another useful reagent, chlorinating anisoles with outstanding para-selectivity in high yield [30]. Reactions require the use of 5 mol equivalents of $CuCl_2$ (as a 50% w/w loading on alumina) to one equivalent of anisole.

3.5.3 Electrophilic bromination

N-Bromosuccinimide on silica gel (BDH chromatography) is an attractive and versatile reagent system for controlled bromination of a wide variety of activated aromatic and heterocyclic compounds (Table 3.8) [31]. In contrast to the traditional procedures for such electrophilic brominations, NBS/SiO_2 often gives high yields and good regioselectivity. In addition, by changing the mole ratio of reagent : substrate it is often possible to introduce bromo substitutents in a discrete and stepwise fashion. Reactions are easy to perform (normally done in dichloromethane) and simple

Table 3.7 — Chlorination of aromatic compounds over supported reagents

Substrate	Product (yield)	Reagent	Reference
	(93%)	^{t}BuOCl/SiO$_2$	[28]
	(65%) + (35%)	^{t}BuOCl/SiO$_2$	[28]
	(44%) + (56%)	^{t}BuOCl/SiO$_2$	[28]
	(76%) + (15%)	ClN(CH$_2$CH$_2$Cl)$_2$/SiO$_2$	[29]
	(87%)	ClN(CH$_2$CH$_2$Cl)$_2$/SiO$_2$	[29]
	(95%)	CuCl$_2$/Al$_2$O$_3$	[30]

filtration gives a solution of essentially pure product (residual NBS and succinimide are retained by the silica).

3.6 ELECTROCYCLIC REACTIONS

3.6.1 Diels–Alder reactions

Several groups have reported improvements of Diels–Alder reactions in the presence of silica or alumina catalysts [32–34]. Thus, silica gel (20 g) promotes the Diels–Alder

Table 3.8 — Bromination of aromatic compounds using NBS/SiO$_2$

Substrate	Products	Yield (%)
OEt (phenyl ether)	OEt, Br (bromo phenyl ether)	100
3-substituted indole (R = H, CN)	2-bromo-R-indole with Br on benzene ring	75 – 95[a]
3-substituted indole	R-indole with Br at 2-position	90 – 96
(CH$_2$)$_n$ bridged diarylamine (n = 1,2)	(CH$_2$)$_n$ bridged diarylamine with Br	86 – 89

[a] Used 2 moles of NBS.

reaction of 1,3-butadiene (13.6 g) and 1,4-naphthoquinone (13.3 g) in xylene (200 ml) at 70°C/3 h (equation (3.12)) [34]. The presence of silica gel results in a 54% yield compared with 9% in the absence of silica gel.

$$\text{naphthoquinone} + \text{butadiene} \xrightarrow[\substack{\text{xylene} \\ 70°C}]{\text{SiO}_2} \text{adduct (54\%)} \qquad (3.12)$$

Parlar and Bauman [33] have studied Diels–Alder reactions over inorganic oxides in the absence of solvents. A solution of cyclopentadiene and acrylic dienophile in diethyl ether was adsorbed onto a range of oxides (e.g. SiO$_2$, Al$_2$O$_3$, sand and montmorillonite clay); the solvent was then evaporated and the heterogeneous phase was heated to 50°C for a few hours. Some interesting observations were made regarding changes in exo/endo ratio for different solids. The technique of 'dry adsorption' Diels–Alder reactions has been further investigated by Smit and coworkers (Table 3.9) [35].

Table 3.9 — Diels–Alder reactions over solids under dry-heterogeneous conditions at 20°C (1–22 h)

Diene	Dienophile	Product	Solid	Yield(%)
			SiO$_2$	73[a]
			SiO$_2$	56[a]
			SiO$_2$	89
			SiO$_2$	79
		(96% endo)	SiO$_2$[b]	90
		(80% exo)	MgO.SiO$_2$	48
			MgO.SiO$_2$	47

[a] 4-Isomer selectivity = 95%. [b] Reaction temperature − 20°C.

It would appear that the promotion of Diels–Alder reactions requires solids that have been dried at elevated temperature (ca. 200°C/6 h), to remove physically bound water. Furthermore, the rate of reaction is directly dependent on the amount of silica (SiO_2 : substrate weight ratio of 10 : 1 preferred) and is improved by solids with a high surface area (preferably greater than 300 m^2 g^{-1}). The acidity of the recommended silicas (e.g. Mallinckrodt) might also be an important factor.

The presence of solvents (especially polar solvents) lowers the rate of conversion. It has been suggested [35] that the role of the silica gel under the 'dry state' conditions is to anchor the reactants within close proximity of one another (within critical bonding distance) and stabilize the pre-reaction complexes of diene with dienophile. Whatever the precise mechanism, it is a fact that under 'dry-heterogeneous' conditions cycloadditions proceed effectively at much lower temperatures (20°C compared with over 100°C under traditional liquid phase conditions) and produce adducts with enhanced regio- and stereoselectivity.

Heterodiene reactions over solids (e.g. last entry of Table 3.9) are also possible. However, the acid lability of vinyl ethers may preclude the use of acidic silica gels. Magnesium silicate (Fluorisil) has poor acid sites and is more suitable [35].

Rideout and Breslow have reported acceleration of the Diels–Alder reaction in aqueous media [36]. The hydrophobic nature of diene and dienophile increases the number of useful collisions and this is an alternative way of forcing the reactants into close proximity.

3.6.2   [3,3]-Sigmatrophic rearrangements

The [3,3]-sigmatrophic Carrol rearrangement of *tert*-allyl acetoacetic esters into γ,δ-unsaturated ketones is also promoted by the use of 'dry-heterogeneous' conditions [37]. In this case, silica gel is not a suitable catalyst, instead promoting acylotropic migration. Alumina, on the other hand, with a mixture of acid and base sites, is a much more effective catalyst (equation (3.13)). Reactions are performed effectively at 60°C, instead of the traditional conditions (160–220°C) and are milder and more convenient than the 'bis-enolate' approach, which requires 2 equivalents of LDA in THF [38].

$$(3.13)$$

$$(74 – 78\%)$$

3.7   NUCLEOPHILIC, SUBSTITUTIONS BY OXYGEN AND NITROGEN NUCLEOPHILES

3.7.1   *O*-Alkylation of alcohols

Alcohols in refluxing cyclohexane give their methyl ethers in high yields using a reagent combination of dimethyl sulphate and alumina [5]. Some diols, when preadsorbed on alumina, react with dimethyl sulphate to produce the corresponding monomethyl ethers selectively (equation (3.14)). This cannot be achieved by use of NaOH or R_2CO_2 with dimethyl sulphate.

It is also possible to use diazomethane as the reagent in the presence of alumina [5] or silica gel [39]. In the latter case, primary, secondary and tertiary alcohols react at room temperature and the silica is thought to participate via formation of methyl diazonium silicate (this slowly decomposes to N_2 and methylated SiO_2). The SiO_2-catalysed method is especially suitable for acid and base labile structures.

More recently, Nishiguchi [83] has demonstrated selective mono-alkylation of 1,n-diols catalysed by aluminium sulphate supported on silica gel.

$$HOCH_2(CH_2)_8 CH_2OH \xrightarrow[Al_2O_3]{Me_2SO_4} HOCH_2(CH_2)_8 CH_2OMe \qquad (3.14)$$
$$(86\%)$$

3.7.2   Ring-opening of epoxides

Nucleophilic ring-opening of epoxides is an important synthetic method for the preparation of *trans*-1,2-difunctionalized molecules. In practice, however, these reactions frequently give mixtures of products, largely as a result of secondary processes such as rearrangement (e.g. to allylic alcohols or aldehydes), elimination, and redox reactions. Furthermore, mixtures of regioisomers are generally obtained from unsymmetrical epoxides. Work by Posner and Rogers [40] represents a significant improvement over standard procedures for the nucleophilic opening of epoxides. In general, the nucleophile (alcohol, amine, carboxylic acid) is supported on alumina (Woelm W-200 neutral) at a loading of 4% by weight. For good results, it is important to remove all the surface water by high-temperature (400°C) vacuum drying (alternatively a freshly opened tin of alumina is used) and thereafter it is recommended that further manipulations are carried out in a glove box. If these conditions are followed, a solution (Et_2O solvent) of epoxide (1 mmol per 7.5 g alumina) can then be applied to the dry alumina, and ring-opening proceeds under exceedingly mild conditions (10 min to 1 h, 25°C). This is particularly useful for acid-sensitive, labile epoxides, giving moderate to excellent yields in most instances of a single product (e.g. equation (3.15)). Moreover, this technique is applicable to polyfunctional molecules (most functional groups are unaffected, e.g. − OH, − CN, ketones, esters and alkenes) and reactions proceed regioselectively (less substituted carbon) and stereospecifically (trans). Particularly impressive are the reactions with medium-sized ring (C_8–C_{12}) epoxides, which proceed cleanly with no products formed via transannular interactions. This appears to be the best method for such epoxides. Similar reactions have been performed with *tert*-butyl hydroperoxide [41].

$$(3.15)$$
$$(95\%)$$

For less sensitive epoxides, it is also possible to perform the reactions with a catalytic amount of W-200-N alumina (10% w/w) on a large scale. For example, cyclohexene oxide (30 g, 0.3 mol) is ring-opened by allyl alcohol (1.5 mol) using only 3 g of alumina (fresh W-200-N) in dibutyl ether at 100°C/1 h to give a 78% yield of β-functionalized alcohol (equation (3.16)) [40].

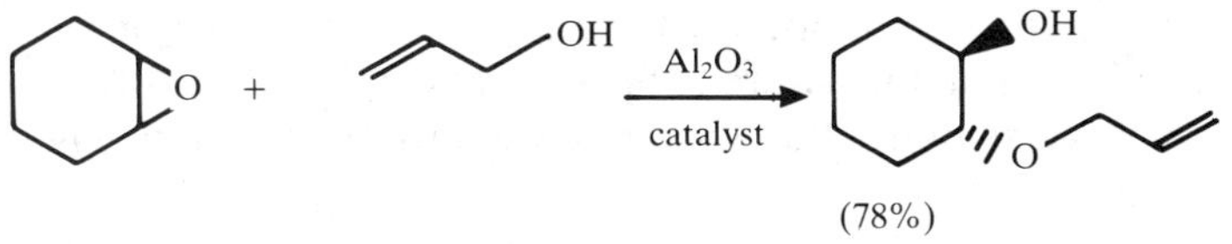

(3.16)

(78%)

Alumina proved ineffective for reactions of 2,3-epoxycarboxylates with nucleophiles, giving products derived from rearrangement. However, recent work by Saa and coworkers [42] illustrates the utility of synthetic $AlPO_4$–Al_2O_3 ring-opening of such compounds by alcohols and carboxylic acids (e.g. equation (3.17)). Amines do not react, but Ritter-type ring-opening with acetonitrile permits regioselective introduction of the acetamido group (equation (3.18)). These reactions are performed in dichloromethane (5 mmol of epoxide in 70 ml CH_2Cl_2) using two equivalents of nucleophile (10 mmol on 7.5 g $AlPO_4$–Al_2O_3). Previous work [43] has shown that electrophilic activation of these epoxides favours attack at the carbon α to the ester functionality. The failure of basic nucleophiles (e.g. $^nBuNH_2$) suggests that Lewis acid sites are responsible for the activity of $AlPO_4$–Al_2O_3. The origin of the regioselectivity is thought [42] to stem from simultaneous coordination by both oxiranic oxygen and carbomethoxy group to the surface Lewis sites of $AlPO_4$–Al_2O_3. Subsequent attack by a nucleophile takes place more rapidly at the harder centre (β-carbon). Furthermore, reaction on a vinylogous ester gave equal amounts of the two regioisomers.

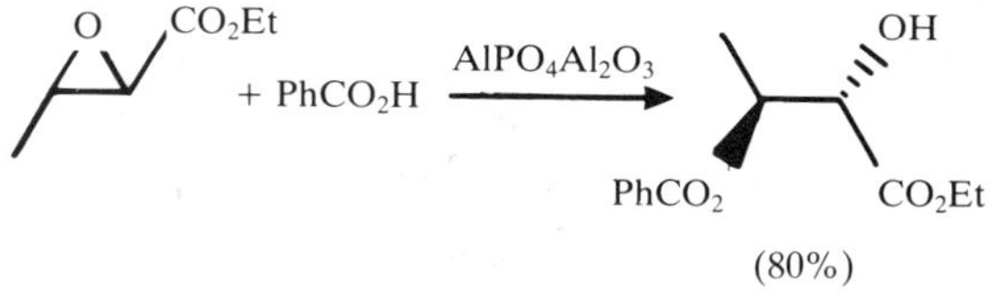

(3.17)

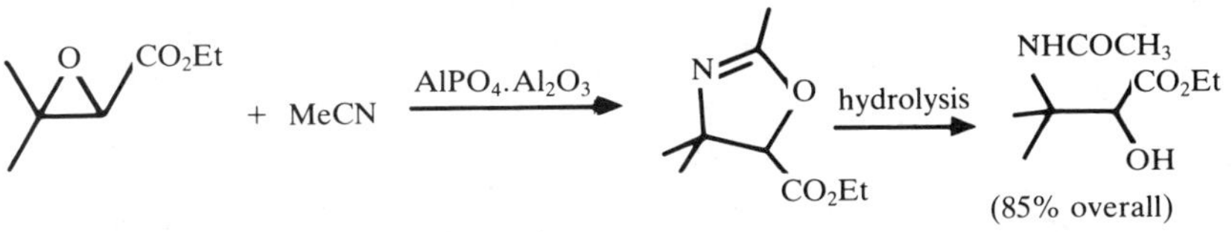

(80%)

(3.18)

(85% overall)

3.7.3 *N*-Substitution reactions

3.7.3.1 N-*Alkylation*
KF/Al_2O_3 promotes smooth *N*-alkylation of carboxamides, lactams and other *N*-heterocycles with alkyl halides or dialkyl sulphates under mild conditions [44]. Conventional methods for such *N*-alkylations require rather drastic conditions (e.g.

NaNH$_2$) and are prone to result in relatively low yields. The method using KF/Al$_2$O$_3$ is useful for *N*-alkylation of secondary amides and *N,N*-dialkylation of primary amides, and often appears to be catalytic in KF/Al$_2$O$_3$. Selective *N*-monoalkylation of primary amides is possible using the stoichiometric base 'KOH on alumina' prepared by grinding a physical mixture of alumina (M-60-B, mean pore diameter = 6 nm) and KOH (powder) in a mortar. Reagents prepared by aqueous impregnation of KOH on alumina are not selective. Furthermore, the selectivity of the reaction was shown to be extremely dependent upon the nature of the solvent employed [45]. Dioxan is the best solvent with few exceptions and is considered to be a stabilizing medium for

Table 3.10 — Reactions of amides and amines with alkylating agents in the presence of solids

Amide/amine	Electrophile	Product	Yield (%)	Solid	Ref.
	MeI		100	KF/Al$_2$O$_3$	[44]
	Me$_2$SO$_4$		89	KF/Al$_2$O$_3$	[44]
	Me$_2$N⌒⌒Cl		77	KF/Al$_2$O$_3$	[46]
	⌒⌒Br	(96.8% monoalkylation)	71	KOH/Al$_2$O$_3$	[45]
	*n*BuBr	(99.5 % monoalkylation)	60	KOH/Al$_2$O$_3$	[45]

potassium enolate-like intermediates in the reaction. Some examples [44-46] of *N*-alkylation reactions are given in Table 3.10.

3.7.3.2 *Transamidation*
Hydrated zirconium oxide is an excellent reusable catalyst for transamidation reactions (e.g. equation (3.19)) [47]. The advantages of this method are high yields, short reaction time (ca. 2 h), easy product isolation, low cost, and stability of the oxide, which can be stored at room temperature in air and used at any time. Hydrated zirconium oxide is a more active catalyst than silicic acid, alumina, or titania.

$$\text{(structure)} + Me_2NCHO \xrightarrow{ZrO_2} \text{(structure)} \quad (3.19)$$

$$(92\%)$$

3.7.3.3 N-*Arylation*
The copper-catalysed *N*-arylation of amides, the Goldberg reaction, usually requires drastic conditions and produces only moderate yields of the desired products. Much better results can be achieved by use of copper on silica gel (RCH 60/35) as a catalyst system for this reaction (Table 3.11) [48]. Potassium acetate can be used as a base, and the acetic acid resulting from the reaction mixture removed by continuous distillation.

3.8   ELIMINATION REACTIONS OVER SOLID SUPPORTS

3.8.1   Dehydration of alcohols
The heterogeneous alumina-catalysed dehydration of alcohols has been known since the end of the eighteenth century. Nevertheless it is only recently that efforts have been made to explain the elementary processes [49]. With the exception of tertiary alcohols, most simple alcohol dehydrations require temperatures well in excess of 250°C. This factor normally means that it is more convenient to derivatize an alcohol (e.g. by tosylation) before elimination to give the desired alkene. Zeolites are anyway generally better dehydration reagents than the simple amorphous solids (e.g. Al_2O_3), and these materials are discussed more fully in Chapter 5. However, amorphous solids do find some utility. For example, ferric chloride on silica gel is a useful stoichiometric reagent for the dehydration of allylic, tertiary and sterically strained secondary alcohols [50]. Reactions are rapid at room temperature and yields of alkenes are excellent; moreover, regioselective monodehydration of various polyhydroxy steroids can readily be effected. The same reagent can be used for pinacol and acyloin rearrangements while a modified reagent, prepared by addition of ca. 2% water, can be used for the epimerization of tertiary alcohols and the conversion of epoxides into

Table 3.11 — Direct *N*-arylation of amides promoted by copper on silica gel with KOAc base

Amide	Haloarene	Product	Yield (%)
			48
			64
	PhBr		72
			73

diols. More recently, the reagent had been used to induce ring enlargement of 1-substituted cyclobutanols and cyclopentanols (e.g. equations (3.20) and (3.21)) [51]. $FeCl_3/SiO_2$ is now commercially available from Fluka.

$$(3.20)$$

$$(3.21)$$

3.8.2   Dehydration of amides

The dehydration of primary amides to the corresponding nitriles has traditionally been carried out with a number of acidic reagents, including $AlCl_3$, P_2O_5, $POCl_3$ and trimethylsilyl polyphosphate. Most of these reagents are used in molar excess and are effectively destroyed during dehydration. Moreover, such materials render the medium corrosive, and cause significant problems during workup of the reaction mixture. Solids (zinc oxide, zeolites) which become strongly acidic at high temperatures (300 to 400°C) are able to promote the reaction, but such conditions are not applicable in many cases. However, certain zirconias are excellent catalysts for the dehydration of primary amides (e.g. equation (3.22)) [52]. Hino and Arata have described [53] a highly active form of zirconia (prepared by treatment of $Zr(OH)_4$ with H_2SO_4, then calcination at 500°C) which has superacid sites with Hammet acidity values (H_0) of -16. Refluxing the amides in diphenylether (259°C) or o-nitrotoluene (223°C) in the presence of zirconia (40% by weight of amide) provides the corresponding nitriles in good yields (61 to 98%) within 8 h [52]. At the end of the reaction, the zirconia can be recovered, washed with n-butanol and dried. In this way the catalyst can be recycled.

$$CH_3(CH_2)_7CH=CH(CH_2)_7CONH_2 \xrightarrow[\Delta]{ZrO_2} CH_3(CH_2)_7CH=CH(CH_2)_7CN \qquad (3.22)$$
$$(94\%)$$

3.8.3   Dehydrohalogenation

Ando and coworkers have found that KF/Al_2O_3 is an excellent basic catalyst for the dehydrohalogenation route to alkenes and acetylenes (Table 3.12) [54]. Even primary halides are converted under mild conditions to give the unsaturated products in high yield. Similarly, KO-^{t}Bu on silica permits controlled vapour phase dehydrohalogenation of small ring halocycloalkanes to form highly strained cycloalkenes (Table 3.12) [55]. Silica gel has also been used to promote mild and selective dehydrobromination of β,γ-dibromoesters in the cepham series (Table 3.12) [56].

3.8.4   Desulphonations

Mildly acidic silica gel (Merck 60) promotes destannylsulphonation of compound (4) at 25°C to afford the corresponding Z-vinylsilane (83% yield, 98% Z isomer) (equation (3.23)) [57]. Alternatively, destannylsulphonation in the presence of benzenesulphinic acid on silica results in a 95% yield of the E-vinyl silane (isomeric purity 97%). Similar eliminations have been reported for β-sulphonylketones, which in the presence of basic alumina give clean formation of the α,β-unsaturated ketones with no rearrangement of the new double bond [58] (equation 3.23)).

3.8.5   Hoffman elimination of N-oxides

Alumina has been used to perform the classical Hoffman elimination of amine N-oxides under very mild conditions [59].

$$(3.23)$$

Table 3.12 — Dehydrobromination reactions effected by solids

Substrate	Product	Yield (%)	Solid	Reference
PhO–CH$_2$CH$_2$Br	PhO–CH=CH$_2$	73	KF/Al$_2$O$_3$	[54]
Ph–C(Br)–CH$_2$Br	Ph–C(Br)=CH$_2$	80	KF/Al$_2$O$_3$	[54]
Ph–CH(Br)–CH$_2$Br	Ph–C≡C–H	70	KF/Al$_2$O$_3$	[54]
spiro cyclopropane–cyclohexane with Br	spiro cyclopropane–cyclohexane	85	KOBu^t/SiO$_2$	[55]
bicyclic dibromide	bicyclic bis-methylene	85	KOBu^t/SiO$_2$	[56]
β-lactam thiazoline with CH$_2$Br, CHBr, RO$_2$C, H	β-lactam thiazoline with =C, CH$_2$Br, CO$_2$R	'high'	SiO$_2$	[56]

3.9 SELECTIVE REDUCTIONS OVER SOLIDS

3.9.1 Conjugate reductions of αβ-unsaturated carbonyl and nitro compounds

There are numerous supported reagents available that catalyse or bring about the reduction of organic substrates. Sodium borohydride either on silica gel or on alumina is now commercially available from Fluka. The immediate advantages of these reagents are that they can be used in non-polar solvents (e.g. benzene) and that alkali-sensitive groups are not affected. For example, NaBH$_4$/SiO$_2$ is an excellent selective reagent for the reduction of β-nitrostyrenes to the corresponding nitroal-kanes (equation (3.24)) [60]. The method avoids Michael addition to the βnitrostyrene starting material because the mildly acidic silica surface brings about rapid protonation of the α-carbanion intermediate before Michael addition can take place.

$$\text{Ar}\diagdown\diagup^{\text{NO}_2}_{\text{R}} \quad \xrightarrow{\text{NaBH}_4/\text{SiO}_2} \quad \text{Ar}\diagdown\diagup^{\text{NO}_2}_{\text{R}} \qquad (3.24)$$

$$(92 - 98\%)$$

Chemoselective reduction of α, β-unsaturated carbonyl or nitro compounds to the corresponding ketones, aldehydes or nitroalkanes is possible using the Hantzsch ester, 3,5-bis(ethoxycarbonyl)-1,4-dihydro-2,6-dimethylpyridine (HEH), in the presence of silica gel (e.g. equations (3.25) and (3.26)) [61, 62]. It has been revealed [62] that the silanol group on the surface of silica gel acts as a general-acid catalyst in the reduction mechanism. Moreover, the silica gel draws the materials subjected to the reduction closer to its surface. Thus HEH/silica gel is a mimetic system (NAD(P)$^+$-NAD(P)H, enzyme-like) which exerts dual ability — binding and catalysis.

$$\xrightarrow{\text{HEH/SiO}_2} \qquad \text{CHO} \qquad (3.25)$$

$$(67\%)$$

$$\xrightarrow{\text{HEH/SiO}_2} \qquad (100\%) \qquad (3.26)$$

NaBH$_4$/alumina has been used for the reduction of ketones in the presence of enol acetates (equation (3.27)) [63]. Under the normal conditions used for NaBH$_4$ reductions, enol acetates undergo rapid hydrolysis.

$$\xrightarrow{\text{NaBH}_4/\text{Al}_2\text{O}_3} \qquad (3.27)$$

3.9.2 Reductions of carbonyl groups

Tributyltin hydride normally reduces aldehydes and ketones only if they bear powerful electon-withdrawing groups. In the presence of silica gel, however, it reduces simple aldehydes and ketones to the corresponding alcohols [64]. Sulphoxides, nitro groups, esters, nitriles and halides are unaffected under these conditions. The rate of reduction of carbonyl groups is in the order aldehydes > dialkyl ketones > alkyl aryl ketones > diaryl ketones, and treatment of mixtures of aldehydes and ketones with the tributyltin hyhdride/SiO_2 reagent results in almost exclusive reduction of the aldehyde.

A truly remarkable selective reagent for the reduction of aldehydes has been reported by Babler and Sarussi [65]. Under homogeneous conditions, pyridine–borane complex reduces carbonyl functions only at elevated temperatures: however, in the presence of alumina (activated, chromatographic grade) it reduces aldehydes at 20°C without affecting ketones, to give the corresponding primary alcohols in high yields. A wide variety of solvents are tolerated and do not significantly affect the observed aldehyde-selective reductions. It is suggested that the role of the alumina (possessing active Lewis sites) is to polarize the pyridine–borane complex so as to provide a more nucleophilic source of hydride [65]. Although borane–THF readily reduces carboxylic acid functions, pyridine–borane complex on alumina failed to reduce lauric acid (over 2 h), and the unreacted borane could be recovered in over 80% yield.

Meerwein–Ponndorf–Verley (MPV) type carbonyl reduction at room temperature using 2-propanol over W-200-N vacuum dehydrated alumina [66] has several useful advantages over alternative methods: (a) α, β-unsaturated aldehydes can be reduced cleanly to the corresponding allylic (i.e. not saturated) alcohols; (b) aldehydes can be reduced even in the presence of some ketones; and (c) many reducible functionalities (e.g. nitro, ester, cyano, bromo and iodo) survive. Although troublesome to prepare (alumina Woelm W-200 calcined at 400°/0.6 torr/24 h), the 2-propanol-alumina reagent can be stored in a sealed vial for at least 8 months without loss of reducing activity.

A clear demonstration of the usefulness of the reagent is given in the reduction of the chromone (5) (equation (3.28)), which led selectively and in 80% yield to the allylic alcohol (6) [66]. The substrate suffered reduction of both the aldehyde group and the α, β-double bond with the following reagents: $NaBH_4$, $NaBH_4 \cdot AlCl_3$, DIBAL and $NaBH_3CN$/acid. Moreover, under homogeneous MPV conditions the chromone failed to react.

$$(3.28)$$

(5)   (6)

A further interesting and very useful feature of the reagent system is that the ease of ketone reduction appears to be largely dependent on steric factors, which in turn dictate the nature of the adsorption of the substrate onto the alumina surface. Thus, 2-substituted cyclohexanones or similarly hindered ketones are not reduced. Further discrimination can be achieved by using the more hindered reagent: 2,4-dimethyl-3-

pentanol on alumina, but even with 2-propanol, significant selectivities can be observed (e.g. equation (3.29)).

Hydrated zirconium oxide is also an attractive catalyst for the MPV reduction of carbonyl compounds [67]. A disadvantage in the use of $Al(O^iPr)_3$ as a homogeneous catalyst is the need for a strong acid workup together with tedious handling of the voluminous aluminate salts. Hydrated zirconium oxide allows easy product isolation, provides high yields under neutral conditions and is a stable oxide which can be stored at room temperature in air and used at any time. Matsushita and Ishiguro have recently described [67] its application for the MPV reduction of quinones to the corresponding hydroquinones (equation (3.30)).

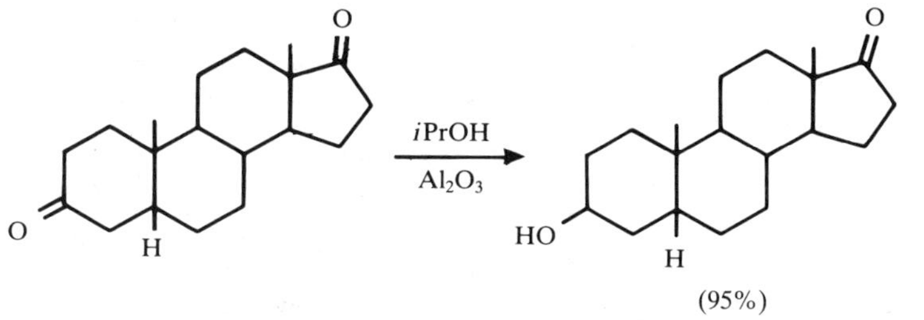

(3.29)

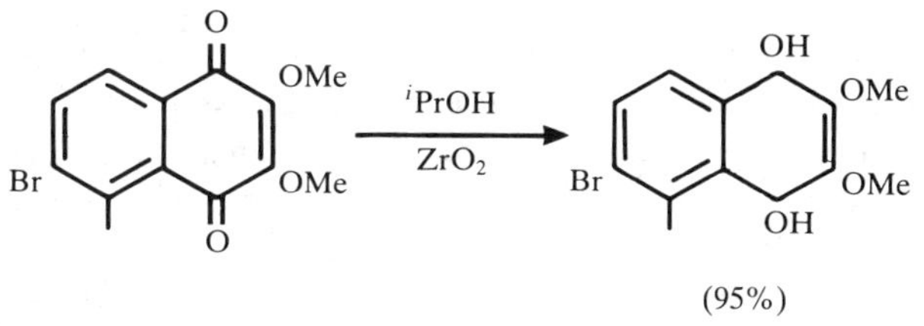

(95%)

(3.30)

(95%)

3.10   OXIDATION OVER SOLIDS

3.10.1   Reagents

A variety of supported reagents have been developed for the purpose of oxidation, and these have been reviewed by McKillop and Young [1]. All of these supported oxidants have the usual advantages of simple workup. The following account therefore focuses on particular supported oxidants which are efficient and fulfil a chemo- or regioselective purpose. Several supported oxidants are commercially available, including chromyl chloride on silica-alumina (Fluka), potassium permanganate on silica gel (Fluka) and sodium (meta)periodate on alumina (Fluka). Spectroscopic studies for the $KMnO_4$ [68] and $NaIO_4$ [69] reagents suggest that the most active loading is a discrete physisorbed salt monolayer. Because of their finely divided state, these inorganic-supported oxidants can be employed in non-protic solvents and therefore overcome potential workup problems (some products might be difficult to extract from aqueous media).

3.10.2   Oxidation of alcohols

The 'Fetizon reagent' [70] (Ag_2CO_3/Celite) is prepared by mixing aq. $AgNO_3$ with Celite (purified) and aq. Na_2CO_3; the green precipitate is dried in a rotary evaporator for several hours. Reactions are normally performed in toluene or dichloromethane.

3.10.2.1   Use of Fetizon's reagent

In the oxidation of an alcohol using Ag_2CO_3/Celite, α-hydrogen abstraction is the rate-determining step (Scheme 3) [71]. Sometimes, dramatic changes in product distribution are observed on changing reaction solvent, and it has been suggested that this may be due to the fate of water (produced as a reaction byproduct), which is only removed from the system if a Dean and Stark trap is employed.

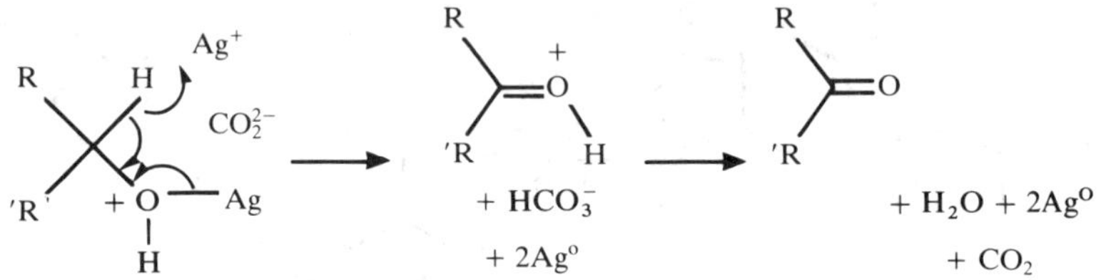

Scheme 3 — Oxidation of a secondary alcohol by silver carbonate on Celite.

An important feature of the Fetizon reagent is that different types of alcohols are oxidized at different rates. This fact is of considerable synthetic utility for selective oxidation of one hydroxyl group in a diol or triol. In general, Ag_2CO_3/Celite oxidizes secondary more readily than primary alcohols. The outcome of oxidation also depends on the relative disposition of neighbouring hydroxyl groups and the presence of conjugating functionalities. The most important factor, however, is the stereochemistry of adsorption of the diol or triol onto the reagent surface.

1,2-Diols linked to unsaturated groups (e.g. alkenes or arenes) tend to give diol cleavage to form aldehydes. Aliphatic 1,3-, 1,7- and 1,8-diols undergo monooxidation, with secondary alcohols being oxidized in preference to primary (e.g. equation (3.31)). Aliphatic 1,4-, 1,5- and 1,6-diols are converted to lactones (e.g. equations (3.32) and (3.33)), or sometimes lactols. A number of other functionalities are unaffected (e.g. epoxide and alkene in equations (3.32) and (3.33)). More extensive discussion and literature references are available in reference [13].

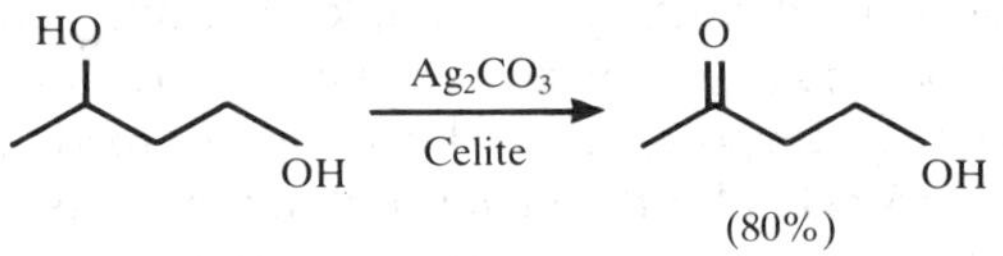

(3.31)

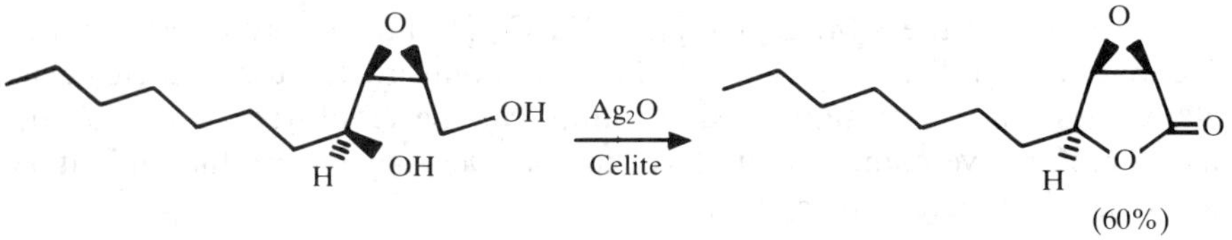

(3.32)

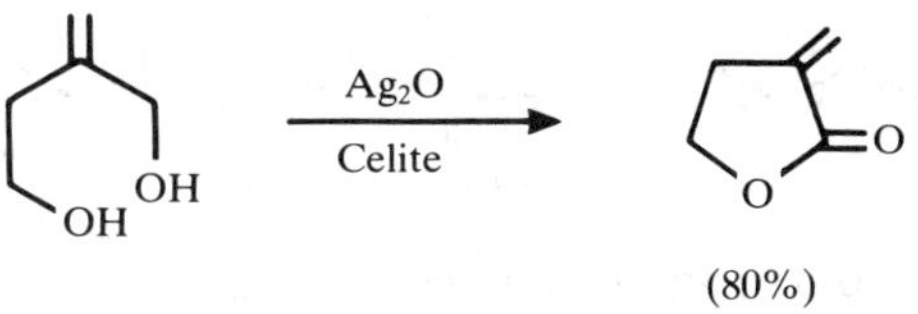

$$(3.33)$$

(80%)

The Fetizon reagent has found extensive application in steroid chemistry [72]. The nature of the products formed on oxidation of steroidal diols and triols depends critically on the stereochemistry of the hydroxyl groups and the manner in which the steroid is adsorbed onto the surface of the reagent. Oxidation of androstane-$3\beta,5\alpha,6\alpha$-triol with Ag_2CO_3/Celite in benzene gives a particular ketodiol as the main product (equation (3.34)). It has been suggested [72] that the selectivity of this transformation is due to the accessibility of the 3α-hydrogen to the flat reagent surface, whereas the 6β-hydrogen is more remote. Equation (3.35) shows an even more selective example.

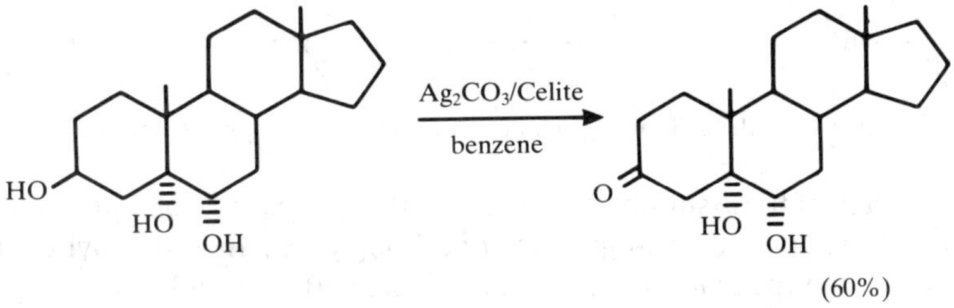

$$(3.34)$$

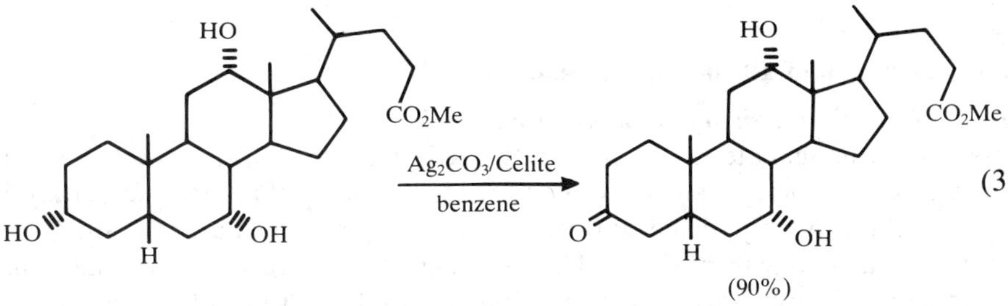

(60%)

$$(3.35)$$

(90%)

3.10.2.2 *Use of supported nitrates*
$Cu(NO_3)_2$ and $Zn(NO_3)_2$, supported on silica gel, are cheap, simple and convenient reagents for the efficient oxidation of alcohols to the corresponding carbonyl compounds (e.g. equation (3.36)) [73]. These reagents also have the advantage of being stable in dry air, although insufficient or excessive drying can cause a decline in oxidizing activity (ideal conditions for activation are 130°C at 7×10^2 Pa). Various oxide supports have been investigated, and silica gel (Merck type 60 or Fuji Davison BW300) is considered to be the most effective. A critical loading value of 2 mmol g^{-1} corresponds to monolayer coverage and maximum rate of oxidation. It would appear from the effect of excess moisture, that acidic surface silanol groups are necessary to activate the nitrate salts.

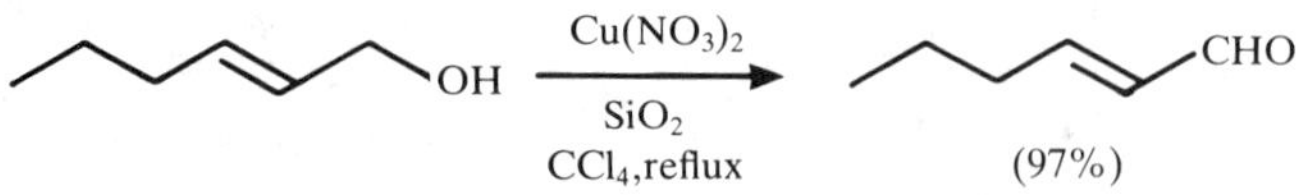

$$(3.36)$$

Nishiguchi and Asano suggest that the mechanism of oxidation involves formation of the radical nitrogen dioxide which then abstracts a hydrogen atom from the alcohol (Scheme 4) [73]. The presence of radical species is evident from the detrimental affect of CO_2 or radical scavengers like 2,4,6-tri-*tert*-butylphenol and the fact that rates of oxidation of para-substituted benzyl alcohols were not affected by electron release or withdrawal (ρ value of the Hammett equation was zero).

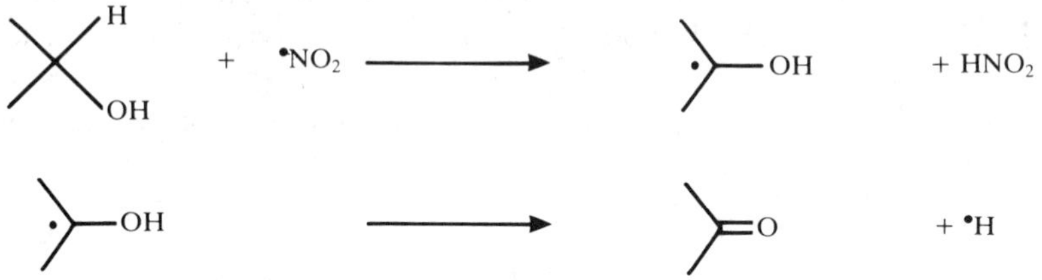

Scheme 4 — Mechanism of oxidation of alcohols with supported nitrate.

A disadvantage of these supported nitrates is the competitive oxidative cleavage of secondary ether groups. For example, acetone (57% yield) and 2-propyl nitrite (25% yield) were formed in the reaction between bis(2-propyl)ether and $Cu(NO_3)_2$ on silica gel in refluxing CCl_4 for 30 min [73].

3.10.3   Oxidation of alkenes and alkanes

3.10.3.1   *Use of potassium permanganate on silica*
Potassium permanganate has been widely used as an oxidant in aqueous solution. The versatility of KMO_4 can be extended with the use of phase transfer catalysis or the preparation of supported forms. Potassium permanganate on silica gel is commercially available from Fluka. The silica-supported reagent is particularly useful for mild and selective carbon–carbon double bond cleavage reactions [73]. The reaction occurs at room temperature over a few minutes and is equally effective for cleaving double bonds having one, two or three substituents, including electron-withdrawing groups. The reactions are conveniently performed by elution of a benzene solution through a short column of $KMnO_4/SiO_2$. In the example given in equation (3.37), all the traditional methods failed (e.g. aq. $KMnO_4$/acid or base; $KMnO_4/MgSO_4$; $KMnO_4/NaIO_4$).

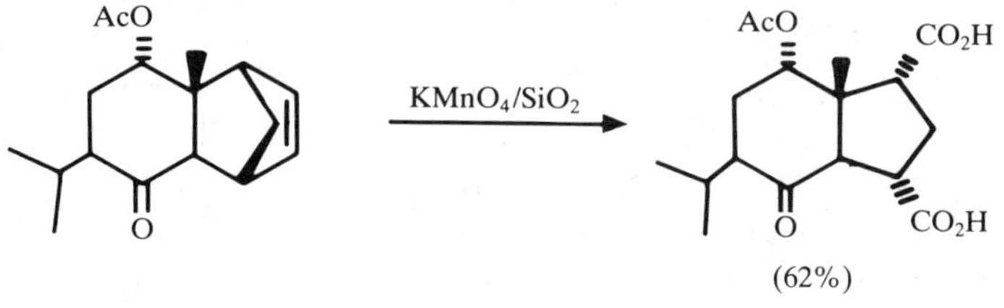

$$(3.37)$$

(62%)

3.10.3.2 Use of ozone and silica

Ozone is adsorbed by silica gel at low temperatures (4.5% w/w at $-78°C$) and the resulting reagent is ideal for direct hydroxylation of hydrocarbons [74]. The oxygen insertion occurs preferentially at tertiary carbon atoms and proceeds with retention of configuration (e.g. equation (3.38)). In certain cases, direct oxidation of a methylene group to a ketone can take place (equation (3.39)) [75]. The ozone/SiO_2 reagent oxidatively destroys aromatic compounds that do not contain electron-withdrawing groups. In general, ozonation of extensive monofunctional structures (such as cyclododecyl acetates) adsorbed on silica gel affords oxidation remote from the binding sites (associated with polar functional groups). Beckwith and Duong have developed this technique so that it can be adopted more easily in synthesis [76]. The ozone/silica reagent can be more conveniently used in Freon 11 solvent (inert to attack by ozone). Remarkably, regioselective hydroxylations can be achieved for certain alcohols by first preparing the hydrogen succinate ester and reacting it with a solution of ozone in Freon 11 in the presence of silica gel. For example, oxidation of the hydrogen succinate ester of 3,7-dimethyloctanol gave the 1,7 diol in 70% yield with virtually no attack at the alternative tertiary position (equation (3.40)). Thus, a strong association between the silica and the hydrogen succinate functionality protects adjacent centres from attack.

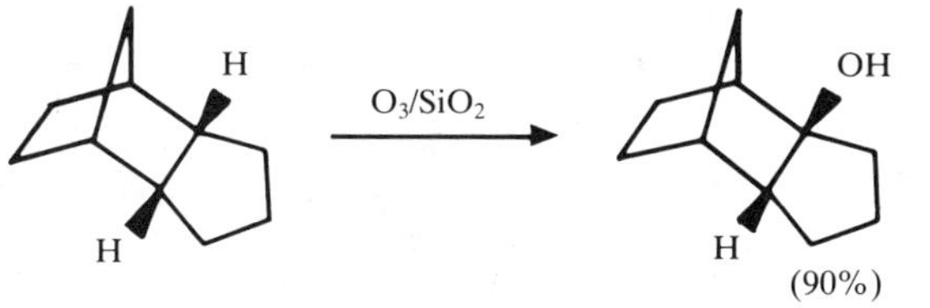

(3.38)

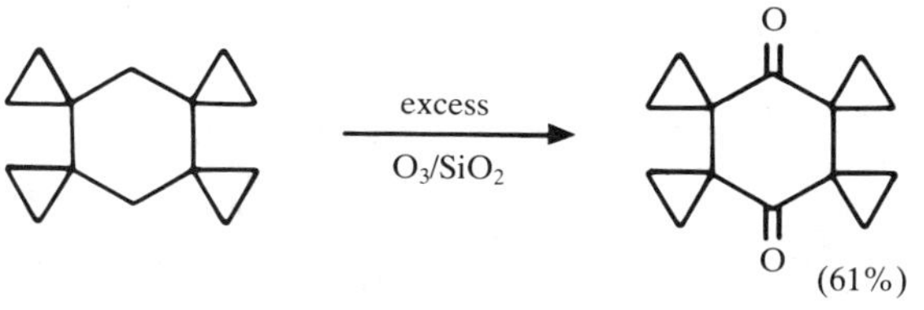

(3.39)

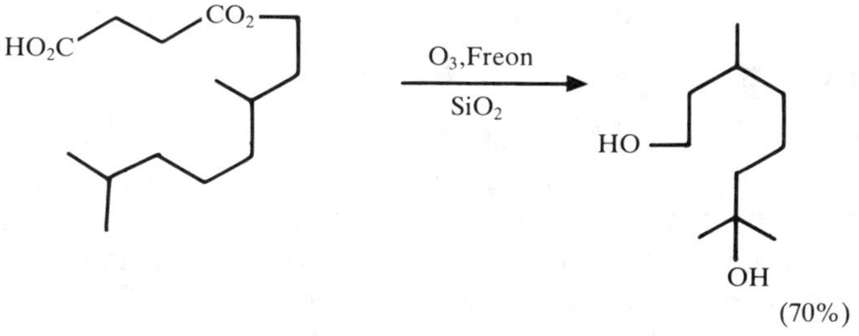

(3.40)

3.10.3.3 Use of tert-butyl hydroperoxide and selenium dioxide on silica

Allylic methyl groups in medium ring compounds are selectively oxidized to primary alcohols and α,β-unsaturated aldehydes using tert-butyl hydroperoxide and selenium dioxide supported on silica gel [77]. This method is highly selective for exocyclic allylic methyl groups, with no trace of methylene oxidation. For example, 6,7-diacetoxybicyclohumulene reacts with 70% *tert*-butyl hydroperoxide (3 equivalents) over SeO_2/SiO_2 (5% loading) to give 6,10-diacetoxybicyclohumulene-12-ol in 80% yield with 20% of unchanged substrate (equation (3.41)) [77]. Another feature of this method is that the intermediate allyl selenide can isomerize to give the most stable geometry and subsequently react to give the corresponding allylic alcohol. For example, caryophyllene (with a *trans*-6,7-double bond) gives *cis*-caryophyllen-13-ol in 90% yield (following reductive workup of the so-formed α,β-unsaturated aldehyde) (equation (3.42)).

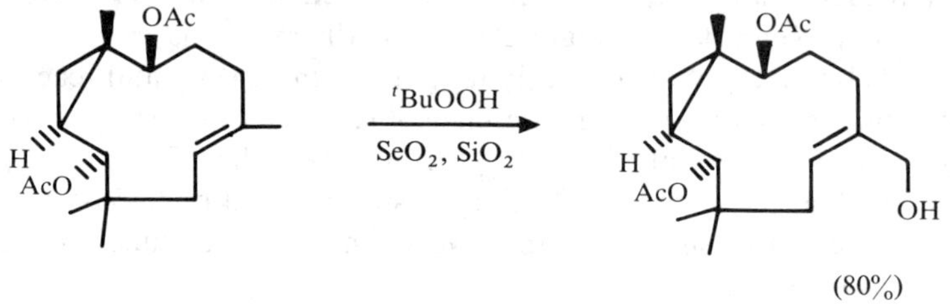

$$(3.41)$$

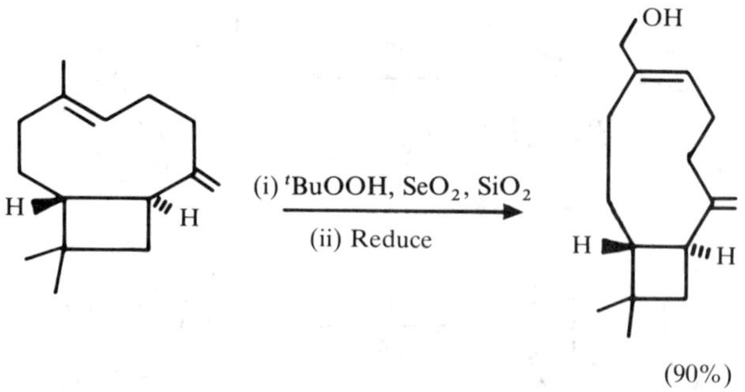

(80%)

$$(3.42)$$

(90%)

3.11   ALLYLIC CHLORINATION OVER SILICA

Excellent regioselective allylic chlorinations are achieved using *tert*-butyl hypochlorite on silica gel [78]. Non-polar solvents (e.g. hexane) with acidic silica gel (Kieselgel 60) promote formation of allylic chlorides from simple olefins in high yields (70–85%). However, changing solvent to diethyl ether results in a more reactive reagent system capable of double allylic chlorination. Geranyl acetate may be chlorinated regioselectively at the terminal alkene (remote from the acetate functionality) in 70% yield (equation (3.43)). This synthetic method has been successfully applied in the synthesis of San Jose Scale pheromones [79]. Furthermore, the reagent *tert*-$BuOCl/SiO_2$ has been used to prepare 1,2-epoxy-3-methyl-3-butene, an intermediate for the synthesis of (*E*)-phytol (equation (3.44)) [80].

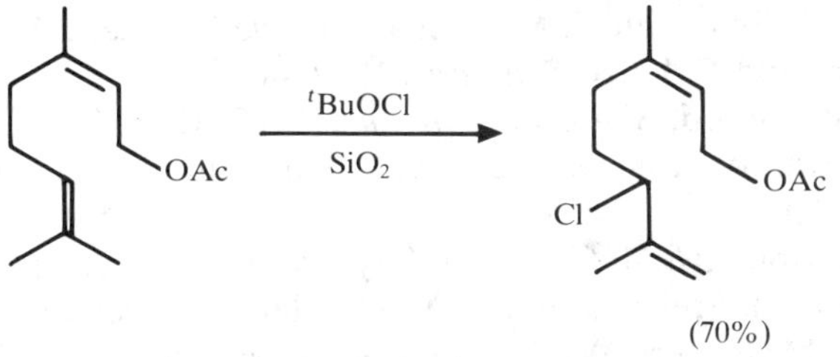

$$(3.43)$$

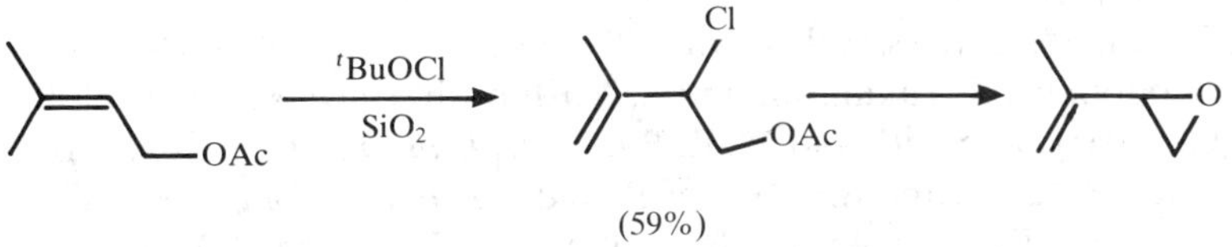

(70%)

$$(3.44)$$

(59%)

REFERENCES

[1] A. McKillop and D. W. Young, *Synthesis*, 1979, 401 and 481.

[2] G. H. Posner, *Angew. Chem. Int. Ed. Engl.*, 1978, **17**, 487.

[3] G. Posner and M. Oda, *Tetrahedron Lett.*, 1981, **22**, 5003.

[4] H. Brockmann and H. Schodder, *Chem. Ber.*, 1941, **74**, 73.

[5] H. Ogawa, Y. Ichimura, T. Chihara, S. Teratani and K. Taya, *Bull. Chem. Soc. Jpn*, 1986, **59**, 2481, and references therein.

[6] J. Bertin, H. B. Kagan, J.-L. Luche and R. Settan, *J. Am. Chem. Soc.*, 1974, **96**, 8113.

[7] D. Villemin and M. Ricard, *Synthetic Commun.*, 1987, **17**, 283, and references therein.

[8] J. H. Clark, D. G. Cork and M. S. Robertson, *Chem. Lett.*, 1983, 1145.

[9] D. Villemin, *J. Chem. Soc. Chem. Commun.*, *1983*, **19**, 1092.

[10] J. Yamawakie, T. Kawate, T. Ando and T. Hanafusa, *Bull. Chem. Soc. Jpn*, 1983, **56**, 1885.

[11] D. Villemin, *Chem. Ind. (London)*, 1985, 166.

[12] D. E. Bergbreiter and J. Lalonde, *J. Org. Chem.*, 1987, **52**, 1601.

[13] J. M. Mèlot, F. Taxier-Boullet and A. Foucaud, *Tetrahedron*, 1988, **44**, 2215.

[14] M. E. Alonso and A. Morales, *J. Org. Chem.*, 1980, **45**, 4530.

[15] B. M. Trost, *Accounts. Chem. Res.*, 1980, **13**, 385.

[16] D. Ferroud, J. P. Genet and J. Muzart, *Tetrahedron Lett.*, 1984, **25**, 4379.

[17] D. Villemin, *J. Chem. Soc. Chem. Commun.*, 1985, 870.

[18] A. B. Alloum and D. Villemin, *Synth. Commun.*, 1989, **19**, 2567.

[19] C. Gagnieu and A. Grouiller, *Carbohyd. Res.*, 1980, **84**, 61.

[20] Y. Kamitori, M. Hojo, R. Masuda, T. Kimura and T. Yoshida, *J. Org. Chem.*, 1986, **51**, 1427.

[21] M. Hojo and R. Masuda, *Synthesis*, 1976, 678.

[22] F. Huet, M. Pellet and J. M. Conia, *Tetrahedron Lett.*, 1977, 3505.

[23] P. Lescure and F. Huet, *Synthesis*, 1987, 404.

[24] K. S. J. Stapleford, *Synthetic Commun.*, 1982, **12**, 651.

[25] Y. Kamitori, M. Hojo, R. Masuda, T. Izumi and S. Tsukamoto, *J. Org. Chem.*, 1984, **49**, 4161.

[26] T. H. Johnson, U.S. 4463207 (31 July 1984); *Chem. Abstr.*, **101**, 151564s.

[27] S.-H. Baek, M. Srebnick and R. Mechoulam, *Tetrahedron Lett.*, 1985, **26**, 1083.

[28] K. Smith, M. Butters. W. E. Paget and B. Nay, *Synthesis*, 1985, 1155; K. Smith, *Stud. Surf. Sci. Catal.*, 1991, **59**, 55.

[29] K. Smith, M. Butters and B. Nay, *Tetrahedron Lett.*, 1988, **29**, 1319.

[30] M. Kodomari, S. Takahashi and S. Yoshitonii, *Chemistry Lett.*, 1987, 1901.

[31] A. G. Mistry, K. Smith and M. R. Bye, *Tetrahedron Lett.*, 1986, **27**, 1051; *idem.* in D. Price, B. Iddon and B. J. Wakefield (eds), *Bromine Compounds: Chemistry and Applications*, Elsevier, Amsterdam, 1988, p. 277; see also H. Konishi, K. Aritomi, T. Okano and J. Kiji, *Bull. Chem. Soc. Jpn*, 1989, **62**, 591.

[32] M. Hudlicky, *J. Org. Chem.*, 1974, **39**, 3460.

[33] H. Parlar and R. Bauman, *Angew. Chem. Int. Ed. Engl.*, 1981, **20**, 1014; see also Patent: DE30355549 Al (29 April 1982).

[34] M. Tanoka, T. Hotta and S. Takeo, *Jpn Kokai Tokkyo, Koho.*, JP 61,122,241 (10 June 1986); *Chem. Abstr.*, **106**, 49780a.

[35] V. V. Veselovsky, A. S. Gybin, A. V. Lozanova, A. M. Moiseenkov, W. A. Smit and R. Caple, *Tetrahedron Lett.*, 1988, **29**, 175.

[36] D. C. Rideout and R. Breslow, *J. Am. Chem. Soc.*, 1980, **102**, 7816.

[37] S. I. Pogrebnoi, Y. B. Kalyan, M. Z. Krimer and W. A. Smit, *Tetrahedron Lett.*, 1987, **28**, 4893.

[38] S. R. Wilson and M. F. Price, *J. Org. Chem.*, 1984, **49**, 722.

[39] K. Ohno, H. Nishiyama and H. Nagase, *Tetrahedron Lett.*, 1979, 4405.

[40] G. H. Posner and D. Z. Rogers, *J. Am. Chem. Soc.*, 1977, **99**, 8208 and 8214.

[41] H. Kropf, H. M. Amirabadi, M. Mosebach, A. Torler and H. von Wallis, *Synthesis*, 1983, 587.

[42] J. Riego, A. Costa and J. M. Saa, *Chemistry Lett.*, 1986, 1565.

[43] D. R. Burfield, S. Gan and R. M. Smithers, *J. Chem. Soc., Perkin Trans. I*, 1977, 666.

[44] J. Yamawaki, T. Ando and T. HanaFusa, *Chemistry Lett.*, 1981, 1143.

[45] K. Sukata, *Bull. Chem. Soc. Jpn*, 1985, **58**, 838.

[46] Y. Nakamoto, Y. Ishizuka, O. Futsukaichi and Y. Ohira, JP. 63,307,866 (15 Dec. 1988); *Chem. Abstr.*, **110**, 192868y.

[47] K. Takahashi, M. Shibagaki and H. Matsushita, *Agric. Biol. Chem.*, 1988, **52**, 853.

[48] B. Renger, *Synthesis*, 1985, 856.

[49] B. H. Davis, *J. Org. Chem.*, 1982, **47**, 900.

[50] E. Keinan and Y. Mazur, *J. Org. Chem.*, 1978, **43**, 1020.

[51] A. Fadel and J. Salaun, *Tetrahedron*, 1985, **41**, 413 and 1267.

[52] G. W. Joshi and R. A. Rajadhyaksha, *Chem. Ind. (London)*, 1986, 876.

[53] M. Hino and K. Arata, *J. Chem. Soc. Chem. Commun.*, 1980, 851.

[54] J. Yamawakie, T. Kaware, T. Ando and T. HanaFusa, *Bull. Chem. Soc. Jpn*, 1983, **56**, 1885.

[55] J. M. Denis, R. Niamayoua, M. Vàta and A. Lablache-Combeir, *Tetrahedron Lett.*, 1980, **21**, 515.

[56] A. Vigevani, M. Foglio, G. Franceschi, P. Masi, A. Suarato, E. Gandini and G. Palamidessi, *J. Chem. Soc., Perkin Trans. I*, 1979, 504.

[57] M. Ochiai, T. Ukita and E. Fujita, *Chemistry Lett.*, 1983, 1457.

[58] J. Vidal and F. Huet, *Tetrahedron Lett.*, 1986, **27**, 3733.

[59] M. Brand and S. E. Drewes, *Synthetic, Commun.*, *1987*, **17**, 795, and references therein.

[60] A. K. Sinhababu and R. T. Borchardt, *Tetrahedron Lett.*, 1983, **24**, 227.

[61] K. Nakamura, M. Fujii, A. Ohno and S. Oka, *Tetrahedron Lett.*, 1984, **25**, 3983.

[62] S. Yasui, M. Fuji and A. Ohno, *Bull. Chem. Soc. Jpn*, 1987, **60**, 4019.

[63] F. Hodosan and N. Serban, *Rev. Roumaine Chim.*, 1969, **14**, 121.

[64] N. Y. M. Fung, P. DeMayo, J. H. Schauble and A. C. Weedan, *J. Org. Chem.*, 1978, **43**, 3977.

[65] J. H. Babler and S. J. Sarussi, *J. Org. Chem.*, 1983, **48**, 4416.

[66] G. H. Posner, A. W. Runquist and M. J. Chapelaine, *J. Org. Chem.*, 1977, **42**, 1202.

[67] H. Matsushita and S. Ishiguro, *Agric. Biol. Chem.*, *1985*, **49**, 3071.

[68] A. Al Jazzaa, J. H. Clark and M. S. Robertson, *Chemistry Lett.*, 1982, 405.

[69] D. N. Gupta, P. Hodge and J. E. Davies, *J. Chem. Soc., Perkin Trans I*, 1981, 2970.

[70] V. Balogh, M. Fetizon and M. Golfier, *J. Org. Chem.*, 1971, **36**, 1339.

[71] F. J. Kakis, M. Fetizon, N. Douchkine, M. Golfier, P. Mourgues and T. Prange, *J. Org. Chem.*, 1974, **39**, 523.

[72] M. Fetizon and P. Mourgues, *Tetrahedron.*, 1974, **30**, 327.

[73] T. Nishiguchi and F. Asano, *J. Org. Chem.*, 1989, **54**, 1531.

[74] Z. Cohen, E. Keinan, Y. Mazur and T. H. Varkony, *J. Org. Chem.*, 1975, **40**, 2141.

[75] E. Proksch and A. de Meijere, *Tetrahedron Lett.*, 1976, 4851.

[76] A. L. H. Beckwith and T. Duong, *J. Chem. Soc., Chem. Commun.*, 1978, 413.

[77] B. R. Chhabra, K. Hayano, T. Ohtsuka, H. Shirahama and T. Matsumoto, *Chemistry Lett.*, 1981, 1703.

[78] W. Sato, N. Ikeda and H. Yamamoto, *Chemistry Lett.*, 1982, 141.

[79] L. Novak, L. Poppe and C. Szàntay, *Synthesis*, 1985, 939.

[80] Y. Kobayashi, F. Sato, S. Suzuki and Y. Fujita, JP. 6,226,279 (26 June 85) *Chem. Abstr.*, **107**, 154224r.

[81] C. Agami, F. Meynier and T. Rizk, *Synthetic. Commun.*, 1987, **17**, 241.

[82] R. K. Razdan, H. C. Dalzell and G. R. Handrick, *J. Am. Chem. Soc.*, 1974, **96**, 5860.

[83] T. Nishiguchi and K. Kawamine, *J. Chem. Soc. Chem. Commun.*, 1990, 1766.

4

Reactions assisted by clays and other lamellar solids — a survey

J. A. Ballantine

4.1 INTRODUCTION

The history of the use of clays as heterogeneous catalysts is interesting, in that it is not the expected story of gradual development to the important position that they hold today, but that a bright start was followed by disappointment, after which these catalysts were almost ignored for 30 years.

In the early 1930s acid-treated natural clays were the pre-eminent solid catalysts used in petroleum refining, as they had the ability to induce vapour phase isomerization and cracking of the large paraffin molecules in crude oil to provide useful fuels for the petrol engine. However, they had one serious drawback: their lack of thermostability under the high-temperature fluidized catalytic cracking (FCC) conditions seriously affected their catalyst lifetime, and the search was on for similar solids that had greater stability. The more stable synthetic amorphous silica-aluminas completely replaced clays in the late 1930s in the high-temperature catalytic crackers of the petroleum industry, and these catalysts were in turn replaced in the 1960s by the new shape-selective synthetic zeolites. However, following a burst of activity in the late 1970s, pillared clays were developed that had increased thermostability, and these are now being considered as specialist FCC catalysts to deal in particular with heavy crudes.

The organic chemist's interest in the use of clays was also kindled with the burst of activity started in the 1970s when the potential of clays as solid acid catalysts and the utility of clay-supported reagents were described. Since that time there have been numerous reports of the catalytic activity of clay-based materials involving a bewildering variety of organic reactions.

Graphite intercalates can also provide some interesting reactions, though the range of applications is more restricted.

4.2   ACIDITY OF CLAYS AND TYPES OF CLAY CATALYSTS

The vast majority of clay catalysts used by organic chemists are based on the naturally occurring smectite clay, montmorillonite (also known as bentonite). This clay has an aluminosilicate structure which can be compared to a multi-decker ham sandwich. The bread layers represent the extended aluminosilicate sheets where two external tetrahedral silica groups surround internal octahedral alumina groups in a tetrahedral–octahedral–tetrahedral ('TOT') structure. The ham represents an interlamellar water layer containing dissolved cations.

The montmorillonites are a group of TOT-type clays in which isomorphous substitution of some of the octahedral aluminium(III) atoms by magnesium(II) or iron(II) atoms has taken place, with the result that the sheet retains a residual negative charge. In the naturally occurring form, this charge is balanced by the introduction into the water layer of interlamellar cations such as Na^+ or Ca^{2+}, some cations also occupying broken edge sites. Different deposits of montmorillonites can be found in which there are between 25 and 125 mequiv. of exchangeable cation per 100 g of clay. Such smectite clays have the added property of swelling in the presence of either water or a host of organic molecules, when the interlamellar distance between the sheets increases to accommodate the guest molecules [1,2].

Natural montmorillonite clays have almost no catalytic activity, but it is relatively easy to convert them into useful catalysts by either (a) acid treatment or (b) cation exchange with polyvalent ions such as Al^{3+} or Cr^{3+}. These cations can polarize their coordinated water molecules to yield protons in the interlamellar zone (equation (4.1)) [3,4].

$$[Al(OH_2)_6]^{3+} \rightarrow [Al(OH_2)_5(OH)]^{2+} + H^+ \qquad (4.1)$$

However, even mild acid ion-exchange of the natural clays can cause partial leaching of aluminium from the octahedral layer, resulting in de-lamination of the aluminosilicate sheets [5], providing a less crystalline aluminium-exchanged clay. Treatment of natural montmorillonites with strong mineral acids causes considerable de-lamination of the structure, producing an increase in surface area, particularly at the sheet edges, as well as adsorption of quantities of acid onto both external and internal surfaces. The tetrahedrally substituted smectite clay, beidellite, can also be used to prepare acid catalysts with high acidity by cation exchange [6,7].

Clay catalysts have been shown to contain both Brønsted and Lewis acid sites [8,9], with the Brønsted sites mainly associated with the interlamellar region and the Lewis sites mainly associated with edge sites. The acidity of the ion-exchanged clays is very much influenced by the quantity of water between the sheets. If the clay is heated (to around 100°C) so as to remove most of the interlamellar water until only 'one layer' of water remains, at about 5% total water level, the Brønsted acidity increases markedly [10,11] to that of a very strong acid indeed. Heating to a higher temperature (at around 200–300°C) results in collapse of the clay interlayer structure as the water is driven out, resulting in a decrease in Brønsted acidity but an increase in Lewis acidity. Further heating (to around 450°C and above) results eventually in complete dehydroxylation of the aluminosilicate lattice, producing a completely amorphous solid that retains Lewis acidity.

The swelling properties of montmorillonite and beidellites can be employed to manufacture 'pillared clays'. Large polyatomic inorganic ions can be intercalated into the swelled clay and then modified by heat treatment, so as to produce inorganic pillars that hold the roof up at a larger than normal interlamellar distance [6,12,13]. Pillared clays can be prepared that have pore sizes larger than most zeolites, as well as having increased thermostability over normal layered clays, and much activity has been concentrated in incorporating transition metal elements into pillared clays to provide useful inorganic catalysts [13–16].

However, organic chemists with synthesis in mind have so far confined their interest to the swellable montmorillonite clays, and almost all of their clay catalysts have been either (a) acid-treated clays such as K-10 [17], or ion-exchanged clays such as Al^{3+}, Cr^{3+} or H^+ exchanged Wyoming or Texas bentonites [18]. Further details of clay structures are given in Chapter 1.

The acid-treated and cation exchanged clays can be simply regarded as solid acids and act as heterogeneous catalysts, with all of the advantages of easy removal of the catalyst from the products. Acid-treated clays, because of their increased surface area and swelling properties, have also been widely used as solid supports for inorganic reagents such as potassium permanganate [19], thallium(III) nitrate [20] and both copper(II) and iron(III) nitrates [21].

The ion-exchanged clays have mostly Brønsted acidity in the interlamellar zone and so are characterized by promoting acid-catalysed reactions often of a bimolecular type between protonated and neighbouring unprotonated reactants [22].

4.3 PREPARATION OF CLAY CATALYSTS

Organic chemists can purchase ready-made acid-treated montmorillonite catalysts from a variety of commercial sources. Süd-Chemie AG, of Munich, West Germany, produce K-10, KSF and Tonsil acid-treated clays, of which K-10 can be obtained in Europe from Fluka AG and in America from Aldrich Chemical Co., and this easy availability has meant that organic chemists have made most use of this material. Additionally, Filtrol clays are manufactured by the Filtrol Corporation in Los Angeles, USA, and a series of clays (Fulmont, Fulbent and Fulcat) are produced by Laporte Industries, Widnes, England.

Ion-exchanged clays, e.g. Al^{3+}-exchanged montmorillonite [18], are simply prepared by suspending the clay in water containing low concentrations (ca. 0.5 M) of salts of the cation to be introduced. The clay is then washed by decantation or centrifugation to get rid of most of the exchanged sodium cations and then dialysed until the supernatant is free of sodium ion.

Pillared clays have an important role to play as industrial catalysts, but to date have not been used much by practising organic chemists for synthesis purposes and hence their preparation will not be detailed here.

Clay-supported reagents have mainly been prepared by depositing inorganic materials on clay supports. Potassium permanganate/montmorillonite is prepared by grinding potassium permanganate and montmorillonite together as dry powders in approximately equal quantities. Reactions are carried out by refluxing the resultant solid in dichloromethane with the appropriate substrate [19].

Thallium(III) nitrate/K-10 (TTN/K-10) is prepared by stirring K-10 montmorillonite with a solution of thallium(III) nitrate in a mixture of methanol and trimethyl orthoformate, then evaporating to dryness [23]. The resultant free-flowing solid catalyst is stable in a well-stoppered bottle for months.

Claycop [copper(II) nitrate/K-10] [24] and Clayfen [iron(III) nitrate/K-10] [25] are prepared by adding K-10 montmorillonite to the acetone solvate of the inorganic nitrates and removing the solvent under vacuum.

4.4 CLAY-CATALYSED ORGANIC REACTIONS WHERE THE CATALYST IS ACTING ESSENTIALLY AS A SOLID ACID

4.4.1 Additions to alkenes
Acid-activated and cation-exchanged clays can act as catalysts for a whole range of addition reactions of alkenes. The first step involves protonation of the alkene to give a carbocation intermediate, which may or may not then rearrange to a more stable carbocation species, followed by reaction with a wide variety of available nucleophiles.

4.4.1.1 *Ether or alcohol formation by addition of alcohols or water to alkenes*
1-Alkenes react with alcohols to produce a mixture of 2-alkyl and 3-alkyl ethers in the presence of Al^{3+}-exchanged montmorillonite at 102–200°C for 3 h in a sealed reactor. The reaction is rather slow with hex-l-ene and ethanol, yielding 2-ethoxyethane (14%) and 3-ethoxyethane (3%) (Scheme 1), with most of the reactants recovered unchanged [26].

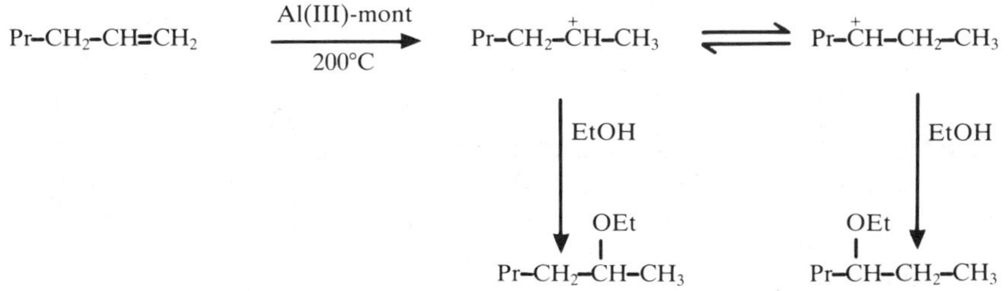

Scheme 1 — Clay-catalysed addition of ethanol to 1-hexene.

However, alcohol additions to substituted alkenes, which have stabilized tertiary carbocations as intermediates, give much improved yields at lower temperatures with a variety of clay catalysts. For example, methyl *t*-butyl ether (MTBE) can be produced in 65–90% yield from the reaction of 2-methylpropene with methanol with

H^+-exchanged montmorillonite at temperatures below 100°C [26–32]. This preparation has also been adapted for flow-rig operation using a new type of lithiated ion-exchange catalyst [33].

One very general synthetic use of this addition reaction is the protection of alcohols as their tetrahydropyranyl ethers in excellent yields using K-10 montmorillonite in dichloromethane solvent at room temperature (equation (4.2)) [34]. The major advantage of this technique over conventional acids is the very much easier workup, which simply requires filtration to remove the solid catalyst.

$$ \text{(4.2)} $$

The addition of water to alkenes under the influence of clay catalysts also works best with substituted alkenes, a 90% yield of *t*-butanol being obtained at 40–60°C from the reaction of 2-methylpropene with water [29,35]. Water can also be added to ethene using ion-exchanged montmorillonites in a flow system [36]. Long chain-1-alkenes react at reflux temperatures with the interlamellar water contained in ion-exchanged montmorillonites to give bis(2-alkyl) ethers, in quantitative yield based on the small amount of interlamellar water present, but the reaction cannot be sustained by the addition of extra water [37,38].

4.4.1.2   Thioether formation by the addition of thiols to alkenes
Both hydrogen sulphide and thiols add to alkenes in the presence of ion-exchanged montmorillonites at 200°C to give good yield of thiols [39] and dialkyl sulphides [40–42]. For example, propene reacts with H_2S at 174°C in a sealed reactor with Filtrol 70 to give a 35% conversion to 2-propanethiol and diisopropyl sulphide in a 4 h reaction (equation (4.3)) [39].

$$ CH_3 - CH = CH_2 + H_2S \xrightarrow[\substack{174°C \\ 4\,h}]{\text{Filtrol 70}} $$

$$ CH_3 - CH(SH) - CH_3 + (CH_3)_2CH - S - CH(CH_3)_2 \quad (4.3) $$

1-Hexene reacts with 1-butanethiol using Al^{3+}-exchanged montmorillonite at 200°C in a sealed reactor for 4 h to give three dialkyl sulphides: 1-hexyl 1-butyl sulphide (12%), 2-hexyl 1-butyl sulphide (30%) and 3-hexyl 1-butyl sulphide (25%) [41].

4.4.1.3   Ester formation by the addition of carboxylic acids to alkenes
Esters can be formed by the direct addition of carboxylic acids to alkenes using ion-exchanged or acid-activated clay catalysts [18,43,44]. The reaction between ethene and acetic acid requires 200°C to produce an efficient yield (96%) of ethyl acetate, whereas substituted alkenes such as 2,2-dialkylalkenes give high yields of *t*-alkyl esters with acids at temperatures below 30°C. Terminal alkenes such as 1-hexene give mixed ester products, the yields of which reflect the relative stability of their intermediate

carbocations. The esterification reaction can also occur intramolecularly with cyclooctene-5-carboxylic acid to produce internal lactones [45].

4.4.1.4 *Amine production by addition of ammonia to alkenes*

It is much more difficult to add ammonia to alkenes using clay catalysts, as the basicity of the ammonia makes this the preferential site of protonation, so effectively making the production of a carbocation unlikely. However, alkenes can be induced to give amines (2% conversion of propene) at high temperatures and pressures over an acid-treated clay [46].

4.4.1.5 Alkene dimerization and oligomerization

Alkenes undergo easy oligomerization with either acid-treated or ion-exchanged clays. Protonation of the alkene yields a carbocation intermediate that can react with a neighbouring non-protonated alkene, either in the interlamellar zone or on the clay surface [26,47–50]. A typical example of an industrial application is a Gulf research oligomerization process whereby a C_{2-5} alkene stream is treated with a calcined montmorillonite to give useful C_{5-24} hydrocarbons [51]. Acidic clays can also be used for catalysis of vinyl polymerization to give, for example, polystyrene [52], polypropylene [53] or polytrichloroethene [54].

Substituted alkenes oligomerize easily at relatively low temperatures (e.g. 2-methyl-propene reacts below 60°C), whereas 1-alkenes require a higher temperature (e.g. 1-hexene requires 120°C).

Clay-catalysed dimerizations of α-substituted styrenes produce both linear dimers [55–57] and cyclic indane dimers [45,58,59]. Long-chain unsaturated fatty acids also give dimers in up to 70% yield [60–63].

4.4.2 Arene dimerization

Many arenes are dimerized by heating with transition metal ion-exchanged (mainly Cu^{2+}, Fe^{3+} or Ru^{3+})montmorillonites. The mechanism is thought to proceed via a cation-radical species [64] to give either 4,4'-disubstituted biphenyls or polymeric products. The reactions of benzene [65], toluene [66,67], anisole [67] and phenols [67–70] have been studied. Chlorophenols, in addition to forming dimers [71], have been shown to produce chlorinated hydroxy-biphenyls, -diphenyl ethers and -dibenzofurans under the influence of clays [72].

4.4.3 Diels–Alder and ene-reactions

4.4.3.1 *Diels–Alder cycloadditions*

It was shown in 1978 that Cu^{2+}-exchanged montmorillonite was an effective catalyst for the Diels–Alder dimerization reaction of butadiene, when 60% of the cycloadduct was obtained from a reaction at 60°C [73]. It was considered that the ability of the transition metal exchanged clays to generate radical cations from unactivated dienes would be advantageous in Diels–Alder-type reactions, and Laszlo, *et al.* examined the use of Fe(III)/K-10 (Clayfen) under mild conditions in dichloromethane solvent with a small amount of added 4-*t*-butylphenol. Under these conditions, 1,3-cyclohexadiene gave a 77% yield of the adduct at 0°C in one hour [74] and 2,4 dimethylpenta-1,3-diene gave a 90% yield of the corresponding adduct [74]. It was later shown that the addition of the phenol was unnecessary, and cyclopentadiene reacted with methyl vinyl ketone at -24°C to give a 96% yield of adduct (endo/exo = 21/1) in 4 h at -24°C (equation (4.4)) [75]. Furan reacted with methyl vinyl ketone to give a 60% yield of adduct after 4 h at -43°C [76].

It was shown that clays exchanged with transition metal cations, such as Cr^{3+} and Fe^{3+}, were as effective as Clayfen in catalyzing Diels–Alder reactions with unactivated dienes although Al^{3+} was not [77,78]. However, the cycloaddition of acryloferrocene and 1-phenyl-1,3-butadiene worked better with the acidic Al^{3+}-exchanged montmorillonite than with the Cr^{3+}-exchanged material [79].

4.4.3.2   *Ene-reactions*
Foucaud *et al.* investigated the potential of clay catalysts in ene-reactions and found that acid-treated K-10 montmorillonite was an effective catalyst for the intramolecular cyclization of $(+)$-citronellal **(1)** to $(-)$-isopulegol **(2)** (56% yield) and $(+)$-neoisopulegol **(3)** (18% yield) at 15°C in dichloromethane (equation (4.5)) [80].

Acid-treated K-10-based catalysts have also proved effective in the intermolecular ene-reaction of diethyl oxomalonate with 2-methyl-2-butene, in which the intermediate adduct cyclizes to give high yields of lactones (Scheme 2) [80].

4.4.4   Additions to epoxides and other cyclic ethers

4.4.4.1   *Oligomerization of epoxides*
Treatment of ethylene and propylene oxides with ion-exchanged or acid-activated clays at moderate temperatures results in extensive polymerization, but cyclic dimers such as 1,4-dioxan can also be isolated [81,82].

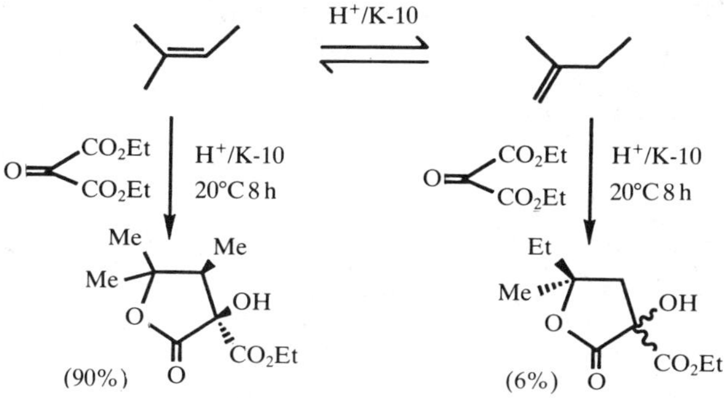

Scheme 2 — Lactone formation via a clay-catalysed intermolecular ene-reaction.

4.4.4.2 Hydroxyether production by addition of alcohols

Alcohols add to ethylene and propylene oxides in the presence of ion-exchanged or acid-activated montmorillonites at moderate temperatures in both liquid (40–70°C) and gas phase (120°C) reactions to give good yields of the 1-alkoxy-2-hydroxyalkane derivatives (equation (4.6)) [47,82–85].

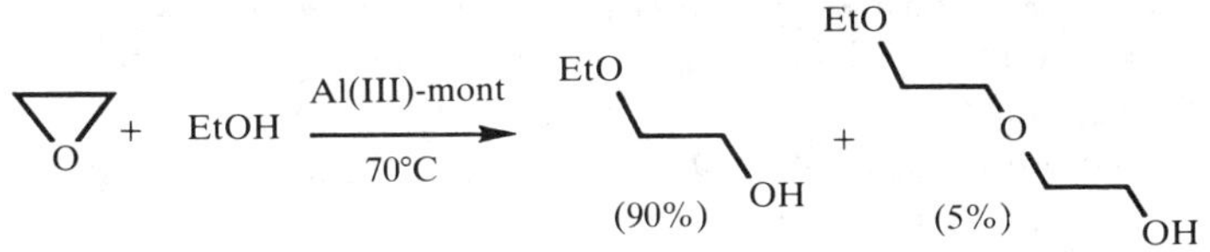

$$\tag{4.6}$$

Epichlorohydrin can also be used to add a variety of alcohols in 30–60% yields by refluxing for 2.5 h in tetrachloromethane with K-10 as catalyst [86].

4.4.4.3 Ester production by addition of carboxylic acids

Ethylene and propylene oxides add acetic acid to give hydroxyalkyl acetates using acid-activated montmorillonites [47,48].

4.4.4.4 Thiol production by addition of hydrogen sulphide

Hydrogen sulphide can be added to 2,3-epoxy-1-propanol using montmorillonite catalyst at 30°C to provide 1,2-dihydroxy-3-propanethiol in 90% yield [88].

4.4.4.5 DABCO production by addition of piperazine

The reaction of piperazine with ethylene oxide at high temperature with acid-treated montmorillonite gives high yields of 1,4-diazabicyclo[2.2.2]octane (DABCO) (equation (4.7)) [89].

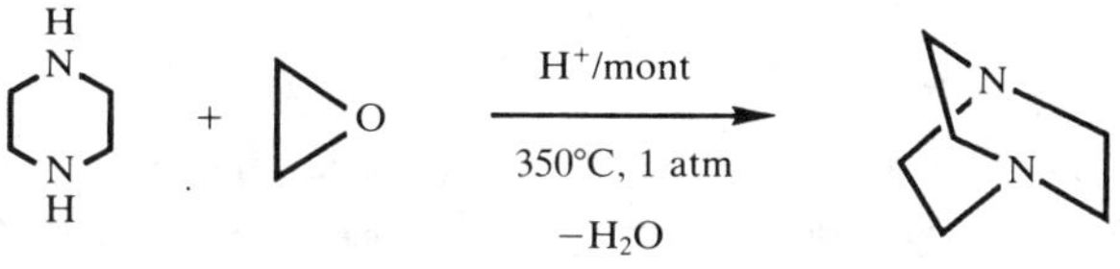

$$\tag{4.7}$$

4.4.4.6   *Dioxolane production by reaction with aldehydes*
Epoxides react with aldehydes under the influence of acid-activated K-10 or KSF montmorillonites to give good yields of dioxolanes at temperatures between 30 and 80°C (e.g. equation (4.8)) [90].

$$\text{(4.8)}$$

4.4.4.7   *Formation of polytetramethylene glycol polymers from tetrahydrofuran*
Tetrahydrofuran reacts with acid-activated montmorillonites at 30°C for 6 h to give polytetramethylene glycol polymer of 1600–2000 molecular weight [91,92].

4.4.5   Reactions of carbonyl compounds and acetals
There are a host of acid-catalysed reactions of carbonyl compounds that can be catalysed efficiently by montmorillonite catalysts to give clean products in high yields.

4.4.5.1   *Formation of acetals from carbonyl compounds and enol ethers*
Cyclic acetals are readily formed by the action of 1,2-diols [93] or ethylene oxide [94] with carbonyl compounds, or from the reaction of enol ethers with 1,2-diols (equation (4.9)) [95], under the influence of ion-exchanged or acid-activated montmorillonites. The yields are very high under mild conditions.

Montmorillonites also catalyse the formation of dimethyl acetals from methanol and carbonyl compounds under mild conditions [96], but these compounds are even more easily prepared in quantitative yield at room temperature in 1 min by the action of trimethyl orthoformate with the appropriate carbonyl compounds (equation (4.10)) [97–99]. As this latter reaction fails totally with natural montmorillonite (Na^+), but is complete in 1 min with Al^{3+}-exchanged montmorillonite, it is likely that the reaction takes place in the interlamellar region where the cation exchange has taken place and not on the outside surface of the clay.

$$\text{(4.9)}$$

$$\text{(4.10)}$$

Protection of sensitive alcohols during synthesis can be effected by acetal exchange with formaldehyde diethyl acetal under the influence of acid-treated montmorillonite under very mild conditions to yield a mixed formaldehyde acetal (equation (4.11)) [100].

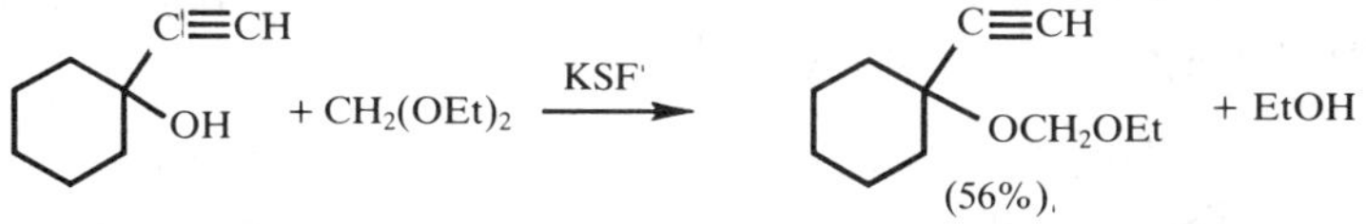

(4.11)

(56%).

Formaldehyde acetals can easily be prepared from the corresponding alcohols by use of a quaternary ammonium-exchanged montmorillonite as a phase transfer catalyst [190]. Diacyl acetals can be prepared from acetic anhydride and a carbonyl compound using acid-treated montmorillonite as catalysts [102].

4.4.5.2   *Formation of thioacetals and thioenol ethers*
The acid-treated montmorillonite KSF has been used to synthesize a wide variety of thioacetals from the reaction of aldehydes and acyclic ketones with thiols and dithiols under mild conditions [103]. Cyclic ketones under similar conditions yield thioenol ethers upon reaction with thiophenol or butane-1-thiol (equation (4.12)) [104].

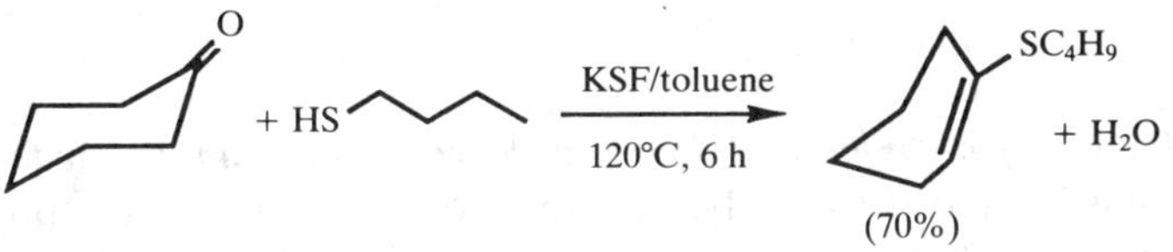

(4.12)

(70%)

4.4.5.3   *Formation of enamines*
K-10 montmorillonite is an excellent catalyst for the preparation of enamines from cyclohexanone and a wide variety of amines, yields of up to 80% being obtained by refluxing in benzene for 3–4 h [105].

4.4.5.4   *Aldol condensation and related reactions*
Ethanal undergoes self-condensation in the presence of ion-exchanged montmorillonites at 100°C to give the α,β-unsaturated aldehyde, 2-butenal, in 24% yield after 8 h [43]. α,β-Unsaturated aldehydes are also produced in high yield by the addition of acetals to enol ethers using K-10 montmorillonite under mild conditions, followed by hydrolysis and elimination of the intermediate acetal (equation (4.13)) [106].

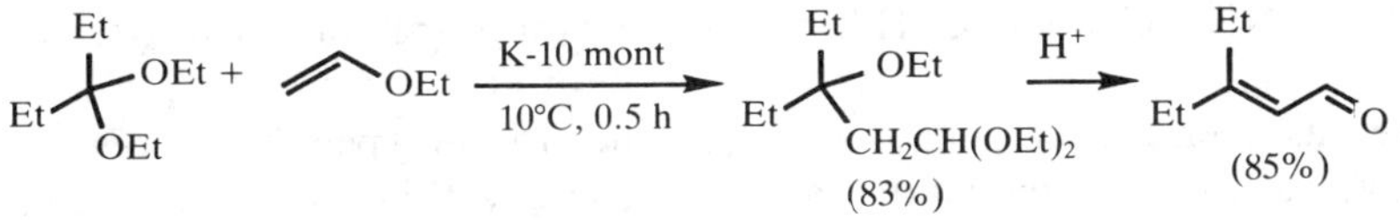

(4.13)

(83%)   (85%)

Glyceraldehyde undergoes self-condensation at 40°C in the presence of raw Na^+-montmorillonite to give mixtures of C_5 and C_6 monosaccharides in overall conversions in the 90% region [107,108].

Al^{3+}-exchanged montmorillonite has proved to be a remarkable catalyst for the crossed aldol reactions between silyl enol ethers and a wide range of aldehydes or acetals (e.g. equation (4.14)) [109–111]. The diastereoselectivity in the reaction is significantly sensitive to the solvent used.

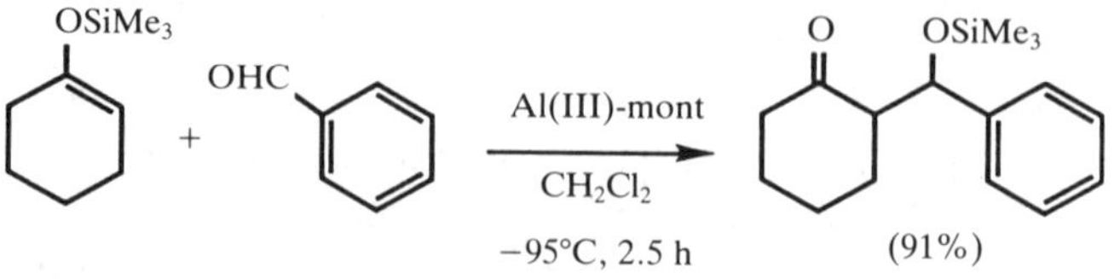

(4.14)

The study has been extended to include Michael additions (e.g. equation (4.15)) [112,113].

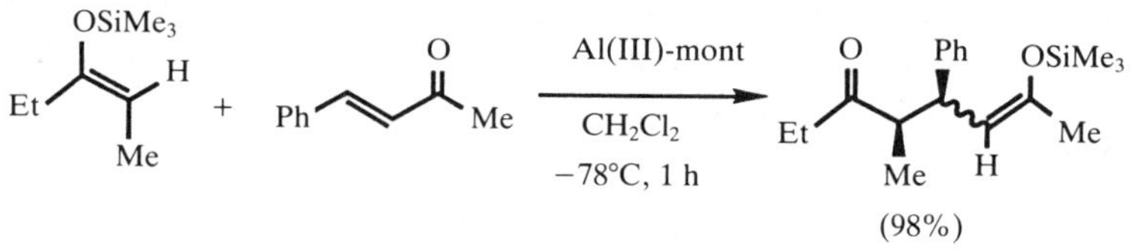

(4.15)

4.4.5.5   *Production of a conjugated diene by the reaction of an aldehyde and alkene*
A natural clay has been used to catalyse the reaction between 2-methylpropene and 2-methylpropanal at 250°C in a flow reactor, when a yield of 58% of 2,5-dimethylhexa-2,4-diene was obtained [114].

4.4.6   Reactions of carboxylic acids and their derivatives

4.4.6.1   *Esterification*
It has long been established that acid-activated clays are useful catalysts for the esterification of acids and alcohols in the gas phase [115], and this has recently been re-examined [116]. Mono-esters of diols can also be prepared using montmorillonite catalysts by ester interchange with ethyl acetate in refluxing toluene [117].

4.4.6.2   *Lactam and amino acid polymerizations in relation to prebiotic synthesis*
ε-Caprolactam can be polymerized over calcined ion-exchanged montmorillonites at 200°C. This reaction is considered to be a surface reaction and not to occur in the interlamellar zone [118]. High conversion (ca. 92%) was achieved at 260°C when a mixed $Fe^{3+}/NH_3^+(CH_2)_{11}CO_2H$-exchanged montmorillonite catalyst was used [119].

There has been a long debate as to whether the polypeptides necessary for life could have been formed by condensation reactions within the interlayers of natural clay minerals [120, 121]. Although the production of simple amides by clay-catalysed condensations of acids and amines has proved to be a difficult high-temperature reaction [122], it was considered that activation of the amino acids as their aminoacyl adenylates might facilitate the reaction (equation (4.16)) to take place at room temperature [121]. Polymerization of the aminoacyl adenylates was found to be much easier if polypeptides were pre-adsorbed onto the natural montmorillonite prior to the reaction [123].

$$
\begin{array}{c}
\xrightarrow{\text{Na}^+\text{-montmorillonite}}
\end{array}
$$

(aa)$_n$-AMP + (aa)$_m$-AMP $\xrightarrow{\text{Na}^+\text{-montmorillonite}}$

$$
\begin{array}{c}
\text{(aa)}_{n+m}\text{-AMP} \\
\text{polypeptide adenylate} \\
+ \\
\text{AMP} \\
\text{adenosine monophosphate}
\end{array}
\qquad (4.16)
$$

aminoacyl adenylates

Natural montmorillonite clays have also been found to catalyse the production of 2′,3′-cyclic AMP in aqueous solution from diiminosuccinonitrile and 3′-AMP, and so strengthen the argument that natural clays could have been involved in the prebiotic synthesis of phosphate esters [124].

4.4.6.3   *Decarboxylation of β-ketoacids*
Ion-exchanged montmorillonites have been studied for the easy decarboxylation of oxaloacetic acid to pyruvic acid. It was considered that the Al atoms at the crystal edges were causing metalloenzyme-like behaviour [125].

4.4.7   Elimination/condensation reactions

4.4.7.1   *Formation of ethers and alkenes by elimination of water from alcohols*
Secondary and tertiary alcohols undergo facile dehydration to alkenes by the use of ion-exchanged and acid-treated montmorillonites [26,96,126,127]. For example, cyclohexanol gives 88% of cyclohexene on treatment with Al^{3+}-exchanged montmorillonite at 200°C (equation (4.17)) [45]. The dehydration of tert-alcohols takes place at much lower temperatures. Primary alcohols give predominantly bis(1-alkyl) ethers, with little alkene produced (equation (4.18)) [26]. 2-Propanol yields mostly bis(2-propyl) ether with very little propene, behaving more like primary alcohols in this respect [26,127].

$$
\text{(4.17)}
$$

(4%)   (88%)

$$
\text{(4.18)}
$$

(R=C$_4$H$_9$ 63%)   (R=C$_4$H$_9$ 3%)

As primary alcohols give 1-alkyl ether products with very little of the corresponding 2-alkyl products, this has been taken as an indication that the protonated alcohols react in an SN$_2$-type nucleophilic substitution mechanism involving a neighbouring unprotonated alcohol molecule as nucleophile, without the development of carbocation intermediates which would have yielded the isomeric products [26].

Ethylene glycol undergoes polymerization with Al^{3+}-exchanged montmorillonite while other α,ω-diols undergo preferential intramolecular cyclodehydration to give high yields of the corresponding cyclic ethers [26]. For example, diethylene glycol gives 60% of 1,4-dioxan at 200°C [26], and butane-1,4-diol gives a quantitative yield of tetrahydrofuran at 200°C [128, 129].

K-10 montmorillonite has been used to effect the ether coupling of two different alkanols. For example, 1-phenylethanol has been reacted with a series of alkanols, including chiral compounds, to yield the corresponding 1-alkoxy-1-phenylethanols [130,131], and a stereocontrolled glycosidation with benzyl alcohol has been performed using K-10 as catalyst (equation (4.19)) [132].

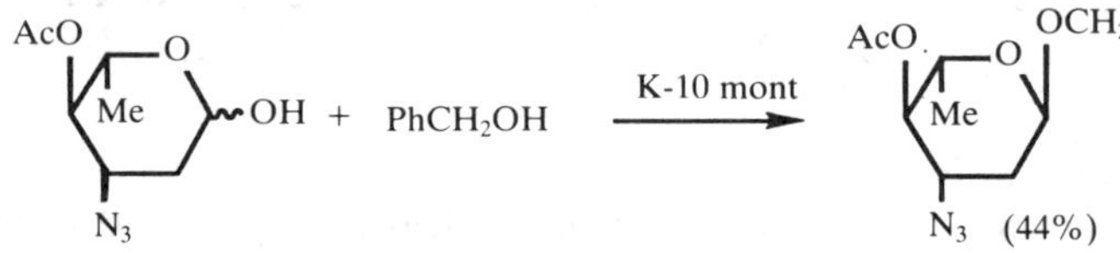

$$\text{(structure: AcO, O, Me, OH, N}_3\text{)} + \text{PhCH}_2\text{OH} \xrightarrow{\text{K-10 mont}} \text{(structure: AcO, O, OCH}_2\text{Ph, Me, N}_3\text{ (44\%))} \qquad (4.19)$$

Benzyl alcohol undergoes an interesting dehydration reaction on heating in the presence of Al^{3+}-exchanged montmorillonite to yield poly(phenylenemethylene) of molecular weight up to 250,000 daltons. The production of this polymer can be rationalized by substitution on the protonated species by the activated ring system of an unprotonated neighbour (Scheme 3) [26].

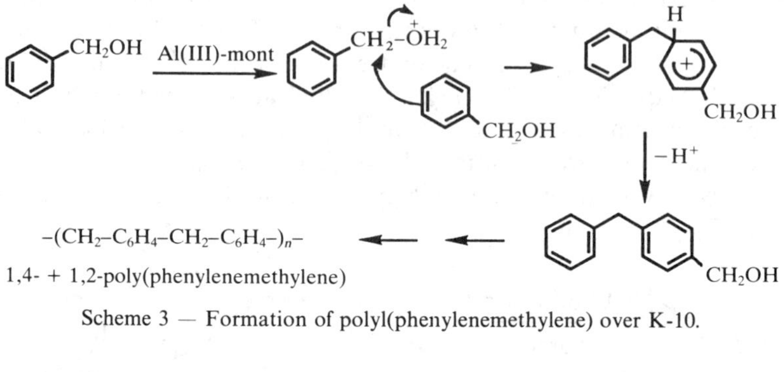

$-(CH_2-C_6H_4-CH_2-C_6H_4-)_n-$
1,4- + 1,2-poly(phenylenemethylene)

Scheme 3 — Formation of polyl(phenylenemethylene) over K-10.

4.4.7.2   *Formation of thioethers by elimination of hydrogen sulphide from thiols*

Primary aliphatic thiols eliminate H_2S at 220°C to give 10–30% of the bis(1-alkyl) sulphides on treatment with Al^{3+}-exchanged montmorillonite [40,41]. The reaction is slow and much of the thiol is recovered unchanged. Secondary thiols react much faster to give up to 80% yield of the bis(2-alkyl) sulphides under the same conditions (e.g. equation (4.20)) [40,41]. 1-Phenylethanol also reacts with a variety of thiols using K-10 as catalyst to give good yields of 1-alkylthio-1-phenylethanes through the preferential elimination of water [133]. Thiophenol undergoes an unusual reaction with Al^{3+}-exchanged montmorillonite at 200°C, giving diphenyl sulphide and benzene [41]. The latter compound corresponds to the formal loss of sulphur!

α,ω-Dithiols cyclize readily with loss of H$_2$S over Al^{3+}-exchanged montmorillonite. For example, butane-1,4-dithiol gives tetrahydrothiophene in good yield [129]. Benzyl thiol eliminates H$_2$S to give the same poly(phenylenemethylene) polymer which was obtained from benzyl alcohol [41].

$$\text{(4.20)}$$

4.4.7.3 *Formation of secondary and tertiary amines by elimination of ammonia from amines*

Although the bimolecular elimination of ammonia from amines is unknown in solution chemistry, this reaction can readily be catalysed by Al^{3+}-exchanged montmorillonite. The reaction with 1-alkylamines is rather slow at 220°C although it yields exclusively bis(1-alkyl)amines. However, with 2-alkylamines and cycloalkylamines, the yields of bis(2-alkyl)amines and dicycloalkylamines are very much better [134,135]. In all cases the reactions are clean, with very little material other than product and starting material being isolated. The reaction is very temperature dependent, the yield of dicyclohexylamine rising from 5% at 160°C to 60% at 220°C [134,135]. Benzylamine, unlike benzyl alcohol, does not produce a polymer but gives high yields of N,N-dibenzylamine [134,135].

Nucleophilic displacement of a protonated amino group by a neighbouring unprotonated species has been proposed for these reactions, and this is supported by an analysis of the products formed during crossover experiments between cyclohexylamine and benzylamine at different molar ratios [99]. However, some experimental evidence for an alternative, oxidation-reduction mechanism for the benzylamine coupling has recently been provided [99a]. Pyrrolidine undergoes a remarkable reaction (equation (4.21)) in which a single molecule of ammonia is expelled from three molecules of pyrrolidine in a ring-opening reaction which has no analogy in solution chemistry [134, 135].

$$\text{(4.21)}$$

4.4.8 Electrophilic aromatic substitutions

4.4.8.1 *Alkylations of aromatic compounds*

Acid-treated and ion-exchanged montmorillonites are effective catalysts for Friedel-Crafts-type alkylation of aromatic compounds. The alkylation with alkenes has been an industrial process for many years: e.g. benzene with propene at 300°C [136,137], phenol with propene at 130–170°C [138–140], aniline with propene at 300°C [138,141], and diphenylamine with styrene at 220°C [142]. Currently, these catalysts are being replaced by the more stable pillared clays for these high-temperature reactions [143]. A recent report has shown that transition metal-exchanged montmorillonites such as Zr^{4+} and Ti^{4+}-exchanged K-10 are effective catalysts for the alkylation of aromatics using halides, alcohols and alkenes [144].

Alkylations of phenols do not require elevated temperatures. For example, α-methylstyrene reacts with phenol at 80°C to give 93% of *p*-cumylphenol using Silton CS-1, an acid-treated montmorillonite (equation (4.22)). Phenol also reacts with acetone to give Bisphenol A in 50% yield at 75°C with acid-activated clays (equation (4.23)) [146].

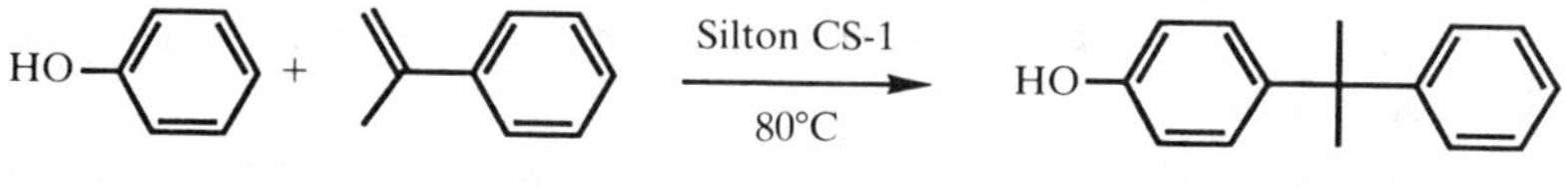

(4.22)

(93%)

(4.23)

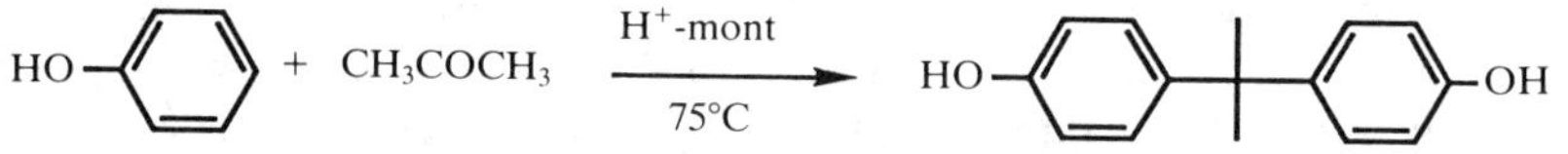

(Bisphenol A 50%)

Formaldehyde reacts with phenols and aniline derivatives to yield diphenylmethane derivatives in high yields at moderate temperatures using acid-activated clays [147–150]. Indoles are easily alkylated at the 3-position in 40–80% yields with methyl vinyl ketone in a Michael-type addition under the influence of Al^{3+}-exchanged montmorillonite [151].

Allylic alcohols can sometimes be used in Friedel–Crafts reactions in the presence of K-10, giving regioselective phenylation at the carbon terminus (e.g. equation (4.24)) [152].

$$PhCH(OH)C(CO_2Et)=CH_2 + PhH \xrightarrow{K-10} PhC(OH)=C(CO_2Et)CH_2Ph \quad (4.24)$$
$$E/Z = 9.1$$

K-10 montmorillonite has also been used to catalyse the coupling of an α-acetoxymethylpyrrole with an α-free pyrrole with remarkable efficiency (equation (4.25)). This electrophilic substitution step provides a route to dipyrromethanes, which are vital to the strategy of the synthesis of the unsymmetrical porphyrins of nature [153].

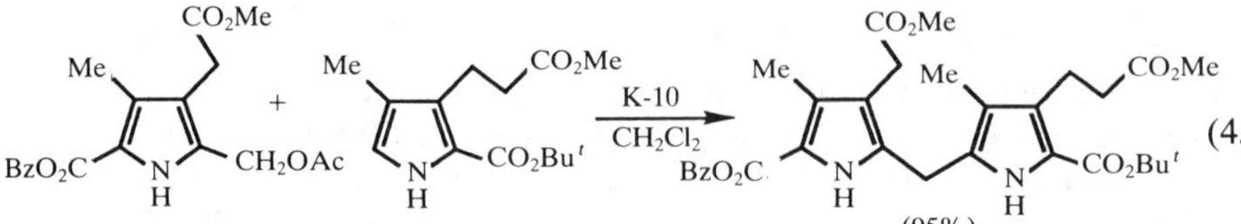

(4.25)

(95%)

4.4.8.2   Acylation of aromatic compounds

Activated aromatic ring systems such as toluenes [154], xylenes [155] and phenols [156] can be acylated with a variety of carboxylic acids in the presence of Al^{3+}-exchanged montmorillonite at 120°C to give long chain aromatic ketones in yields of up to 80%.

4.4.8.3   Heterocyclic ring-closure reactions on aromatic systems

Montmorillonite has been used as catalyst for the Fischer indole synthesis by ring closure of a variety of phenylhydrazones [157], and the cycloaddition ring closure of benzylideneanilines with vinyl ethers produces both tetrahydroquinoline and azetidine derivatives (equation (4.26)) [158].

$$R^1 = Et, R^2 = H \qquad\qquad (95\%) \qquad\qquad (0\%)$$

$$R^1, R^2 = (CH_2)_3 \qquad\qquad (47\%) \qquad\qquad (47\%)$$

4.4.8.4.   Nitration of aromatics

Toluene has been nitrated using nitric acid over K-306 montmorillonite in vapour phase reactions at 300°C, with very short contact times of the order of seconds, to give 65% of mononitration products among which the *para*-isomer predominates [159]. The lower temperature liquid phase nitration of toluene over a modified montmorillonite using nitric acid treated with acetic anyhyride in tetrachloromethane can give up to 80% of mononitrotoluenes; also with high *para*-selectivity [160].

Activated aromatic compounds can be nitrated with iron(III) nitrate deposited on K-10 montmorillonite (Clayfen) at room temperature in solvent to give mononitrated compounds, with very high regioselectivity for *ortho*-nitration (e.g. equation (4.27)) [161].

Aromatic hydrocarbons are also nitrated with good regioselectivities over montmorillonite-supported copper(II) nitrate in the presence of acetic anhydride to give up to 80% yields of nitrated aromatics [162].

4.4.8.5   *Halogenation of aromatics*

Aromatic compounds can be chlorinated under vapour flow conditions at 150°C using acid-activated montmorillonite (K-306). Toluene yields 80% of monochlorotoluenes with a normal isomer distribution in the remarkably short contact time of 3 s [163].

4.5   CLAY-SUPPORTED INORGANIC REAGENTS

4.5.1   Clay-supported potassium permanganate

A clay-moderated potassium permanganate can be used for the oxidation of alcohols to the corresponding carbonyl compounds [19]. The reagent is prepared by grinding the potassium permanganate with an equal quantity of montmorillonite as a dry solid and reaction is performed by a gentle warming with the reactant in dichloromethane with efficient stirring. Although primary alcohols, other than benzyl alcohol, give poor yields of the aldehydes, the reaction works well with secondary alcohols when up to 95% of the corresponding ketones are obtained. Allylic secondary alcohols react extremely easily to give α,β-unsaturated ketones in yields up to 90% (e.g. equation (4.28)) [19].

$$\text{(4.28)}$$

4.5.2   Clay-supported iron(III) nitrate (Clayfen) and copper(II) nitrate (Claycop)

Clayfen and Claycop reagents are simply prepared by treating K-10 montmorillonite with the acetone solvate of the metal nitrate and evaporating the solvent under vacuum. These oxidizing agents often produce high yields at room temperature and have achieved wide use [21].

4.5.2.1   *Oxidation of secondary alcohols*

Secondary alcohols can readily be oxidized in high yields at moderate temperatures by simply adding Clayfen to a solution in hexane. Gentle evaporation of the solvent induces the reaction, which can be monitored by the evolution of brown fumes. The product is simply isolated by filtration of the spent reagent and evaporation of the solvent. Yields in the region 74–100% are reported [25]. The reaction also works well for the oxidation of benzylic alcohols and for benzoins to benzils [164]. Aromatic primary alcohols give excellent yields of the aromatic aldehydes, but aliphatic primary alcohols give complex mixtures of products and consequently poor yields of aldehydes.

4.5.2.2   *Oxidative coupling of thiols*

Clayfen and Claycop are remarkably effective agents for the oxidative coupling of thiols to disulphides. Reaction is usually complete within 1 min at room temperature, giving yields up to 97% [165].

4.5.2.3   Cleavage of thioacetals to carbonyl compounds

Clayfen and Claycop can be used for the oxidative cleavage of linear or cyclic thioacetals to give the corresponding ketones, in very high yields (e.g. equation (4.29)). This process is the method of choice for the deprotection of carbonyl groups at room temperature [166].

$$
\xrightarrow[\text{20°C, 5h}]{\text{Clayfen/hexane}} \tag{4.29}
$$

(100%)

4.5.2.4   Cleavage of hydrazones to carbonyl compounds

Clayfen catalyses the oxidative cleavage of *N,N*-dimethylhydrazones at low temperature to give the corresponding carbonyl compounds in yields of 60–90% [167]. The reaction can be extended to include the cleavage of tosylhydrazones, phenylhydrazones and 2,4-dinitrophenylhydrazones in high yield (equation (4.30)) [168].

$$
\xrightarrow{\text{Clayfen CH}_2\text{Cl}_2} \tag{4.30}
$$

(76 – 89%)

4.5.2.5   Oxidative nitrosation of hydrazines to azides

Clayfen effects the oxidative conversion of hydrazines at moderate temperatures in refluxing dichloromethane to give 55–83% yields of the corresponding azides [169].

4.5.2.6   Oxidative conversion of thiobenzophenones to benzophenones

Clayfen has been used to catalyse the oxidative conversion of thiobenzophenones to the corresponding benzophenones at room temperature in yields of 77–100% [21].

4.5.2.7   Dehydrogenation of 1,4-dihydropyridines

Both Clayfen and Claycop can be used to promote the dehydrogenation of 1,4-dihydropyridine derivatives to the corresponding pyridines in refluxing chloroform in yields of 59–91% [170].

4.5.3   Clay-supported thallium(III) nitrate

Thallium(III) nitrate supported on K-10 montmorillonite (TTN/K-10) is an oxidizing agent that is prepared by mixing a solution of thallium(III) nitrate in methanol and trimethyl orthoformate (TMOF) with K-10, then removing the solvents to produce a stable free-flowing powder.

4.5.3.1   Oxidative rearrangement of aryl alkyl ketones

Aryl alkyl ketones are easily oxidized by TTN/K-10, with rearrangement to the corresponding methyl arylacetates in excellent yields at room temperature (equation (4.31)) [20].

$$\text{(4.31)}$$

This reaction has been extended into a useful method of preparing 3-methoxycar-bonylmethylpyrrole derivatives, which are required for the synthesis of the natural unsymmetrical porphyrins, via TTN/K-10 oxidative rearrangement from the readily available 3-acetylpyrroles [171].

4.5.3.2   *Oxidative rearrangement of alkenes*

Unconjugated cyclic alkenes undergo rapid oxidative rearrangement at room temperature with TTN/K-10 to give a high yield of a ring-contracted aldehyde which is isolated as its dimethyl acetal (equation (4.32)) [20,23].

$$\text{(4.32)}$$

The reaction can be extended to styrenes (equation (4.33)) [20,23].

$$\text{(4.33)}$$

4.5.4   Clay-supported sodium hypochlorite

Epoxides can be prepared in 80–98% yield from alkenes which are *gem*-disubstituted with two electron withdrawing groups using sodium hypochlorite in the presence of a montmorillonite [172]. The alkenes can conveniently be prepared *in situ* from the reactions of aldehydes with compounds like methyl cyanoacetate.

4.5.5   Clay-supported hydrogen peroxide

Phenol and phenyl ethers are hydroxylated in the ring by the action of hydrogen peroxide in the presence of an acid-activated montmorillonite (e.g. phenol gives a mixture of 30% catechol and 11% hydroquinone by stirring at 80°C for 2.5 h [173]. With neutral clay, the yields are negligible.

4.5.6   Clay-supported transition metal complexes as hydrogenation catalysts

A diphenylphosphinepalladium(II) complex has been anchored in the interlamellar zone of a montmorillonite and has proved to be an effective catalyst for the hydrogenation of styrene [174]. This Pd(II) complex also effects stereoselective hydrogenation of alkynes, enynes and dienes with *cis*-selectivity always better than 85% [175]. A montmorillonite-bipyridinepalladium(II) diacetate complex behaves similarly [176]. Montmorillonite combined with a silylphosphinepalladium(II) complex also shows almost complete *cis*-selectivity in the hydrogenation of alkynes [177]. A montmorillonite-ruthenium(IV) complex has been used for the hydrogenation of natural oils under ambient conditions [178], and a montmorillonite-silylaminepalladium(II) complex is effective for both the selective dehydrobromination of aryl bromides [179] and the selective and sequential reduction of nitroaromatics [180].

4.6   GRAPHITE INTERCALATION COMPOUNDS

4.6.1   General considerations

True graphite intercalation compounds (GICs) involve nothing more than physical occupation by intercalated molecules of sites within the interlamellar space of the graphite with no chemical interaction between the guest and the host. In many cases, however, some sort of chemical interaction also occurs. For example, alkali metal intercalates involve electron transfer from the metal to the polyaromatic graphite sheets not unlike that occurring in sodium naphthalene ($NaC_{10}H_8$). In other cases the π-cloud of the graphite sheets may be complexed. For the purposes of this brief account, all types are considered as GICs irrespective of the nature of any interaction between host and guest.

GICs often exhibit reactivity patterns similar to those of the free reagent. However, intercalation into graphite may provide a means of rendering a potentially dangerous or highly reactive material easier to handle or store. For example, XeF_4 and related substances can be handled as solids, are less corrosive and do not produce explosive byproducts when intercalated into graphite. Although the preparation of such GICs requires the handling of the free reagents [181,182], their use is quite straightforward [183–185].

Intercalated Grignard reagents can be prepared using graphite-intercalated magnesium. Such reagents can be stored in dry conditions for months without loss of activity [186].

Simple intercalation compounds may react via slow release of the active reagent into the surrounding medium. In other cases, however, reaction may take place within the interlamellar space or the graphite may be specifically involved. In such cases, reactivity patterns can be quite different from those of the corresponding free reagents. For example, catalytic quantities of $C_{24}^+HSO_4^-.2H_2SO_4$, formed on electrolysis of sulphuric acid with a graphite anode, allow quantitative formation of ethyl esters from the parent carboxylic acids and triethyl orthoformate (equation (4.34)), whereas similar quantities of sulphuric acid itself do not [187].

$$RCO_2H + (EtO)_3CH \xrightarrow{\;C_{24}^+HSO_4^-.2H_2SO_4\;} RCO_2Et + EtOH + HCO_2Et \quad (4.34)$$

GICs are non-stoichiometric compounds. Nevertheless, GICs with relatively reproducible proportions can be obtained, depending on whether almost all available intercalation sites are occupied (as in KC_8, for example) or whether only, say, about one in three sites are occupied (e.g. in KC_{24}).

There are a number of reviews [188–193] of the use of GICs, and treatment here is necessarily brief and non-comprehensive.

4.6.2   Activated metal intercalation compounds

4.6.2.1   Preparation

Potassium-graphite is easily obtained by heating a mixture of the two elements at 70–200°C in the absence of air [194–195]. Adjustment of the stoichiometry allows the formation of C_8K, $C_{24}K$, $C_{36}K$, etc., corresponding to intercalates with different numbers of graphite layers between the metal layers.

Other metal intercalates can be prepared either by treating an appropriate metal halide with potassium-graphite or by reacting a metal derivative, graphite and a reducing agent directly. GICs of Mg, Hg, Cd, Zn and a range of transition metals have been obtained [186,191,196–202]. The nature of the compounds has been discussed [203].

4.6.2.2   Catalytic applications

Potassium-graphite, C_8K, acts as a catalyst for the polymerization of ethene [204] and hydrogenation of benzene [205–206]. Polymerization of lactones has been achieved [207] with the GIC, KC_{24}.

4.6.2.3   Reactions with organic halides

In general, the reactions of C_8K with organic halides are similar to those of potassium dispersions, giving coupling, reduction and dehydrohalogenation reactions [208]. Dechlorination of polychlorinated biphenyls can be achieved [191] with KC_8 or NaC_{16}, whilst both KC_8 and Fe-in-graphite have been used to dehalogenate *vic*-dibromides [209,210].

Mg-graphite reacts readily with alkyl, aryl or allyl halides to form the corresponding Grignard reagent GICs, which react in the expected manner [186]. Similarly, Zn-graphite can be used to carry out Reformatsky and related reactions (e.g. equation (4.35)) [211]. The yields are exceptionally good (75–90%).

$$(4.35)$$

The Fe GIC is less successful in Reformatsky-type reactions, but α,α'-dibromoketones react with Fe-graphite to give intermediates that can undergo useful cyclization reactions (e.g. equations (4.36), (4.37)) [210].

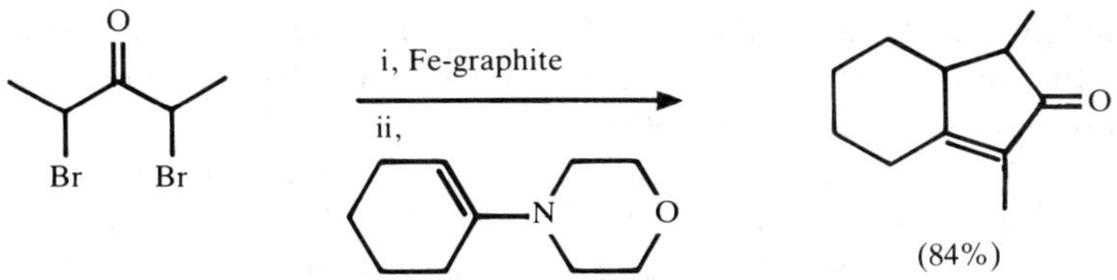

(4.36)

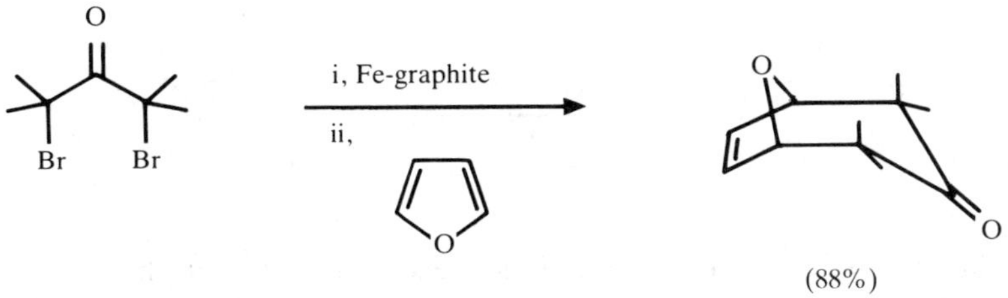

(84%)

(4.37)

(88%)

4.6.2.4   Use in reductions

Benzophenone undergoes pinacol-type coupling reduction with KC_8 in aprotic conditions, but simple reduction in proton-donating solvents [194, 212]. Interestingly, reduction of camphor gives the *exo*-alcohol predominantly (equation (4.38)) [194]. α,β-Unsaturated carbonyl compounds are also readily reduced by KC_8 [213,214].

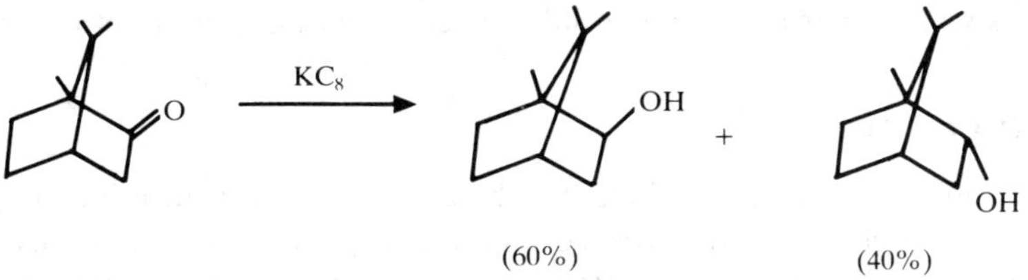

(4.38)

(60%)   (40%)

4.6.2.5   Miscellaneous reactions

Metallation can occur with KC_8, which is a fairly powerful base. The reaction is more useful with nitriles and esters having α-hydrogen atoms than with the corresponding ketones, where reduction is predominant. In cases where metallation occurs, α-alkylation can be achieved [191,215]. Other applications are listed in reviews [191,192].

4.6.3   Other graphite intercalation compounds

Chromium trioxide intercalated into graphite apparently does not function as an oxidant [193]. However, a reagent prepared from CrO_3 and graphite, apparently consisting of Cr_2O_8 deposited on graphite [216], is a useful selective oxidant for conversion of primary alcohols into aldehydes [194].

The compound $C_{24}^+HSO_4^- \cdot 2H_2SO_4$ is an excellent catalyst for a range of esterification reactions [187]. The reactions of primary and secondary alcohols with aliphatic carboxylic acids take place at room temperature, utilize stoichiometric amounts of reactants and give nearly quantitative yields [187]. Triethyl orthoformate can be used instead of ethanol (section 4.6.1). A related nitric acid GIC has been used in both oxidation and nitration reactions [193].

GICs of $XeOF_4$, XeF_4 and XeF_6 can be used to fluorinate aromatic hydrocarbons in modest yields [183–185]. The XeF_6 GIC is also capable of fluorinating β-diketones, and gives a 90% yield of 5-fluorouracil from uracil (equation (4.39)) [217].

$$(4.39)$$

Intercalated $SbCl_5$ is useful for halogen exchange with aliphatic bromides or iodides [218]. The products can be quite different from those obtained with free $SbCl_5$. The $SbCl_5$ reagent is also capable of chlorinating p-benzoquinone [219].

Graphite has been modified to provide a carboxyl group to which appropriate functionalities can be attached. In this way it is possible to attach, for example, a chiral diphosphine ligand that can be used to complex transition metal catalysts. Chiral inductions obtained from reactions of such species are, however, so far disappointing [191,220].

Other applications of GIC's are known [191] and it may be expected that new applications will appear as the potential of such species becomes better understood.

4.7   CONCLUSION

The wide range of organic reactions in which clays participate is remarkable. The ability of smectite clays to swell to accommodate organic guest molecules, as well as inorganic guest molecules as supported reagents, have resulted in these aluminosilicates becoming virtuosos of catalysis. In most cases, such inorganic reagents are much less effective in the absence of the aluminosilicate support, and the easy removal of the catalyst by simple filtration facilitates the isolation of the products. The added ability of the montmorillonites to develop extremely high acidity in the interlamellar zone by gentle heating provides a micro-environment in which bimolecular substitution reactions can take place, which may be either unfavourable or unknown in solution chemistry. It seems inevitable that the use of clays as catalysts or supports will increase markedly as organic chemists become more aware of the benefits that they can provide.

Although graphite intercalation compounds do not generally have the convenience or catalytic potential of clays and clay-supported reagents, they nevertheless offer some interesting reactions. Some compounds can exhibit catalytic activity.

REFERENCES

[1] B. K. G. Theng, *Formation and Properties of Clay–Polymer Complexes*, Elsevier, Amsterdam, 1979.

[2] B. K. G. Theng, *The Chemistry of Clay–Organic Reactions*, Adam Hilger, London, 1974.

[3] M. M. Mortland and K. V. Raman, *Clays Clay Mineral.*, 1968, **16**, 393.

[4] J. J. Fripiat and M. I. Cruz-Cumplido, *Annu. Rev. Earth Planet Sci.*, 1974, **2**, 239.

[5] H. van Olphen, *An Introduction to Clay Colloid Chemistry*, Wiley, New York, 1977.

[6] D. Plee, A. Schultz, G. Poncelet and J. J. Fripiat, in *Catalysis by Acids and Bases*, B. Emelik *et al.* (eds), Elsevier, Amsterdam, 1985, p. 343.

[7] P. A. Diddams, J. M. Thomas, W. Jones, J. A. Ballantine and J. H. Purnell, *J. Chem. Soc. Chem. Commun.*, 1984, 1340.

[8] L. P. Aldridge, J. R. McLaughlin and C. G. Pope, *J. Catal.*, 1973, **30**, 409.

[9] L. Forni, *Catal. Rev.*, 1973, **8**, 65.

[10] J. J. Fripiat, M. C. Gastuche and R. B. Richard, *J. Phys. Chem.*, 1962, **66**, 806.

[11] B. K. G. Theng, *Dev. Sedimentol.*, 1982, **35**, 197.

[12] N. Lahav, U. Shani and J. Shabtai, *Clays Clay Mineral.*, 1978, **26**, 107.

[13] D. E. W. Vaughan, R. J. Lussier and J. S. Magee, U.S. Pat. 4,176,090, 1979.

[14] S. Yamanaka and G. W. Brindley, *Clays Clay Mineral.*, 1979, **26**, 21.

[15] S. Yamanaka and G. W. Brindley, *Clays Clay Mineral.*, 1979, **27**, 119.

[16] T. J. Pinnaaia, M. S. Tzou and S. D. Landau, *J. Am. Chem. Soc.*, 1985, **107**, 4783.

[17] Süd-Chemie AG, Munich, West Germany.

[18] J. A. Ballantine,, M. Davies, R. M. O'Neil, I. Patel, J. H. Purnell, K. J. Williams and J. M. Thomas, *J. Mol. Catal.*, 1984, **26**, 57.

[19] D. G. Lee and N. A. Noureldin, *Tetrahedron Lett.*, 1981, **22**, 4889.

[20] C.-S. Chiang, A. McKillop, E. C. Taylor and J. F. White, *J. Am. Chem. Soc.*, 1976, **98**, 6750.

[21] A. Cornélis and P. Laszlo, *Synthesis*, 1985, 909.

[22] J. A. Ballantine, J. H. Purnell and J. M. Thomas, *J. Mol. Catal.*, 1984, **26** 157.

[23] A. McKillop and D. W. Young, *Synthesis*, 1979, 481.

[24] M. Balogh, A. Cornélis and P. Laszlo, *Tetrahedron Lett.*, 1984, **25**, 3313.

[25] A. Cornélis and P. Laszlo, *Synthesis*, 1980, 849.

[26] J. A. Ballantine, M. Davies, J. H. Purnell, M. Rayanakorn, K. J. Williams and J. M. Thomas, *J. Mol. Catal.*, 1984, **26**, 37.

[27] Gulf Oil Canada, U.S. Pat. 4,042,633, 1980.

[28] A. Bylina, J. M. Adams, S. H. Graham and J. M. Thomas, *J. Chem. Soc. Chem. Commun.*, 1980, 1003.

[29] J. M. Adams, D. E. Clement and S. H. Graham, *Clays Clay Mineral.*, 1982, **30**, 129.

[30] M. P. Atkins, Br. Pat. Appl. 22,151,603, 1985.

[31] J. M. Adams, T. Clapp and D. E. Clement, *Clay Miner.*, 1983, **18**, 411.

[32] J. M. Adams, K. Martin, R. W. McCabe and S. Murray, *Clays Clay Mineral.*, 1986, **34**, 597.

[33] M. P. Atkins, J. Williams, J. A., Ballantine and J. H. Purnell, Eur. Pat. Appl. EP 284397, 1988; J. Williams, J. H. Purnell and J. A. Ballantine, *Catal. Lett.*, 1991, in press.

[34] S. Hoyer, P. Laszlo, M. Orlovic and E. Polla, *Synthesis*, 1986, 655.

[35] J. A. Ballantine, W. Jones, J. H. Purnell, D. T. B. Tennakoon and J. M. Thomas, *Chem. Lett.*, 1985, 763.

[36] M. P. Atkins, D. J. H. Smith and D. J. Westlake, *Clay Miner.*, 1983, **18**, 423.

[37] J. M. Adams, J. A. Ballantine, S. H. Graham, R. J. Laub, J. H. Purnell, P. I. Reid, W. Y. M. Shaman and J. M. Thomas, *Angew, Chem. Int. Ed. Engl.*, 1978, **17**, 282.

[38] J. M. Adams, T. V. Clapp, D. E. Clement and P.I. Reid, *J. Mol. Catal.*, 1984, **27**, 179.

[39] Phillips Petroleum Co., Belg. Pat. 886,261, 1981.

[40] J. A. Ballantine, R. P. Galvin, R. O'Neil, J. H. Purnell, M. Rayanakorn and J. M. Thomas, *J. Chem. Soc. Chem. Commun.*, 1981, 695.

[41] R. P. Galvin, Ph.D. Thesis, University of Wales, Swansea, 1983.

[42] Phillips Petroleum Co., U.S. Pat. 2,610,981, 1947.

[43] J. A. Ballantine, M. Davies, J. H. Purnell, M. Rayanakorn, J. M. Thomas and K. J. Williams, *J. Chem. Soc. Chem. Commun.*, 1981, 8.

[44] R. Gregory, D. J. H. Smith and D. J. Westlake, *Clay Miner.*, 1983, **18**, 431.

[45] J. M. Adams, S. E. Davies, S. H. Graham and J. M. Thomas, *J. Catal.*, 1982, **78**, 197.

[46] J. O. H. Peterson and H. S. Fales, U.S. Pat. 4,375,002, 1983.

[47] M. P. Atkins, Eur. Pat. Appl. EP 0,130,055, 1985.

[48] Idemitsu Petroleum Co., *Japan Kokai Tokkyo Toho*, JP 58,147,496, 1983.

[49] R. E. Reusser, E. A. Clayton and B. E. Jones, U.S. Pat. 4,351,980, 1982.

[50] J. M. Adams, A. Bylina and S. H. Graham, *Clay Miner.*, 1981, **16**, 238.

[51] Gulf Research & Dev. Co., U.S. Pat. 3,947,483, 1976.

[52] D. Njopwound, G. Roques and R. Wandji, *Clay Miner.*, 1987, **22**, 145.

[53] C. T. O'Connor, L. L. Jacobs and M. Kojima, *Appl. Catal.*, 1988, **40**, 277.

[54] M. M. Mortland and S. A. Boyd, *Environ. Sci. Technol.*, 1989, **23**, 223.

[55] Monsanto Co., U.S. Pat. 691, 925, 1976.

[56] B. Chaudhuri and M. M.. Sharma, *Ind. Eng. Chem. Res.*, 1989, **28**, 1757.

[57] J. C. Wygant, Ger Pat. 2,724,491, 1977.

[58] Chem.-Werke Huls AG, Ger. Pat. 2,906,284, 1979.

[59] J. M. Adams, S. H. Graham, P. I. Reid and J. M. Thomas, *J. Chem. Soc. Chem. Commun.*, 1977, 67.

[60] C. Burba and E. Griebsch, Ger. Pat. 2,226,397, 1973.

[61] K. Pirner, E. Griebsch and E. Niemann, Ger. Pat. 2,118,702, 1972.

[62] A. G. Schering, Fr. Pat. 2,199,754, 1974.

[63] K. Arimoto, M. Miyamoto, H. Shibata, K. Kawaguchi, M. Fujiwara and A. Totsuka, *Japan Kokai Tokkyo Toho*, JP 62,022,742, 1987.

[64] M. M. Mortland and L. J. Halloran, *Soil Sci. Soc. Am. J.*, 1976, **40**, 367.

[65] Y. Soma, M. Soma and I. Harada, *Chem. Phys. Lett.*, 1983, **99**, 153.

[66] M. J. Tricker, D. T. B. Tennakoon, J. M. Thomas and S. H. Graham, *Nature*, 1975, **253**, 110.

[67] Y. Soma, M. Soma and I. Harada, *J. Phys. Chem.*, 1985, **89**, 738.

[68] B. L. Sawhney, R. K. Kozloski, P. J. Isaacson and M. P. N. Gent, *Clays Clay Mineral.*, 1984, **32**, 108.

[69] Y. Soma, M. Soma and I. Harada, *J. Contam. Hydrol.*, 1986, **1**, 95.

[70] B. L. Sawhney, *Clays Clay Mineral.*, 1985, **33**, 123.

[71] S. A. Boyd and M. M. Mortland, *Environ. Sci. Technol.*, 1986, **20**, 1056.

[72] Y. Soma and M. Soma, *Chemosphere*, 1989, **19**, 1895.

[73] R. S. Downing, J. van Austel and A. H. Joustra, U.S. Pat. 4,125,483, 1978.

[74] P. Laszlo and J. Lucchetti, *Tetrahedron Lett.*, 1984, **25**, 1567.

[75] P. Laszlo and J. Lucchetti, *Tetrahedron Lett.*, 1984, **25**, 2147.

[76] P. Laszlo and J. Lucchetti, *Tetrahedron Lett.*, 1984, **25**, 4387.

[77] J. M. Adams, *Applied Clay Sci.*, 1987, **2**, 331.

[78] J. M. Adams, K. Martin and R. W. McCabe, *Proceedings of the International Clay Conference, Denver*, 1985. L. G. Schultz *et al.* (eds), The Clay Minerals Society, Bloomington, Indiana, U.S.A., pp. 324–328.

[79] S. Toma, P. Elecko, J. Gazova and E. Solcaniova, *Collect. Czeck, Chem. Commun.*, 1987, **52**, 391.

[80] J. F. Roudier and A. Foucaud, *Proceedings of the NATO Advanced Workshop on Chemical Reactions in Organic and Inorganic Constrained Systems, Les Bezards*, R. Setton (ed.), Reidel, Dordrecht, 1985, pp. 229–235.

[81] Farbenfabriken Bayer, Ger. Pat. 1,049,103, 1959.

[82] M. E. Davies, Ph.D. Thesis, University of Wales, Swansea, 1982.

[83] W. Herold, Ger. Pat., DE 2,406,293, 1975.

[84] R. Gregory, Eur. Pat. Appl., EP 0,073,141, 1983.

[85] K. Fujita, Y. Ishida and J. Suezana, *Japan Kokai Tokkyo Toho*, JP 62,289,537, 1987.

[86] T. Vu Moc, H. Petit and P. Maitte, *Bull. Soc. Chim. Belg.*, 1982, **91**, 261.

[87] M. P. Atkins, Brit. Pat. Appl., 2,151,603, 1985.

[88] S. Kleeman, M. Nygren and R. Wagner, Ger. Pat. DE 3,014,165, 1985.

[89] M. D. Oakes, L. L. Upson and M. H. Ziv, U.S. Pat. 3,772,293, 1973.

[90] T. Vu Moc, H. Petit and P. Maitte, *Bull. Soc. Chim. Belg.*, 1980, **89**, 759.

[91] H. Müller, O. Huchler and H. Hoffman, U.S. Pat. 4,243,799, 1981.

[92] H. Müller, Ger. Pat. DE 3,236,432, 1984.

[93] J. Y. Conan, A. Natat and D. Privolet, *Bull. Soc. Chim. Fr.*, 1976, **11-2**, 1935.

[94] Süd-Chemie AG, Ger. Pat. 1,086,241, 1961.

[95] T. Vu Moc, H. Petit and P. Maitte, *Bull. Soc. Chim. Belg.*, 1979, **15**, 264.

[96] J. A. Ballantine, J. H. Purnell and J. M. Thomas, *Clay Miner.*, 1983, **18**, 347.

[97] E. C. Taylor and C. Chaing, *Synthesis*, 1977, 467.

[98] M. McInnes, Ph.D. Thesis, University of Wales, Swansea, 1986.

[99] J. H. Purnell and J. A. Ballantine, *J. Mol. Catal.*, 1984, **27**, 169.

[99a] S. Bank and R. Jewettt, *Tetrahedron Lett.*, 1991, **32**, 303.

[100] U. Schaper, *Synthesis*, 1981, 794.

[101] A. Cornélis and P. Laszlo, *Synthesis*, 1982, 162.

[102] Farbwerke Hoechst AG, Ger. Pat. 1,146,871, 1963.

[103] B. Labiad and D. Villemin, *Synth. Commun.*, 1989, **19**, 31.

[104] B. Labiad and D. Villemin, *Synthesis*, 1989, 143.

[105] S. Hunig, K. Huber and E. Benzig, *Chem. Ber.*, 1962, **95**, 926.

[106] D. Fishman, J. T. Klug and A. Shani, *Synthesis*, 1981, 137.

[107] N. Evole Martil and F. Aragon de la Cruz, *An. Quim.*, 1985, **Ser B, 81**, 22.

[108] N. Evole Martil and F. Aragon de la Cruz, *An. Quim.*, 1986, **Ser B, 82**, 256.

[109] M. Kawai, M. Onaka and Y. Izumi, *Chem. Lett.*, 1986, 1581.

[110] M. Onaka, R. Ono, M. Kawai and Y. Izumi, *Bull. Chem. Soc. Jpn.*, 1987, **60**, 2689.

[111] M. Kawai, M. Onaka and Y. Izumi, *Bull. Chem. Soc. Jpn.*, 1988, **61**, 1237.

[112] M. Kawai, M. Onaka and Y. Izumi, *J. Chem. Soc. Chem. Commun.*, 1987, 1203.

[113] M. Kawai, M. Onaka and Y. Izumi, *Bull. Chem. Soc. Jpn.*, 1988, **61**, 2157.

[114] Y. Hagashio, K. Takashashi, Eur. Pat. Appl., 215,567, 1987.

[115] BASF AG, Ger. Pat. 1,211,643, 1966.

[116] H. Slosiarkova and S. Micik, *Conf. Clay Mineral Petrol., (Proc.)*, 1984, 67.

[117] Mitsui Petroleum Industries, *Japan Kokai Tokkyo Toho*, JP 60,025,946, 1985.

[118] M. T. Bryk, A. Sh. Goikhman, I. E. Skobets and F. D. Ovcharenko, *Kolloidin.*, 1983, **45**, 1043.

[119] Y. Fukushima, S. Inagaki and A. Usuki, *Japan Kokai Tokkyo Toho*, JP 62,064,827, 1987.

[120] A. Katchshima and A. Ailam, *Biochem. Biophys. Acta*, 1967, **140**, 1.

[121] M. Paecht-Horowitz, *Biosystems*, 1977, **9**, 93.

[122] S. Kessaissai, B. Siffert and J. B. Donnet, *Clay Miner.*, 1980, **15**, 383.

[123] M. Paecht-Horowitz and R. Frederick, *Origins Life Evol. Biosphere*, 1988, **18**, 359.

[124] J. P. Ferris, C. H. Huang and W. J. Hagan, *Origins Life Evol. Biosphere*, 1988, **18**, 121.

[125] B. Siffert and A. Naidja, *Clay Miner.*, 1987, **22**, 435.

[126] W. Franz, P. Gunther and C. E. Hofstadt, *Proceedings of 5th World Petroleum Congress, New York*, 1959.

[127] Gulf Oil Canada Ltd., U.S. Pat. 1,211,643, 1966.

[128] D. Kotkar and P. K. Ghosh, *J. Chem. Soc. Chem. Commun.*, 1986, 650.

[129] D. Kotkar and P. K. Ghosh, *J. Chem. Soc. Perkin Trans.* **1**, 1988, 1749.

[130] Y. Okuda, M. Yoshihara, T. Maeshima, M. Fujii and T. Aida, *Yokagaku*, 1989, **38**, 153.

[131] Y. Okuda, M. Yoshihara, T. Maeshima, M. Fujii and T. Aida, *Chem. Express*, 1989, **4**, 17.

[132] J. C. Florent and C. Monneret, *J. Chem. Soc. Chem. Commun.*, 1987, 1171.

[133] R. C. Michaelson, U.S. Pat. 4,691,073, 1987.

[134] J. A. Ballantine, J. H. Purnell, M. Rayanakorn, K. J. Williams and J. M. Thomas, *J. Mol. Catal*, 1985, **30**, 373.

[135] J. A. Ballantine, J. H. Purnell, M. Rayanakorn, K. J. Williams and J. M. Thomas, *J. Chem. Soc. Chem. Commun.*, 1981, 9.

[136] R. S. Aries, U.S. Pat., 2,930,820, 1960.

[137] G. C. Joris, U.S. Pat., 2,945,072, 1960.

[138] R. Stroh, J. Ebersberger, H. Haberland and W. Hahn, Ger. Pat. DE 1,051,271, 1959.

[139] Süd-Chemie AG, Ger. Pat. 1,051,271, 1961.

[140] E. Polla, *Bull. Soc. Chim. Belg.*, 1985, **94**, 81.

[141] General Aniline and Film Co., U.S. Pat. 3,287,422, 1966.

[142] E. A. Rogers, Ger. Pat. DE 2,748,749, 1978.

[143] R. Gregory and D. J. Westlake, Eur. Pat. Appl., EP 0,083,970, 1983.

[144] P. Laszlo and A. Mathy, *Helv. Chim. Acta*, 1987, **70**, 577.

[145] M. Imanari, H. Iwane, T. Sugawara and S. Ootaka, *Japan Kokai Tokkyo Toho*, JP 63,208,545, 1988.

[146] K. K. Sun, U.S. Pat. 4,052,466, 1977.

[147] E. Herdieckerhoff and W. Sutter, Ger. Pat. DE 1,051,864, 1959.

[148] Süd-Chemie AG, Ger. Pat. 1,089,108, 1961.

[149] K. Wada, Eur. Pat. Appl., EP 3,107,518, 1981.

[150] F. E. Bentley, U.S. Pat. 4,071,558, 1978.

[151] Z. Iqubal, A. H. Jackson, K. Rao and K. R. Nagarata, *Tetrahedron Lett.*, 1988, **29**, 2577.

[152] D. Saib and A. Foucaud, *J. Chem. Res. Synop.*, 1987, 372.

[153] A. H. Jackson, R. Pandey, R. K. Rao and E. Roberto, *Tetrahedron Lett.*, 1985, **26**, 793.

[154] B. Chiche, A. Finiels, C. Gauthier, P. Geneste, J. Graille and D. Pioch, *J. Mol. Catal.*, 1987, **47**, 229.

[155] B. H. Chiche, P. Geneste, C. Gauthier, F. Figueras, A. Finiels, J. Graille and D. Pioch, Fr. Pat. 2,599,275, 1987.

[156] T. Fujita, M. Ishiguro, K. Takahata and K. Saeki, *Japan Kokai Tokkyo Toho*, JP 60,252,436, 1985.

[157] P. Bhattacharyya and S. S. Jash, *Indian J. Chem., Sect. B*, 1987, **26**, 1177.

[158] J. Cabral, P. Laszlo and M. T. Montaufier, *Tetrahedron Lett.*, 1988, **29**, 547.

[159] J. M. Bakke and J. Liaskan, Ger. Pat. DE 2,826,433, 1979.

[160] A. Cornélis, A. Gerstmans and P. Laszlo, *Chem. Lett.*, 1988, 1839.

[161] A. Cornélis, P. Laszlo and P. Pennelneau, *Bull. Soc. Chim. Belg.*, 1984, **93**, 961.

[162] P. Laszlo and J. Vandormael, *Chem. Lett.*, 1988, 1843.

[163] J. M. Bakke, J. Liaskan and G. B. Loretzen, *J. Prakt. Chem.*, 1982, **324**, 488.

[164] M. Basemann, A. Cornélis and P. Laszlo, *C.R. Acad. Sci. Ser. 2*, 1984, **299**, 427.

[165] A. Cornélis, N. Depaye, A. Gerstmans and P. Laszlo, *Tetrahedron Lett.*, 1983, **24**, 3103.

[166] M. Balogh, I. Hermecz, Z. Meszaros and P. Laszlo, *Helv. Chim. Acta*, 1984, **67**, 2270.

[167] P. Laszlo and E. Polla, *Tetrahedron Lett.*, 1984, **25**, 3309.

[168] P. Laszlo and E. Polla, *Synthesis.*, 1985, 439.

[169] P. Laszlo and E. Polla, *Tetrahedron Lett.*, 1984, **25**, 3701.

[170] M. Balogh, I. Hermecz, P. Laszlo and Z. Mészáros, *Helv. Chim. Acta*, 1984, **67**, 2270.

[171] E. Adelakun, A. H. Jackson, N. S. Ooi and K. R. N. Rao, *Tetrahedron Lett.*, 1984, **25**, 6049.

[172] A. Foucaud and M. Bakouetila, *Synthesis*, 1987, 854.

[173] M. Constantini, J. M. Popa and M. Gubelmann, Eur. Pat. Appl., EP 314583, 1989.

[174] R. K. Kumar, B. M. Choudary, Z. Jamil and G. Thyagarajan, *J. Chem. Soc. Chem. Commun.*, 1986, 130.

[175] G. V. M. Sharma, B. M. Choudary, R. M. Sharma and K. K. Rao, *J. Org. Chem.*, 1989, **54**, 2997.

[176] B. M. Choudary, G. V. M. Sharma and P. Bharathi, *Angew. Chem.*, 1989, **101**, 506.

[177] B. M. Choudary, K. Mukkanti and Y. V. Subba Rao, *J. Mol. Catal.*, 1988, **48**, 151.

[178] M. M. Taqui Kahn, S. A. Samad and M. R. H. Siddiqui, *J. Mol. Catal.*, 1989, **50**, 97.

[179] Y. V. Subba Rao, K. Mukkanti and B. M. Choudary, *J. Organomet. Chem.*, 1989, **367**, C29.

[180] K. Mukkanti, Y. V. Subba Rao and B. M. Choudary, *Tetrahedron Lett.*, 1989, **30**, 251.

[181] H. Selig and O. Gani, *Inorg. Nucl. Chem. Lett.*, 1972, **11**, 75.

[182] H. Selig, M. Rabinowitz, I. Agranat, C. H. Lin and L. B. Ebert, *J. Am. Chem. Soc.*, 1976, **98**, 1601.

[183] I. Agranat, M. Rabinowitz, H. Selig and C. H. Lin, *Synthesis*, 1977, 267.

[184] M. Rabinowitz, I. Agranat, C. H. Lin and L. B. Ebert, *J. Am. Chem. Soc.*, 1977, **99**, 953.

[185] M. Rabinowitz, I. Agranat, H. Selig, C. H. Lin and L. B. Ebert, *J. Chem. Res. (M)*, 1977, 2350.

[186] C. Ungurenasu and M. Palie, *Synth. React. Inorg. Met. Org. Chem.*, 1977, **7**, 581.

[187] J. Bertin, H. B. Kagan, J. L. Luche and R. Setton, *J. Am. Chem. Soc.*, 1974, **96**, 8113.

[188] H. Selig and L. B. Ebert, *Inorg. Chem. Radiochem.*, 1980, **23**, 2281.

[189] L. B. Ebert, *J. Mol. Catal.*, 1982, **15**, 275.

[190] J. Golé, G. Merle and J. P. Pascault, *Synth. Metals*, 1982, **4**, 269; R. Setton and F. Béguin, *ibid.*, p. 299.

[191] R. Setton in *Preparative Chemistry Using Supported Reagents*, P. Laszlo (ed.), Academic Press, London, 1987, p. 255.

[192] A.. McKillop and D. W. Young, *Synthesis*, 1979, 481.

[193] H. B. Kagan, *Chemtech*, 1976, **6**, 510.

[194] J.-M. Lalancette, G. Rollin and P. Dumas, *Can. J. Chem.*, 1972, **50**, 3058.

[195] D. Tamarkin, D. Benny and M. Rabinowitz, *Angew. Chem. Int. Ed. Engl.*, 1984, **23**, 642.

[196] G. Bewer, N. Wichmann and H. P. Böhm, *Mat. Sci. Eng.*, 1977, **31**, 73.

[197] D. Braga, A. Ripamonti, D. Savoia, C. Trombini and A. Umani-Ronchi, *J. Chem. Soc. Dalton Trans.*, 1979, 2026.

[198] E. Kikuchi, T. Ino and Y. Morita, *J. Catal.*, 1979, **57**, 27; *ibid.*, **62**, 189.

[199] H. J. Jung, P. L. Walker and M. A. Vannice, *J. Catal.*, 1982, **75**, 416.

[200] S. Parkash and J. G. Hooley, *J. Catal.*, 1980, **62**, 187.

[201] R. Schlögl, W. Jones, H. P. Böhm and W. Breimeir, *Synth. Metals*, 1983, **8**, 323.

[202] M. E. Vol'pin, Y. N. Novikov, N. D. Lapkina, V. I. Kasatochkin, Y. I. Struchkov, M. E. Kasakov, R. A. Stukan, V. A. Povitskij, Y. S. Karimov and A. V. Zvarikina, *J. Am. Chem. Soc.*, 1975, **97**, 3367.

[203] H. P. Böhm, Y.-S. Ko, B. Ruisinger and R. Schögl, in *Chemical Reactions in Organic and Inorganic Constrained Systems*, R. Setton (ed.), Reidel, Dordrecht, 1985, p. 429.

[204] H.Podall, W. E. Foster and A. D. Giriatis, *J. Org. Chem.*, 1958, **23**, 401.

[205] M. Ichikawa, Y. Inoue and K. Tamaru, *J. Chem. Soc. Chem. Commun.*, 1972, 928.

[206] R. Setton, *J. Mol. Catal.*, 1984, **27**, 263.

[207] I. Rashkov, I. Gitsov and I. Panayotov, *J. Polym. Sci., Polym. Chem. Ed.*, 1983, **21**, 923.

[208] F. Glockling and D. Kingston, *Chem. Ind. (London)*, 1961, 1037.

[209] M. Rabinowitz and D. Tamarkin, *Synth. Commun.*, 1984, **14**, 377.

[210] D. Savoia, E. Tagliavini, C. Trombini and A. Umani-Ronchi, *J. Org. Chem.*, 1982, **47**, 876.

[211] G. P. Boldrini, D. Savoia, C. Trombini and A. Umani-Ronchi, *J. Chem. Soc. Chem. Commun.*, 1978, 927.

[212] D. Tamarkin and M. Rabinowitz, *Synth. Metals*, 1984, **9**, 125.

[213] M. Contento, D. Savoia, C. Trombini and A. Umani-Ronchi, *Synthesis*, 1979, 30.

[214] R. Setton, F. Béguin and S. Piroelle, *Synth. Metals*, 1982, **4**, 1982.

[215] D. Savoia, C. Trombini and A. Umani-Ronchi, *Tetrahedron Lett.*, 1977, 653.

[216] L. B. Ebert, R. A. Huggins and J. I. Brauman, *Carbon*, 1974, **12**, 199; L. B. Ebert and L. Matty, *Synth. Metals*, 1982, **4** 345.

[217] S. S. Yemul, H. B. Kagan and R. Setton, *Tetrahedron Lett.*, 1980, **21**, 277.

[218] J. Bertin, J. L. Luche, H. B. Kagan and R. Setton, *Tetrahedron Lett.*, 1974, 763.

[219] M. G. Heineman and H. P. Latscha, *Chem. Ztg*, 1981, **105**, 255.

[220] H. B. Kagan, T. Yamagishi, J. C. Motte and R. Setton, *Isr. J. Chem.*, 1978, **17**, 274.

5

Zeolite-assisted organic synthesis — a survey

M. Butters

5.1 INTRODUCTION

In contrast to amorphous or lamellar inorganic solids, zeolites have a well defined three-dimensional crystalline structure. Although they are widely recognized as effective molecular sieve drying agents (e.g. for solvents), zeolites have received relatively little attention as catalysts or reagent supports in organic synthesis. The characteristic properties of zeolites, such as variable acidity/basicity, shape-selectivity and thermal stability, make them attractive hosts for organic reactions. Indeed, they already fulfil important roles in the petrochemical and oil processing industries. The application of zeolites in the synthesis of fine chemicals on the other hand is underdeveloped. Several reviews [1–3,94,138,139] have appeared that cover some of the developments in this field. It is intended that this survey will cover examples of the use of zeolite reagents to promote organic reactions under liquid phase conditions.

The uses of zeolites as synthetic tools fall into five categories:

(a) as catalysts (or supports for catalysts) that offer the potential for shape-selective control, either by transition-state selectivity or by exclusion of competing reactants on the basis of critical molecular diameter;

(b) as highly selective adsorbants, able to remove small components from a reaction mixture (and thereby influence reaction equilibria) or to act as chromatographic phases for the separation of isomers;

(c) as hosts for reactive reagents (e.g. Br_2), enabling controlled release by diffusion into the solvent phase;

(d) as high-surface-area insoluble supports for stoichiometric reagents (e.g. NaN_3), providing easier workup or improved conversions;

(e) as ion-exchange lattices, providing controlled sources of synthetically useful cations (e.g. Ag^+, Tl^+) or as non-nucleophilic bases (e.g. $K^{+-}OZeolite$).

5.1.1   Advantages of zeolite-supported reagents
- Zeolites have well defined structures and consequently it is possible to make reagents with near identical compositions that behave in a predictable fashion.
- They are cheap and readily available synthetic solids.
- They offer the potential for 'shape-selective' control of chemical reactions and present the organic chemist with new avenues for catalysis and synthetic methodology.
- Zeolites, like other solids, are safe to handle and frequently offer simple experimental procedures and purifications.
- The chemical properties of zeolites can be finely tuned by virtue of cation exchange properties.

5.1.2   Disadvantages of zeolite-supported reagents
- Some zeolite activation procedures require the use of elaborate equipment (e.g. quartz-tube furnace with gas supply, e.g. N_2, CO, H_2).
- Some zeolite reagents are air and moisture sensitive (particularly reduced metal loaded forms) and require storage in air-tight containers.
- Zeolite-based reagents sometimes require low loadings of reactants in order to achieve acceptable reaction under liquid phase conditions.
- Their shape-selective molecular recognition excludes organic molecules of moderately large dimensions. They are therefore not universally applicable to all organic structures.

5.1.3   Specifications of zeolite structures in relation to organic chemistry
Appropriate zeolites can separate molecules based on size and shape, and also discriminate on the basis of polarity or degree of unsaturation. Most zeolites show a preference for polar or unsaturated compounds, the exposed counter cations of the zeolites acting as sites of strong localized positive change that electrostatically attract the negative end of polar molecules. Thus, organic compounds containing heteroatoms (O, S, N, Cl, etc.) are all potential candidates for adsorption.

An important variable in the main group of zeolites is the Si/Al ratio, which has a major influence on several properties. For instance, as the Si/Al ratio decreases, the hydrophilicity and affinity for polar compounds (e.g. MeOH) increases. Conversely, as the Si/Al ratio increases the affinity for non-polar compounds increases. Silicalite-1 (a pure SiO_2 polymorph) is an extreme case, and displays a preference for hydrocarbons over water. The zeolites that have so far found utility in organic synthesis include A, X, Y, mordenite (MOR), L, ZSM-5 and silicalite (and related structures such as the titanosilicate, TS-1). The most prominent applications for each of these structure types are summarized in Table 5.1.

Table 5.1 — Properties and applications of common zeolites

Zeolite type (pore opening)	Properties and applications in organic chemistry
A (300–500 pm) [depending on cation]	Hydrophilic; high Al/Si ($\leqslant 1.0$), useful for dehydration (chemical and physical) or for removal of other small, polar molecules (e.g. MeOH, HCl); a good support for Ag^+ and Tl^+ reagents (by exchange).
ZSM-5 (ca. 560 pm)	Relatively low affinity for water; generally very low Al/Si (1/100–1/1000); the proton form has been used for hydrolysis reactions (e.g. oxiranes to diols); has some utility as a photochemical support.
Silicalite (ca. 560 pm)	Pure SiO_2. Has a high affinity for hydrocarbons and therefore described as hydrophobic; a related titanosilicate is a potent catalyst for H_2O_2 oxidations of organic functionalities.
Mordenite (ca. 700 pm)	Relatively hydrophilic; moderate Al/Si (1/5–1/50); useful catalyst for para-selective aromatic nitrations; has not been used extensively for synthesis.
L (710 pm)	Relatively hydrophilic; typical Al/Si (ca. 1/3); has been applied as a catalyst for para-selective aromatic bromination and chlorination.
Y and X (740 pm)	Available with a full range of Al/Si ratios; versatile zeolites in organic synthesis.

5.2 CHEMICAL DEHYDRATION; FORMATION OF ALKENES, ENAMINES, ACETALS AND CONDENSED HETEROCYCLES

5.2.1 Introduction

Cronstedt discovered in 1756 that crystals of stilbite visibly lost water when heated and named such minerals 'zeolites' from classical Greek (meaning 'boiling stones') [4]. The readily available synthetic zeolites (A, X and Y) can generally adsorb up to 25% of their own weight in water. For this reason, zeolites are cheap and versatile drying agents, and are finding increasing application in organic chemistry.

For example, zeolite NaA proved to be an essential component for activity and selectivity in the asymmetric Sharpless epoxidation [5]. The authors assume that the role of the zeolite is primarily the protection of the Ti-catalyst by removal of water. Similarly, NaA serves to adsorb water in the presence of the enzyme invertase [6]. In other cases, however, the zeolite may promote reactions by providing catalytic sites, as well as possibly adsorbing water produced during a reaction. Examples of such applications are discussed in the following sections.

5.2.2   Dehydration of alcohols; production of alkenes

Reversible addition–elimination reactions are often promoted by zeolites. The position of the equilibrium depends on temperature, but most studies have been carried out at high temperatures, favouring elimination [1]. For example, the dehydration of alcohols to alkenes takes place in markedly better yields than the reverse addition reaction. Conversions of 100% and selectivities over 95% can be achieved using H-ZSM-5 (equation (5.1)) [7,8,13].

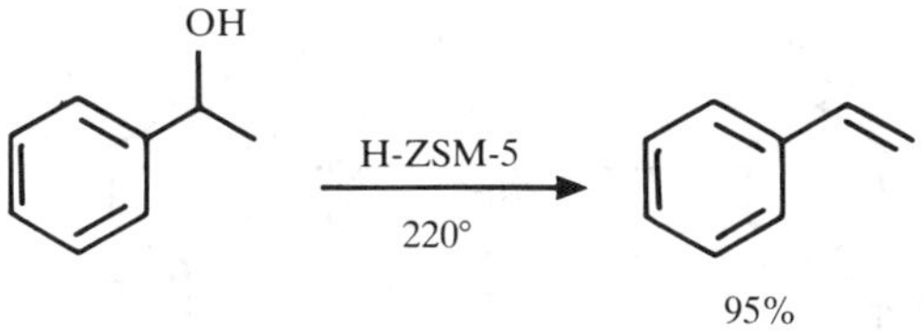

$$\text{(5.1)}$$

Temperatures in excess of 250°C are required in order to dehydrate unactivated primary alcohols under zeolite catalysis. Nevertheless, by making use of the reactant selectivity (see Scheme 1) of CaA, dehydration of 1-butanol and not 2-butanol can be achieved [9]. The 2-butanol is excluded from entrance to the CaA pore system because of its relative bulk compared with the straight chain 1-butanol. Thus, only 1-butanol has access to catalytic sites and only 1-butene is formed as a product. Recently, it has been reported [142] that secondary alcohols can be dehydrated (e.g. 2-methylcyclohexanol gives mainly 1-methylcyclohex-1-ene) over zeolite Y using microwave irradiation.

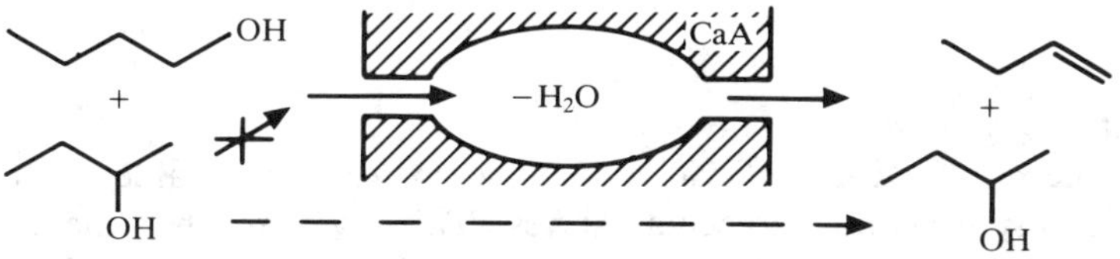

Scheme 1 — Selective dehydration of 1-alcohol (the pore entrance diameter for CaA is ca. 440 pm).

Zeolites are excellent catalysts for the selective dehydration of tertiary-alcohols and similarly activated alcohols. For example, clean dehydration of allylic tertiary alcohols takes place over CaA (zeolite 5A) at room temperature to give conjugated dienes in quantitative yield (equation (5.2)) [10]. It is clear from the range of successful examples of dehydration of tertiary alcohols with critical diameters greater than 500 pm that sites on the external surface of A-type zeolites are responsible for the catalysis e.g. (equations (5.3), (5.4)) [11]. However, the internal pore structure serves as a trap for the chemically formed water.

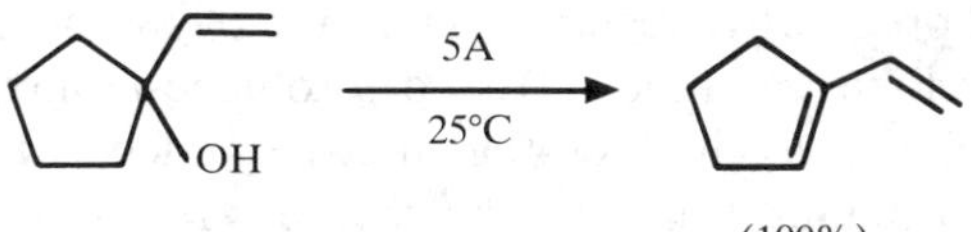

$$\text{(5.2)}$$

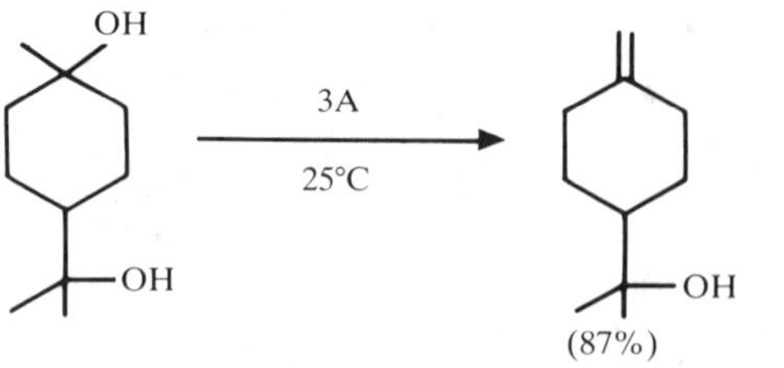

(5.3)

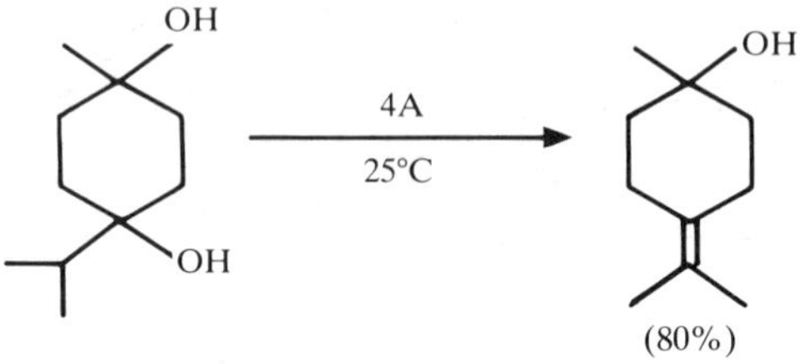

(5.4)

The dehydration of 1,2-tertiary diols proceeds in the presence of faujasite and pentasil type zeolites, but the carbocation intermediate which is formed undergoes a pinacol rearrangement. Yields of up to 83% of ketone are achieved at 105°C using zeolite HY (equation 5.5) [12].

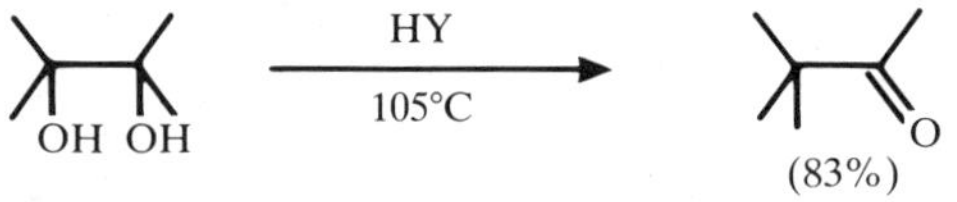

(5.5)

5.2.3   Acetal formation

Acetals can be prepared in high yield from the corresponding aldehydes or ketones and alcohols using zeolite molecular sieves (Table 5.2) to shift the equilibria by adsorption of the water of condensation. In normal acid-catalysed formation of ketone acetals the equilibrium constant is small and a reasonable conversion is possible only by adding a 'chemical trap' (e.g. triethyl orthoformate) for the water of condensation. Azeotropic removal of water is in general inadequate due to temperature dependence of the equilibrium position. Th. M. Wortel *et al.* have reported a series of mild acetalization methods [14–16] that combine an acid catalyst (e.g. *p*-toluenesulphonic acid or zeolite SK-500) with a selective water adsorbent (e.g. zeolite CaA or 3A). Near quantitative yields of acetals are obtained at 0°C in a few minutes.

The problem of cation exchange between the acid catalyst and drying agent can be overcome by using the Ca^{2+} form of zeolite A. The divalent Ca cation is relatively difficult to remove from the zeolite A lattice using monovalent cations (e.g. H^+, Na^+) and exchange of Ca is limited to 20%. In contrast, the rate of proton exchange in KA is very fast. In a typical preparative reaction, a mixture of ketone (or aldehyde) (0.1 mol), alcohol (0.4 mol) cyclohexane (150 ml), and *p*-toluenesulphonic acid monohydrate (3.8 g) is stirred at 0°C and zeolite 5A (40 g, dried) is added. After 2 h, triethylamine (6 ml) is used to quench the reaction, and the zeolite is removed by filtration.

Table 5.2 — Preparation of acetals using molecular sieves [14–16]

Carbonyl	Alcohol	Product	Yield[a]
	MeOH		95%
PhCHO	EtOH		99%
			80%

[a]Based on carbonyl compound; all reactions carried out using a *p*-toluenesulphonic acid catalyst and CaA at 0°C.

5.2.4 Enamine and imine formation

When an amine and a carbonyl compound are condensed under acid catalysis, the water of condensation can be efficiently removed using small pore zeolites such as 3A [17]. Many enamines are sensitive to moisture, and if water is not continuously removed during formation, very poor yields are obtained. Zeolites 3A and 4A have been successfully utilized in a variety of synthetic procedures involving enamine/imine formation (e.g. equations (5.6)–(5.8)) [17,18]. The most efficient method for preparing enamines [17] involves the combination of a silica-alumina 'cracking catalyst' together with zeolite 5A (for water removal). Enamine formation seems limited by the rate of water adsorption by the zeolite. The activity of silica-alumina is ascribed to the presence of acid sites, which enhance the reactivity of the adsorbed carbonyl compound. Strong Brönsted acids such as *p*-toluenesulphonic acid and Amberlyst 15 are less active, and rates are a factor of more than a 100 times slower than with silica-alumina (which is re-usable). The reported procedure [17] involves 32 g 5A per 0.1 mol of ketone and amine together with 8 g of silica-alumina in 80 ml cyclohexane.

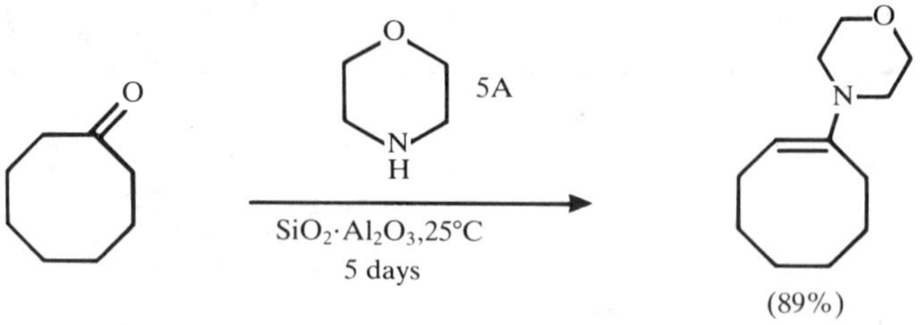

$$(5.6)$$

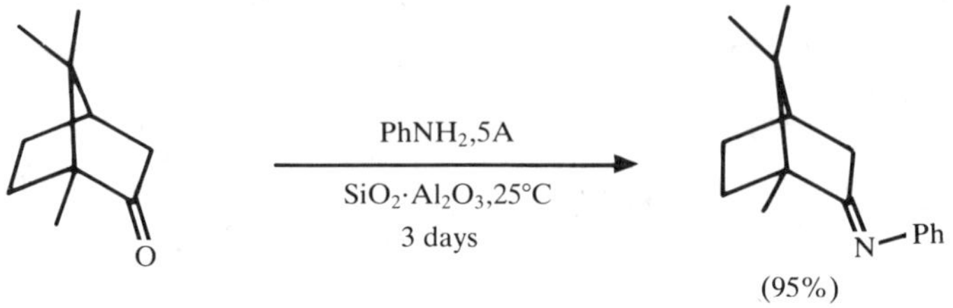

(89%)

$$(5.7)$$

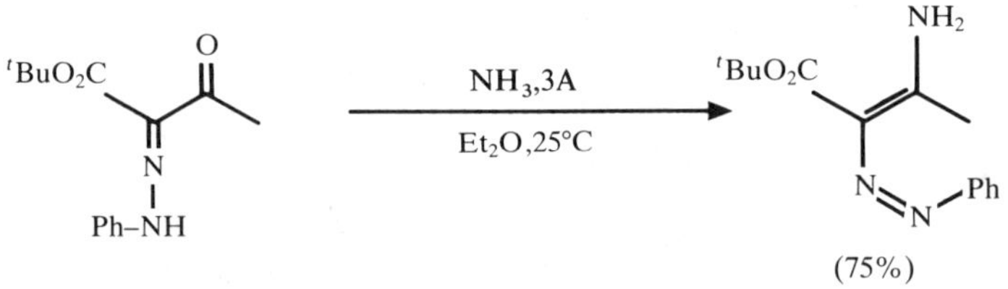

(95%)

$$(5.8)$$

(75%)

A mixture of zeolite 4A and aqueous salicylaldehyde provides an excellent catalyst for the total racemization of optically active amino acids [19]. For example, D-lysine is racemized to DL-lysine with 97% recovery. Presumably, an imine/enamine intermediate is involved. A convenient one-pot procedure [20] for the reductive amination of α-ketolactams using optically active amino acid esters can be effected by the use of zeolite 4A (imine formation) followed by reduction using sodium trihydrocyanoborate.

5.2.5 Formation of heterocycles

D. Boger has applied zeolite 4A to great effect in an elegant synthesis of 3,4-disubstituted pyridines [21]. Pyrrolidine is reacted with a variety of ketones in the presence of 1,2,4-triazine. Zeolite 4A drives the formation of pyrrolidine enamines, which subsequently undergo a regiospecific inverse electron demand Diels–Alder reaction with 1,2,4-triazine. Moreover, the zeolite fortuitously serves as a mild catalyst for the final aromatization step (base-catalysed elimination) involving regeneration of pyrrolidine (Scheme 2). The use of zeolite 4A in this synthetic sequence overcomes two major problems, namely: (a) isolation of unstable enamines is not necessary; and (b) mild catalysis of pyrrolidine elimination (otherwise untrustworthy) is obtained.

Synthetic zeolites are found to be very effective and selective catalysts for the vapour phase ring transformations of oxygen-containing heterocycles into nitrogen or sulphur-containing heterocycles [22]. The essence of these transformations is reaction

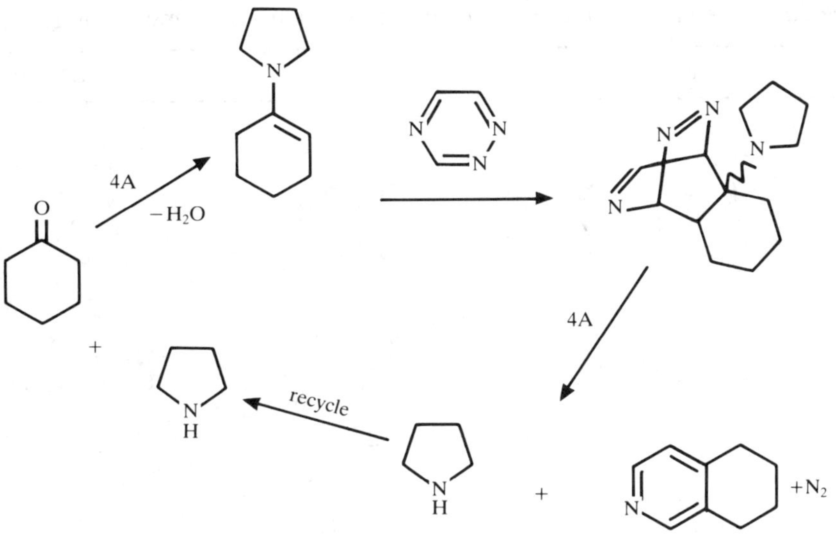

Scheme 2 — Synthesis of 3,4-disubstituted pyridines promoted by zeolite 4A.

of the oxygen-bearing heterocycle with NH_3 or H_2S to produce the new heterocycle, together with one mole of water which is adsorbed by the zeolite.

For example, butyrolactone is converted into 2 pyrrolidone with a selectivity of 80-90% on CuY zeolite [23] at 230°C. Most of the observed heterocycle transformations over zeolites have been in the vapour phase and are therefore of limited value to the synthetic organic chemist. Zeolites and clays have been suggested as catalysts in the prebiotic formation of biomolecules. In this connection the observed formation [24] of adenine and guanine from C_1 precursors using zeolite NaX is interesting.

A-type zeolites have been applied to the synthesis of several heterocyclic ring systems at moderate temperatures. In particular, they have been used to promote condensation reactions leading to thiazoles [25], lactones [26] and quinoxaline-1,4-dioxides [27] (Table 5.3). Zeolites have also been used [140] to catalyse the Claisen–Schmidt condensation of acetophenone with benzaldehyde. Benzofurans are available [141] by cyclodehydration of α-aryloxyketones in the presence of HY.

5.3   PREPARATION OF ESTERS, ETHERS AND AMIDES ASSISTED BY REMOVAL OF SMALL REACTANTS (e.g. ROH, AcOH, HCl) FROM REACTION EQUILIBRIA USING ZEOLITE MOLECULAR SIEVES

5.3.1   Removal of small alcohols

5.3.1.1   Ester formation

The zeolite-catalyzed esterification of acids with alcohols has been described several times [28,29]. For example, acetic acid and alcohols react on wide-pore H-Z-zeolite at 100°C, with yields over 93% [29]. The use of the active centres on the inner surface of zeolites appears to be limited to small molecules. However, zeolite pores can be used

Table 5.3 — Heterocycle-forming condensation reactions promoted by zeolites A

Starting materials	Product	Yield (%)	Conditions
		80	4A, CH$_2$Cl$_2$
		88	3A, dry
		49	5A, Toluene 100°C, 4 h

to adsorb a small reactant selectively from a reaction equilibrium. Thus, the position of an equilibrium can sometimes be moved in favour of a more valuable product. This principle has been applied to transesterification reactions and appears to be advantageous for conversions involving sensitive esters (e.g. β-ketoesters, acrylic esters). High conversions are obtained for a wide range of displacing alcohols (including tertiary, secondary and primary). Phenols and unsaturated alcohols are also tolerated.

Following a comparative study of zeolites 3A, 4A and 5A, van Bekkum and coworkers [30] advised the use of soxhlet extraction for the displaced alcohol, using 3A for reactions of methyl esters with primary and secondary alcohols. 4A or 5A is preferred when higher molecular weight displaced n-aliphatic alcohols are involved and the appropriate alkoxide is also added, as a catalyst. The equilibrium constant for alcoholysis of methyl benzoate using *tert*-butanol is extremely unfavourable ($K_{eq} = 0.008$ at 25°C) and the general preparation of tertiary alkyl esters using conventional ester exchange conditions is often difficult or unsuccessful. The use of a slurry of zeolite 3A or 5A in the presence of potassium butoxide, however, is successful and several new *tert*-butyl esters have been obtained and characterized using this technique [30]. Some problems are encountered with adsorption and/or exchange with metal alkoxide catalysts; however, this can be overcome by judicious choice of zeolite. In a typical reaction, 20 g of methyl benzoate, a solution of 1.2 g potassium in

160 g of *t*-butanol and 60 g of 5A pellets are refluxed for 70 h to give *tert*-butyl benzoate in 100% conversion (equation (5.9)).

$$PhCO_2Me + Bu^tOH \xrightarrow[\text{reflux}]{Bu^tOK,\ 5A} PhCO_2Bu^t \qquad (5.9)$$

Zeolite-assisted transesterification has been extended to the preparation of allylic esters of β-ketoacids. Equimolar amounts of β-ketoester, allylic alcohol and 4-(*N*,*N*-dimethylamino)pyridine (DMAP) are reacted in refluxing toluene and oven-dried zeolite 4A (5 g mmol^{-1} ester) is added to remove the liberated alcohol (MeOH or EtOH). Tertiary alcohols and non-enolizable ketones are unreactive using the DMAP/zeolite method.

5.3.1.2 *Vinyl ether formation*
Transetherification of methyl vinyl ethers is catalysed by palladium and mercuric salts and has wide utility. Using zeolite 4A pellets it is possible to shift the equilibrium towards the desired product by selective adsorption of the methanol produced. A typical procedure involves reaction of methyl vinyl ether (0.1 mol) with isobutyl alcohol (0.1 mol) at 0°C in the presence of Hg(OAc)$_2$ (1.6 mmol, cat.) and dried zeolite 4A (20 g). The reaction is quenched with K$_2$CO$_3$ to give a 76% isolated yield of isobutyl vinyl ether. In view of the gaseous nature of methyl vinyl ether, a more convenient reagent is ethyl vinyl ether (bp 33°C). Other examples of this reaction include the preparation of *p*-chlorobenzyl vinyl ether (57% yield) and *p*-methylbenzyl vinyl ether (43% yield). Some problems are experienced with polymerization of vinyl ether products, but reactions can be performed successfully in diethyl ether as a diluent [33].

5.3.2 Removal of acids (HCl, AcOH)
Zeolites have also been utilized for acylation processes. Two recent examples have appeared in which zeolite 3A has been adopted for scavenging HCl in the reactions of acid chlorides with alcohols [34] and amides [35]. In the latter case, the molecular sieve mediates transacylation in the synthesis of the antibiotic cefoxitin, by selective removal of HCl in the presence of acid chloride and cephalosporin (equation (5.10)).

(Cefoxitin)   (5.10)

Nitrile oxides (R–C≡N—O) are useful intermediates for 1,3-dipolar cycloaddition; these are readily available by 4A-mediated dehydrochlorination of hydroximoyl chlorides [143].

Proton exchanged zeolite Y is both a catalyst and an adsorbent (removal of AcOH or HCl) for the acylation of alcohols by anhydrides [36] and the acylation of uracils by acid chlorides [37].

5.3.3   Removal of thiols (CH₃SH)

Zeolite NaY has been employed [144] to effect the condensation of C–H acids (e.g. CH_3NO_2) with imines of dithiocarbonates. This reaction involves molecular sieve removal of methanethiol and generation of functionalized ketene S,N-acetals in 50% to 85% yield.

5.4   SEPARATION OF ISOMERS AND CLOSELY RELATED COMPOUNDS FROM A MIXTURE

5.4.1   Separation of n-paraffins from branched hydrocarbons

The behaviour of a given zeolite as sorbent, diffusion medium and molecular sieve is regulated by the free dimensions of its channel networks and also by the size, charge and locations of the cations present in its structure. One of the most useful commercial molecular sieve processes involves CaA zeolite in the separation of n-paraffins from iso- and other branched hydrocarbons. On replacing approximately eight Na^+ ions per unit cell by four Ca^{2+} ions in zeolite A, the free diameter of the pore entrance is increased from 400 pm to 500 pm. This allows n-paraffin molecules to be adsorbed selectively in the presence of non-straight chain hydrocarbons. In 1975, the world production of n-paraffins was about 1.6 M tons p.a., obtained largely by the use of CaA zeolite, with about 96% recovery of n-paraffin and purity better than 98%. Other well established separations using zeolites include separations based on shapes of substituted benzenes over medium pore zeolites and separations based on affinity of hexoses or of hexitols over X and Y zeolites [38].

5.4.2   Separation of substituted benzenes

In a recent patent [39] KX zeolite was employed to separate isomers of dichloronitrobenzene. A column was filled with the KX adsorbent and a (50:50) mixture of 2,6-dichloronitrobenzene and 2,4-dichloronitrobenzene was eluted using n-hexanol solvent. This resulted in complete separation of the regioisomers.

Similarly, various cation exchanged forms (Na, Ca, H) of zeolite Y have been used to selectively adsorb 2,3- and 3,4-dinitrotoluene from mixtures containing 2,4- and 2,6-dinitrotoluene [40]. The adsorbed isomers are then desorbed from the zeolite with an alcohol, ketone or ester solvent.

A procedure for the adsorption of phenols, cresols and benzyl alcohol from aqueous solutions using silicalite-1 has been reported [93].

5.4.3   Separation of sugars and related compounds

There is now a widespread application of adsorption processes using zeolite adsorbents for the industrial-scale separation of low-molecular-weight sugars [41,42]. The 'Sarex' process for separation of fructose and glucose uses as the adsorbent a CaY zeolite that shows an equilibrium selectivity for fructose [42]. Separation principles in the glucose/fructose separation are: (a) the preferred Ca-coordination of fructose; and (b) the partition of the mixed solvent yielding a relatively water-rich mixture inside the hydrophilic zeolite, increasing the adsorption capacity. Thus fructose and glucose can be conveniently separated [43] by column chromatography using NaCaX as the stationary phase and eluting with aqueous (1:3) methanol. This system also allows separation of other monosaccharides, of monosaccharides from disaccharides, and of monosaccharides from related alditols. Generally, well-shaped elution peaks, without tailing, are observed. Loadings up to 115 mg of each compound are easily separated over 10 g of zeolite. The order of elution is in agreement with the order of the stability constants of the corresponding Ca(II) complexes.

In view of the known complexation of alditols with lanthanide cations, NaLaX zeolites have been used for separation of these compounds [44]. The role of the cation is further illustrated by the fact that zeolite KX is glucose selective when brought into contact with an aqueous glucose/fructose solution [45]. Thus, column chromatography over suitable X or Y zeolites appears to be a valuable tool in the preparation of pure samples of carbohydrates.

5.5 SELECTIVE PHOTOREARRANGEMENTS WITHIN ZEOLITE CAVITIES

5.5.1 Photolysis of acylbenzenes
Turro and coworkers have made some interesting observations regarding the photochemical transformations of ketones adsorbed on zeolite molecular sieves [46–49]. The channel structures of zeolites create a unique environment for photochemical reactions, and significant deviations from free solution behaviour occur. For example, octanophenone undergoes a Norrish type II photochemical reaction to give a triplet 1,4-biradical. This intermediate reacts to give acetophenone by fragmentation or cyclobutanols by cyclization (Scheme 3). The presence of a zeolite has a marked effect on the proportion of the products (Table 5.4).

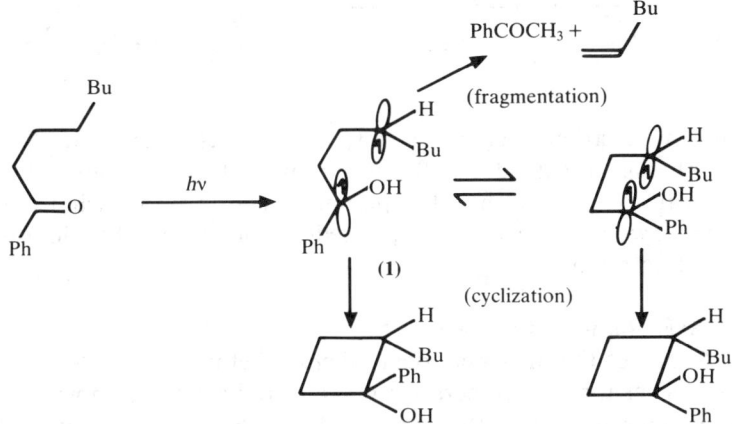

Scheme 3 — Photochemical reactions of octanophenone.

Table 5.4 — Photochemical reactions of octanophenone[a] over zeolites

Zeolite	Ratio of fragmentation: cyclization	Ratio of $Z:E$ cyclobutanols
NaA	2.3	2.2
NaY	0.83	0.72
NaX	3.8	Isomerization
Na mordenite	2.2	2.3
Silicalite-1	50	not available

[a]Similar results were obtained with valerophenone.

Silicate-1 gives outstanding selectivity for fragmentation. It appears that the channel dimensions of silicalite are constraining the reaction pathway of the Norrish type II process. Both fragmentation and cyclization require rotation about σ-bonds in intermediate biradical (1). The transition state leading to ring formation, however, appears to be too bulky to exist within the channels of silicalite-1. Thus, a more extended conformation is favoured, leading to fragmentation. NaY facilitates a much greater degree of molecular mobility, and interestingly (Table 5.4) the ratio of fragmentation to cyclization is less than one with this zeolite. The amount of E-cyclobutanol exceeds the amount of Z with this zeolite too.

When α-alkyldeoxybenzoins are photolysed over LiY or LiX, the normal Norrish type II reactivity is suppressed in favour of the type I process [50]. Moreover, the abnormal p-alkylbenzophenone product is selectively formed (equation (5.11)). The role of the lithium cation is important; the extent of selectivity decreases with Na, K, Rb and Cs as cations. It is possible that the Li^+ cation coordinates with the carbonyl in a conformation that is not suitable for α-hydrogen abstraction.

$$\text{(5.11)}$$

In the narrow restricting channels [53] of silicalite-1, the triplet lifetime of β-phenylpropiophenone is extended to 0.18 ms from the 1 ns occurring in free solution. It would appear that the β-phenylpropiophenone is mainly included within the silicalite channels which restrict degrees of freedom and prevent the adoption of a self-quenching conformation.

5.5.2 Photolysis of dibenzyl ketones

Photolysis [51,52] of the unsymmetrical dibenzyl ketones (2) in 2-propanol in the presence of silica gel at room temperature gives the radical pair combination products (3), (4) and (5) in a ratio of 1:3:1 (Scheme 4). At $-165°C$ there is exclusive formation of (4) and thus only 'cage' products are observed under these conditions.

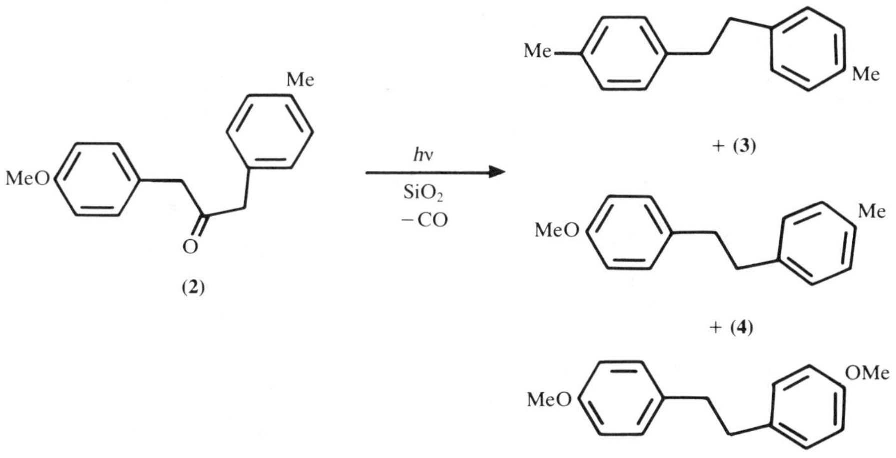

Scheme 4 — Photolysis of an unsymmetrical dibenzyl ketone in the presence of silica.

Extension of this work to zeolites [48] has shown that the 'cage' effect is most pronounced in silicalite-1. For a range of ketone loadings (0.5 to 5% w/w), silicalite-1 consistently gives a higher proportion of the cage combination product than with silica. NaX, Na mordenite, NaY and NaA are less effective in this respect.

The photolytic conversion of dibenzyl ketone to diphenylethane (loss of CO) is 66% in the presence of NaX and 85% in the presence of NaY, respectively [49]. However, if benzene vapour is added to these zeolite-supported reactions, the conversion to diphenylethane falls dramatically in favour of isomerization. Turro suggests that resultant congestion of cavities by benzene favours recombination of the initial radical pair, resulting in isomerization. If water vapour is used as an additive, the radicals appear to have more rapid diffusion, and the reaction products resemble those found in homogeneous systems (equal distribution).

5.6 ELECTROCYCLIC REACTIONS

5.6.1 [2 + 2]Cyclodimerization of cyclopropenes

In the presence of zeolites with pores too small to allow the entry of cyclopropene molecules, [2 + 2]cyclodimerization to tricyclo[3.1.0.0^{2,4}]hexanes is observed (equation (5.12)) [54,55]. Cyclopropene and methyl cyclopropenes can be converted into the corresponding tricyclohexanes over KA or NaA in high yields (ca. 94%), whereas over CaA, NaX or NaY, polymerization is observed. These findings suggest that the sites responsible for catalysis of cyclopropene dimerization are situated on the external surface of KA and NaA.

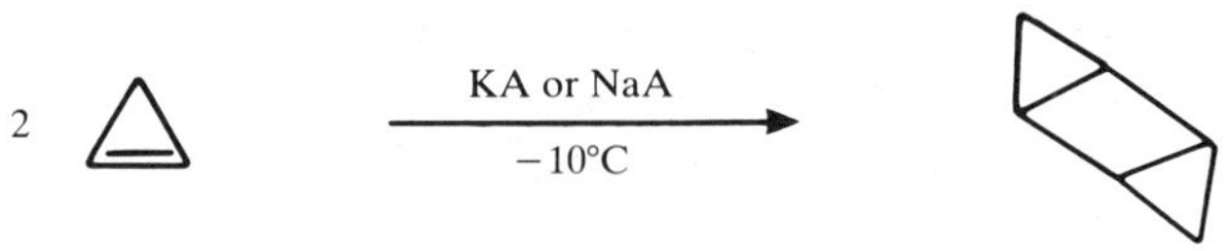

$$2 \quad \triangle \quad \xrightarrow[-10°C]{\text{KA or NaA}} \qquad \qquad (5.12)$$

5.6.2   [4 + 2]Cycloadditions; Diels–Alder reactions

Rate acceleration and selectivity enhancement of the Diels–Alder reaction using water, silica or clay have recently attracted much interest. The possibility of shape-selective control of cycloadditions within the framework of zeolites has also been studied.

Cu^+-containing Y and X zeolites catalyse Diels–Alder cycloadditions. For example, butadiene reacts with acetylene in the presence of Cu^+X to produce 1,4-cyclohexadiene and a small amount of 4-vinylcyclohexene (equation (5.13)) [56,57].

$$\text{(5.13)}$$

(Major)   (Minor)

Highly selective cyclodimerization of butadiene to vinylcyclohexene (VCH) over Cu^+Y has been reported (equation (5.14)) [57,58], but catalyst deactivation due to oligomerization is a problem. This side-reaction is catalysed by acidic sites that are probably introduced during the reduction of Cu^{2+} to Cu^+. However, when unwanted acidic sites are neutralized with ammonia, very stable cycloaddition catalysts can be formed. Maxwell attributes the selectivity of the reaction to transition-state selectivity; the Cu^+-intermediate required to form VCH is less space demanding than the intermediates necessary for the formation of other possible oligomers [58]. Ni and Rh exchanged zeolites, together with 'low acidity' forms of zeolites (e.g. ZSM-20, zeolite beta) have also been used for catalysis of these sorts of processes [59–61].

$$\text{(5.14)}$$

Cu^+ exchanged zeolite Y is an efficient catalyst for a wide range of [4 + 2] cycloaddition reactions [62]. When reactions are performed at 0°C, the catalysis of oligomerization due to Brønsted sites is suppressed and it is not necessary to neutralize the acidic sites, even for reactions of activated dienes/dienophiles. The otherwise sluggish reactions of furan with dienophiles such as methyl vinyl ketone or acrolein are greatly accelerated in the presence of Cu^+Y zeolite. In some cases, better results are obtained using Cu^+Y than are possible by other methods, including high-pressure conditions. The generality of the method is exemplified by the reactions of various dienes with a number of dienophiles given in Table 5.5.

In a typical reaction, a solution of isoprene (10 mmol) and acrolein (10 mmol) in dichloromethane (5 ml) is added slowly (over 30 min) to a slurry of Cu^+Y (1 g) in dichloromethane (5 ml) at 0°C. After 12 h, the CuY is extracted with pentane using a Soxhlet to give 65% yield of 4-methyl-3-cyclohexene-1-carboxaldehyde. Thus Cu^+Y catalyses the Diels–Alder reaction in a 'shape-selective' fashion, giving only the *para*-disubstituted cyclohexene product.

Table 5.5 — Diels–Alder reactions in the presence of CuY (CH_2Cl_2, 0°C, 1–48 h)

Diene	Dienophile	Adduct	Yield (%)	exo/endo
furan	COMe	COMe	73	2.5
furan	CHO	CHO	31	5
cyclopentadiene	COMe	COMe	84	1/24
1,3-cyclohexadiene	CHO	CHO	76	1/33
2,3-dimethyl-1,3-butadiene	CHO	CHO	65	
2,4-dimethyl-1,3-pentadiene	(vinyl ketone)	O	87	

Bauld and coworkers have demonstrated [63,64] that zeolites are able to function as single-electron acceptors to catalyse an electron transfer mechanism of Diels–Alder cycloaddition. Of the zeolites studied, NaX was the most effective but was rather mild in comparison to homogeneous catalysis using aminium salts. For example, irradiation of 1,3-cyclohexadiene in the presence of NaX results in clean [4 + 2]cycloaddition to cyclohexadiene dimers with the characteristic 4:1 endo/exo ratio found for most 'hole catalysed' systems (equation (5.15)). Irradiation in the absence of NaX gives [2 + 2]cycloaddition products.

NaX has also been used to catalyse Brønsted-type [4 + 2]cyclodimerization of 2,4-dimethyl-1,3-pentadiene (equation (5.16)) [63,64].

$$\text{(cyclohexadiene)} \xrightarrow[\text{CH}_2\text{Cl}_2]{\text{NaX, }h\nu} \text{(dimer)} \qquad (5.15)$$

(37%)
(endo:exo = 4:1)

$$\text{(structure)} \xrightarrow[\text{CH}_2\text{Cl}_2]{\text{NaX, 45°C}} \text{(structure)} \qquad (5.16)$$

(90%)

5.6.3 [2 + 2 + 1]Cycloadditions; preparation of cyclopentenones

Interaction of methylenecyclopropane with dicobalthexacarbonyl complexes of various alkynes proceeds smoothly on the surface of solid supports [65] and affords spiro-[2.4]-heptenone derivatives in good yields (e.g. equation (5.17)). Attempts to perform this [2 + 2 + 1] Khand–Pauson reaction homogeneously in hexane at 20°C give only poor yields (10%) of cycloadducts. Several heterogeneous supports, of which NaX is generally the best, are successful under dry-state heterogeneous conditions, in which reactants are adsorbed onto the solid and the solvent is removed before heating to 50°C. The scope of the reaction is quite general, and various derivatives of spiro-[2.4]-heptenones can be prepared. The cycloaddition proceeds with complete regioselectivity with respect to the alkyne component (producing 2-substituted derivatives from terminal alkynes). Regioselectivity of addition across the double bond of methylene-cyclopropene for reactions with terminal alkynes is in favour of the isomer-containing spirosubstituents at the β-position. Conversely, α-spirosubstituted products are formed predominantly in reactions of internal alkynes.

$$\text{(structure)} + \text{MeO} \equiv \xrightarrow[\text{NaX, 50°C}]{\text{Co}_2(\text{CO})_8} \text{(structure)} \quad \text{OMe} + \text{(structure)} \quad \text{OMe} \qquad (5.17)$$

(66%)   (13%)

5.6.4 [2 + 2 + 2]Cycloadditions; synthesis of pyridines

A new procedure for the synthesis of substituted pyridines involves shape-selective cyclotrimerization of phenylacetylene and acetonitrile over Co^+ zeolites [66]. Cobalt^{2+} exchanged forms of zeolite Y and Z are reduced to the Co^+ form using NaBH_4. Phenylacetylene (30 ml) and an excess of acetonitrile (50 ml) are then refluxed in the presence of the Co^+ zeolite (10 g, with 3% w/w Co content) for 2 days (equation (5.18)). In contrast to reactions with CoCl_2 as catalyst [67] which give 80% of isomer (**6**), NaCoY and NaCoZ catalysts yielded 70% and 60% of isomer (**6**) respectively. Such variations of isomer ratio are indicative of the constraints imposed by the zeolite channel system. Zeolite Z has a smaller pore size than does Y, and gives a greater proportion of the more symmetrical isomer (**7**) (equation (5.18)).

$$\text{Ph} \equiv \text{—H} \atop + \atop \text{CH}_3\text{CN} \xrightarrow[\text{100°C}]{\text{Na}^+\text{Co}^+ \text{Zeolite}} \text{(structure)} + \text{(structure)} \qquad (5.18)$$

(**6**)   (**7**)

5.6.5   Claisen rearrangement
Phenolic allyl ethers undergo a cyclic Claisen rearrangement in the presence of zeolite Y and microwave irradiation [142] to give *ortho*-allylphenols in 45% to 53% yields.

5.7   SHAPE-SELECTIVE ELECTROPHILIC ATOMATIC SUBSTITUTION

5.7.1   Friedel–Crafts alkylation
The liquid phase Friedel–Crafts alkylation of arenes using Lewis acid catalysts. (e.g. AlCl$_3$, FeCl$_3$) is carried out on a large industrial scale, e.g. for the preparation of ethylbenzene. The use of such homogeneous catalysts carries problems of corrosivity, toxicity and troublesome workups. In this respect, heterogeneous catalysts offer attractive benefits. Zeolites are very good catalysts for the alkylation of benzenoid compounds and there are numerous examples of *para*-selective alkylations, both in the liquid phase and in the gas phase. For example, zeolite HY has been used to catalyse the alkylation of toluene by styrene with high *para*-selectivity and a conversion of 96% equation (5.19)) [68]. Similarly, in the alkylation of benzene with propene over CeX zeolite [69], a 96% conversion and 96% *para*-selectivity are achieved at 120°C (selectivity is improved by the addition of SO$_2$). Rare earth exchanged Y zeolites are particularly effective catalysts in the liquid phase alkylation of benzenes [69]. In a batch-type (atmospheric pressure) process, reaction is initiated at room temperature and the heats of reactions raise the temperature to reflux (normally about 10 g of CeY catalyst per mole of aromatic substrate will suffice). H mordenite is also a useful catalyst and gives better selectivity for formation of 2-phenylalkanes from 1-alkenes than other zeolites [69a]. Alkyl halides can also be used as alkylating agents. The hydrogen halide gas released during the course of reaction has a limited effect on the catalyst, provided anhydrous conditions are maintained. For example, at atmospheric pressure and room temperature, benzene and *tert*-butyl chloride yield *tert*-butylbenzenes over CeY. Benzene and 1,4-dichlorobutane yield mainly chlorobutylbenzene with some tetralin (no diphenylbutane formed) [69].

(5.19)

(96%)

Phenolic substrates can often be alkylated with a high degree of selectivity and control. Procedures for controlled *tert*-butylation [70] and benzylation [71] have been described. Aniline can be alkylated on faujasite zeolites in the liquid phase [72]. In general, alkylation of alkylbenzenes is much faster than that of phenol or anisole because of the better balance of adsorption between alkenes and alkylbenzenes. For this reason, it is sometimes useful to try different orders of addition when attempting zeolite-catalysed reactions for the first time. Thus it might be important to add polar reactants last of all, so that less polar reactants can take up protic sites within the zeolite framework in readiness for reaction with the strongly adsorbed polar reactant. Zeolite-catalysed nuclear alkylation reactions afford large potential for industrial exploitation as a result of their good selectivity and environmental acceptability. Table 5.6 shows some typical examples. A spectacular example [145] of shape-selective alkylation involves the use of dealuminated H mordenites to catalyse the alkylation of biphenyl with propene. This reaction produces 4,4'-diisopropylbiphenyl in 98% biphenyl conversion and 73.5% selectivity for the 4,4'-isomer.

5.7.2 Friedel–Crafts acylation

In aromatic acylation, present industrial practice involves stoichiometric amounts of metal chloride 'catalyst' together with acid chlorides. Like alkylation, the acylation of arenes can also be carried out using heterogeneous zeolite catalysis. It has been known for some time that zeolites can catalyse useful arene acylation in the gas phase [73]. For example, at 350°C, NaX promotes reaction of benzene and phthalic anhydride to give 98% anthraquinone. Recently, it has been shown that CeY is able to catalyse the *para*-selective acylation of alkylbenzenes in the liquid phase (150°C) using carboxylic acids (equation (5.20)) [74].

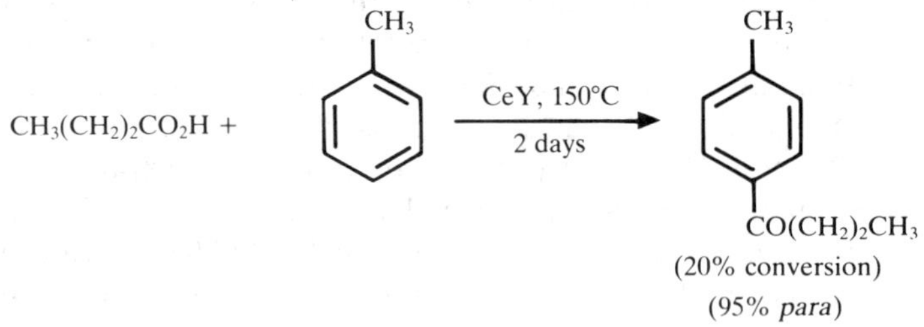

$$CH_3(CH_2)_2CO_2H + \quad \underset{\text{CH}_3}{\bigodot} \quad \xrightarrow[\text{2 days}]{\text{CeY, 150°C}} \quad \underset{\overset{|}{CO(CH_2)_2CH_3}}{\underset{\text{CH}_3}{\bigodot}} \tag{5.20}$$

(20% conversion)
(95% *para*)

5.7.3 Fischer indole syntheses

A number of ketone phenylhydrazones undergo cyclization to indole derivatives (Fischer synthesis) in the presence of zeolite catalysts at 150°C [1]. Good yields of 2-methylindole and tetrahydrocarbazole are formed from the cyclization of the phenyl-hydrazones of acetone and cyclohexanone, respectively, over CaX or rare earth X (e.g. Ce X) catalysts (equations (5.21) and (5.22)).

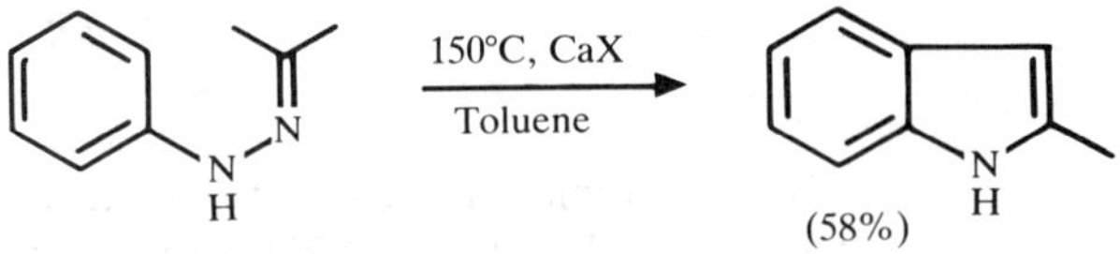

$$\xrightarrow[\text{Toluene}]{\text{150°C, CaX}} \tag{5.21}$$

(58%)

$$\xrightarrow[\text{Toluene}]{\text{150°C, CaX}} \tag{5.22}$$

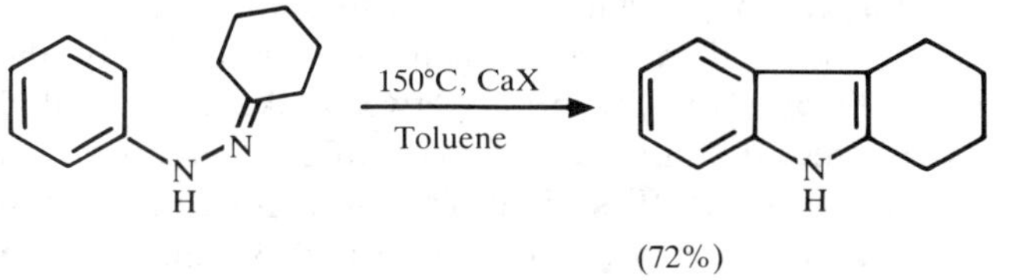

(72%)

Table 5.6 — Alkylation of aromatic substrates over zeolite catalysts

Substrate	Alkylating agent	Product	Conversion (%)	Conditions	Ref.
			90	CeY 150°C 4 h	[69]
		(n = 1,2)			
			90	CeY 130°C 1 h	[69]
			95	CeY 163°C 4 h	[69]
	^tBuOH		74	HY, CCl_4 45°C	[70]
	Ph⌒OH		95	NaY, 200°C 3 h	[71]
			98.5	CaY, 180°C	[72]

5.7.4   Halogenation

5.7.4.1   *Chlorination*

Irreversible damage is caused to zeolites by wet hydrogen halides, so that the use of dihalogen compounds (e.g. FCl, Cl_2, ClBr, Br_2, etc.) generally limits the lifetime of a zeolite catalyst. Nevertheless, some useful systems have been developed for the shape-selective halogenation of aromatic compounds. X, Y and L zeolites can be employed for the chlorination of anisole, toluene and halogenobenzenes. The ratio of *para:ortho* isomers is much higher than with conventional catalysts such as $FeCl_3$. A particularly useful combination is chlorine and KL catalyst, which provides exellent *para*-selectivity and conversions for deactivated benzenes [75]. Thus chlorobenzene and bromobenzene are chlorinated in the liquid phase (70 to 100°C) with *p/o* ratios greater than 95/5 and near quantitative conversion of arene. This reagent is not the most effective option for selective chlorination of moderately activated aromatics, and sensitive functionalities are not tolerated by reactive chlorine at elevated temperatures. Furthermore, the zeolite is not particularly stable to hydrogen chloride. *tert*-Butyl hypochlorite [76] in the presence of HX provides highly *para*-selective monochlorination of various monosubstituted benzenes (equation (5.23)). For example, chlorobenzene in acetonitrile solvent (40°C) gives dichlorobenzene (92% isolated yield) with an isomer ration of 97% *para*/3% *ortho*. Similarly, toluene is chlorinated at room temperature to give a quantitative yield of monochlorotoluenes with a *para:ortho* ratio of 92:8. *t*-BuOCl/HX is most effective for alkyl- phenyl- and halogenobenzenes, giving high *para*-selectivity in all cases (Table 5.7), but highly deactivated benzenes (e.g. benzonitrile, nitrobenzene) are not chlorinated even at elevated temperatures (Table 5.7) [76].

$$\text{(5.23)}$$

No side-chain chlorination is detected with the ^{t}BuOCl/HX reagent and the mild reaction conditions would appear useful for chlorination of benzenes with sensitive functionalities. The role of the zeolite in this reaction is two-fold. HX is a localized source of protic acid that polarizes ^{t}BuOCl into a powerful electrophilic reagent. The network of pores and cavities in the zeolite framework are able to constrain the transition state of the reaction to favour attack at the *para*-position (the transition state for *ortho* attack cannot be easily accommodated). It is likely that much of the *ortho*-chlorinated product in this reaction originates from catalysis on the external surface of the zeolite crystal. Indeed, when larger crystals of HX are employed (ratio of internal sites to external sites increases) the site selectivity for *para* improves [77]. A further advantage of the reagent is that the only byproduct of the reaction is *t*-butanol, and consequently the zeolite lattice is not destroyed.

Table 5.7 — *Para*-selective chlorination of arenes according to equation (5.23)

Substrate	Products para/ortho ratio	Yield (%)	Conditions
Me (toluene)	92:8	100	'BuOCl, HX, 25°C CH_2Cl_2/Et_2O
Et (ethylbenzene)	90:10	100	'BuOCl, HX CH_3CN, 40°C
(t-butylbenzene)	98:2	99	'BuOCl, HX CH_3CN, 40°C
(biphenyl)	86:14	86	'BuOCl, HX CH_3CN, 50°C
Cl (chlorobenzene)		92	'BuOCl, HX CH_3CN, 40°C
Cl (chlorobenzene)	97:3	92	'BuOCl, HX, CH_3CN, 50°C

5.7.4.2 Fluorination
Aromatic fluorination over zeolite catalysts using molecular fluorine has been studied [83].

5.7.4.3 Bromination
Zeolites have also been successfully employed in the *para*-selective monobromination of monosubstituted benzenes under liquid phase conditions. Molecular bromine appears to be the only reagent reported. Useful *para*-bromination reactions can be achieved using various cation exchanged forms of zeolites X, Y and L [78, 79]. Unfortunately, significant problems are encountered because the release of HBr degrades the zeolite structure (low conversions, fall in selectivity, etc.). These problems can be circumvented to some extent by the use of sodium hydrogen carbonate. de la

Vega and Sasson [80] have reported a procedure for the selective *para*-bromination of toluene catalysed by NaY zeolite in the presence of an epoxide. The epoxide serves as an HBr scavenger and dramatically improves the selectivity of the reaction, giving 98/ 2, *para/ortho* selectivity. Unfortunately, only 10% of the toluene is converted to product, and addition of more NaY is necessary (suggesting that the epoxide/ bromohydrin might be strongly adsorbed or polymerizing within the zeolite lattice). It would appear that certain arenes are more suitably brominated in the vapour phase [81]. *para*-Dibromobenzene can be prepared from bromobenzene or benzene using Br_2/LiY zeolite at 200°C. In this case the conversion of benzene is 100% and the *para/ortho* ratio 92:8.

Onaka and Izumi [82] have published a very useful procedure for the *para*-selective bromination of various substituted anilines (equation (5.24)). In this reagent system, it is necessary to pre-adsorb molecular bromine on CaA zeolite (predried 400°C/3 h) in tetrachloromethane and then add the aniline compound in the presence of two equivalents of 2,6-lutidine at 0°C. Remarkable control is possible under these conditions and high conversions (greater than 84%) of clean monobromoanilines are obtained with *para*-selectivities in the region of 94%. Unlike Br_2, aniline is too large to enter the pore cavities of CaA and it would appear that regioselective bromination is achieved through controlled release of bromine, from the pore system. It is not possible to draw any further conclusions about the mechanism of regioselective attack. The lutidine is no doubt useful from neutralization of HBr, which might otherwise adversely affect the reactivity of aniline (via HBr salt).

$$(5.24)$$

5.7.5 Hydroxylation

A further example of aromatic substitution is the direct hydroxylation of benzenes with hydrogen peroxide. This reaction has been reported with catalysis by proton-exchanged and Ti-containing zeolites [84–87]. The *para* isomer is formed preferentially for most monosubstituted benzenes with minimum amounts of side reactions. Aromatic compounds that have been hydroxylated include cresol, anisole, nitrobenzene and toluene. The best results are found when the Ti(IV)-catalyst is a Ti-containing silicalite [85,86] with Si/Ti ratio ∼ 50. The mechanism is tentatively assumed to involve individual lattice Ti species that are able to adopt a coordination number of more than four. Some of these catalyst systems offer attractive improvements over existing indirect hydroxylation procedures (e.g. equation (5.25)).

$$(5.25)$$

5.7.6 Nitration

Nitration of toluene is a notoriously unselective reaction. Few procedures are capable of producing more than 50% of *para*-nitrotoluene. Thus, there is a requirement for a convenient, general *para*-selective method for nitration of alkylbenzenes. A practical laboratory method for the *para*-selective mononitration of alkyl-and aryl-substituted benzenes involves the use of benzoyl nitrate (2.5 mmol) with an equimolar amount of arene in the presence of proton or aluminium exchanged mordenite (1.5 g) [88]. Within 2 h, near quantitative yields of mononitro products are obtained with good *para*-selectivity (e.g. equation (5.26)). Although the reactions are not so selective, zeolites can also help in nitration with nitric acid [89].

$$
\underset{\text{Al}^{3+},\text{H}^+ \text{ mordenite}}{\xrightarrow{\text{PhCO}_2\text{NO}_2}}
$$

(5.26)

(67%)   (1%)   (32%)

5.8 CONTROL OF ALKENE ADDITION REACTIONS

5.8.1 Addition of hydrogen halides

van Bekkum *et al.* have investigated various reports of anti-Markovnikov HBr addition in the presence of CaA and demonstrated that this occurs only by virtue of free radical initiating peroxide impurities [90]. If steps are taken to remove peroxide from starting materials, the normal Markovnikov addition is catalysed by CaA zeolite. The most selective hydrobromination conditions utilize AgA zeolite (68% Ag$^+$ exchange), which gives quantitative hydrobromination of 1-octene. The isomer distribution in this reaction is 90% 2-bromooctane and 10% 3-bromooctane (reflecting a small amount of octene isomerization). Under similar conditions, in the absence of the AgA catalyst, 1-octene gives only 12% conversion with a product distribution of 10% 1-bromo-, 83% 2-bromo- and 8% 3-bromooctane.

5.8.2 Addition of bromine

Several groups have investigated the 1,2-dibromination of alkenes. The combined use of bromine and zeolite 5A was applied to the selective bromination of sterically unhindered olefins. Risbood and Ruthven claimed [92] that a reagent composed of molecular bromine adsorbed on zeolite CaA gave an ionic bromination of unhindered olefins. For example, a mixture of styrene and cyclohexene (1:2 mole ratio) gave a 95% yield of α,β-dibromoethylbenzene and no bromination of cyclohexene. However, further investigations by Dessau [93] indicated that this type of selectivity is quite normal for a free solution pathway. The pore openings of CaA are not large enough to permit the entry of branched alkenes, and it is likely therefore that a small amount of bromine diffuses from the zeolite into the solution phase where reaction with the alkene can take place. Morover, in the absence of styrene, cyclohexene also reacts under the Br_2/CaA conditions, indicating that the selectivity for styrene over cyclohexene might be independent of the CaA zeolite.

In principle, if a zeolite with an appropriate pore structure is used it should be possible to select between a straight chain and a branched or cyclic analogue because only one of the compounds can enter the pores. Thus when an equimolar mixture of 2-octene and cyclohexene is reacted with bromine in the absence of a zeolite, there is little differentiation between the two alkenes. However, when silicalite-1 [94] is used to pre-adsorb octene selectively, leaving cyclohexene in free solution, addition of one mole equivalent of bromine results in almost 100% selective attack on the more bulky cyclohexene. On the other hand, when the bromine is added to a slurry of silicalite-1, followed by addition of the octene/cyclohexene mixture, a reversal in selectivity is observed (Scheme 5).

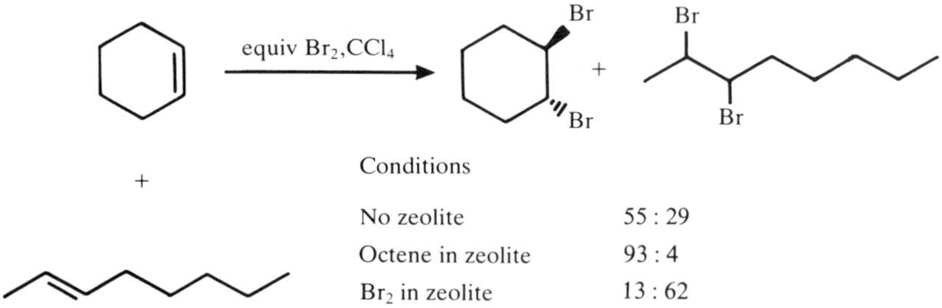

Conditions	
No zeolite	55 : 29
Octene in zeolite	93 : 4
Br_2 in zeolite	13 : 62

Scheme 5 — Selective addition of bromine to alkenes in the presence of silicalite.

5.8.3   Addition of water

Alcohols can be prepared by the acid-catalysed hydration of alkenes. The use of zeolites as catalysts offers marked advantages; they often accelerate the formation of a carbocation but disfavour polymerization, a common side reaction with conventional catalysts. A procedure [95] for selective hydration of cyclohexene involves mixing the alkene with a zeolite ($SiO_2/Al_2O_3 = 64$) in 2:1 water/phenol and heating at 100°C for 8 h. This method provides cycloalkanols (e.g. cyclohexanol, cyclopentanols and cyclooctanols) in moderate yields (ca. 42%). Numerous terpenes have been hydrated over protic zeolites in good yields [96]. For example, the hydration of limonene (**8**), α-pinene (**9**), β-pinene (**10**), and Δ^3-carene (**11**) with wet formic acid in the presence of HY results in high conversions to α-terpineol (**12**) as the main product in high selectivity (scheme 6). Similarly, camphene is hydrated to give isoborneol selectively.

5.8.4   Cyclopropanation

Cyclopropanation of olefins by ethyl diazoacetate (e.g. equation (5.27)) has been carried out with assistance from CuNaX zeolite [97]. The copper cations are responsible for generation of a carbene (by decomposition of the diazo functionality), which gives clean addition reactions with alkene substrates. Although the stereoselec-

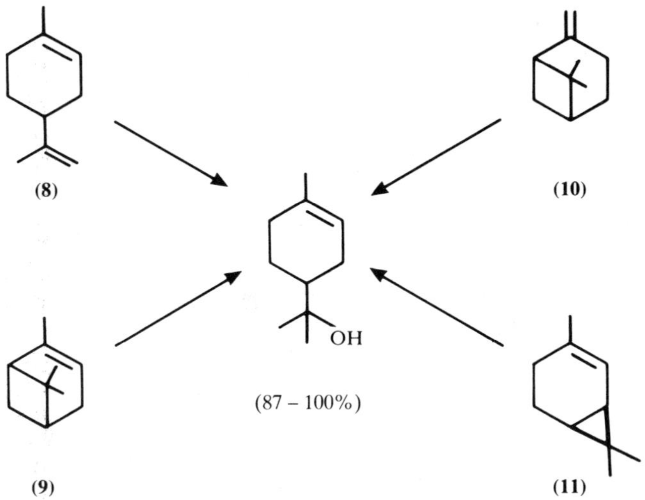

Scheme 6 — Production of α-terpineol by hydration of various terpenes using HY and wet formic acid.

tivity is comparable to conventional homogeneous copper salt catalysis (e.g. CuCN, CuSO$_4$), CuNaX gives relatively low amounts of polymeric side products.

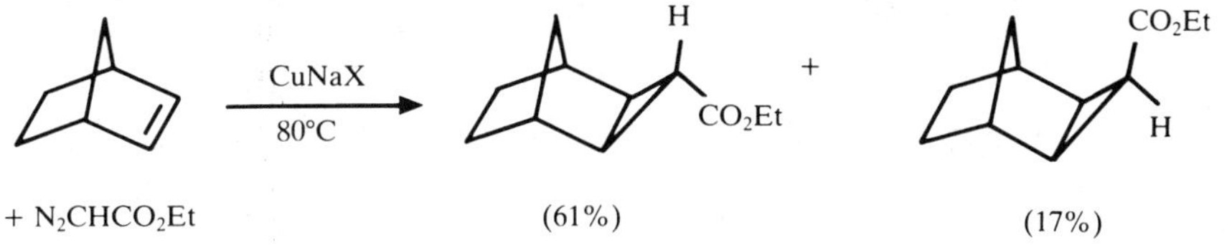

$$(5.27)$$

(Z)-1,2-Diarylethenes are prepared in high yield and high selectivity by the CuNaY-catalysed decomposition of aryldiazomethanes (e.g. equation (5.28)) [98]. Although the best homogeneous catalyst (Cu(ClO$_4$)$_2$) generally gives slightly better yields of stilbenes, the Z/E ratio is much lower. An exchange level of 5% Cu^{2+} appears to be about optimum for gaining Z selectivity. Reactions are best performed at − 70°C in CH$_2$Cl$_2$ using 1 mmol of diazo compound in 7 ml solvent and 10 mol% Cu^{2+} catalyst. Reactions are complete within 7 h. It is assumed by Onaka et al. [98] that copper carbenoid intermediates are formed in the supercages and that the restricted circumstances around the copper ions govern the stability of transition states leading to Z and E isomers.

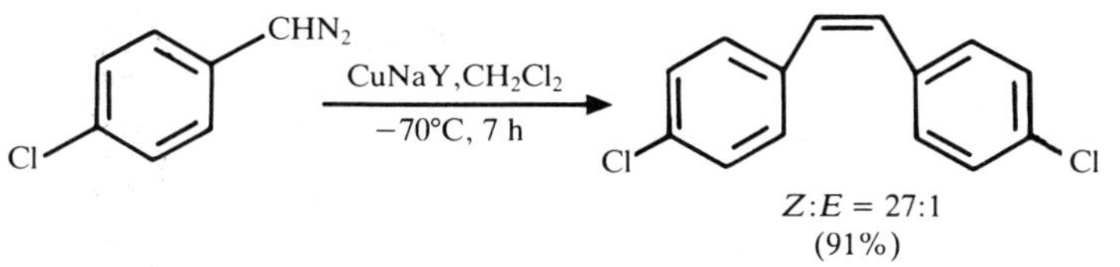

$$(5.28)$$

$$Z{:}E = 27{:}1$$
$$(91\%)$$

5.9 ISOMERIZATION AND OLIGOMERIZATION OF ALKENES

Isomerization and oligomerization of alkenes are important industrial reactions. Many low-temperature examples of oligomerization promoted by ion-exchanged zeolites have been noted. On RhY, NiY and RuY the dimerization of ethene to butenes proceeds selectively at relatively low temperatures (0 to 100°C) [99,100]. For example, ethene is converted into 10% l-butene, 20% Z-2-butene and 70% E-2-butene over RuY at 0°C. The active sites are assumed to be zero valent metal atoms or clusters. Some shape-selective formation of linear oligomers can occur up to 30°C. Low-temperature oligomerization of larger alkenes such as 1-decene and 1-tetradecene occurs over HZSM-5 in the liquid phase [101].

Much of the work on alkene isomerization over zeolites involves high temperatures and gas phase reactions. For example, a sodium–boron zeolite is suitable for contrathermodynamic isomerization of E-2-butene into 1-butene and Z-2-butene in a ratio of 7:5 [102]. More recently, Jacobs *et al.* have developed a series of zeolite-supported sodium clusters, which are excellent low-temperature isomerization catalysts (via formation of carbanions) [103]. The most useful catalyst to date is prepared by impregnation of sodium azide on NaY and calcining at 300°C. Compounds containing allylic or benzylic hydrogen atoms undergo rapid isomerization at 25°C in the presence of Na° on zeolite Y. Compounds without hydrogens α to a double bond (e.g. 3,3-dimethyl-1-butene) fail to react. Alkene isomerizations can be conducted in a bench-scale reactor, and reactions are clean (no coke formation) and show a marked selectivity for Z-alkene formation. Unfortunately, the sodium cluster–zeolite Y catalysts are very sensitive to H_2O, O_2 and CO_2.

5.10 NUCLEOPHILIC SUBSTITUTIONS OF EPOXIDES AND ALKYL HALIDES

5.10.1 Ring opening of epoxides

The use of zeolite Y for catalysis of the reaction between ethylene oxide and a range of alcohols is limited to low-molecular-weight alcohols, since reaction yield decreases for alcohols with a long hydrocarbon chain [104]. The results with acid ion-exchange resins are similar, and zeolites offer little advantage.

Epoxide opening with water is usefully catalysed under mild conditions using proton-form zeolites (e.g. HZSM-5) [105]. The corresponding diols are normally obtained in over 90% yield. For example, if cyclohexene oxide (1.96 g) in dioxan (10 ml) is stirred with a slurry of HZSM-5 (0.5 g) in water (1 ml) for 10 h at 25°C it gives a 92% yield of *trans*-1,2-cyclohexanediol (equation (5.29)).

$$\text{(5.29)}$$

Onaka *et al.* [106] have studied zeolite-catalysed ring opening of epoxides with amines. From a variety of solids investigated, NaY was selected on the grounds of high conversions and high selectivity for attack at the less substituted carbon (equation (5.30)). In the case of styrene oxide, however, better results were obtained with the more basic KY catalyst. It was found that protic catalysts including HY and silica gel (Merck 7734) gave exclusive attack at the benzylic position of styrene oxide.

$$ (5.30) $$

Sodium azide supported on CaY is an efficient and selective nucleophilic source of azide for ring opening of 2,3-epoxyalcohols (equation (5.31)) [107]. It generally gives high regioselectivity for attack at the C_3 carbon of such substrates, compared with selectivity for attack at the less substituted end of simple epoxides. It was deduced that the role of calcium ions in the zeolite is to increase the acid strength of the catalyst, facilitating ring cleavage of epoxide substrates and fixing the conformation of epoxyalcohols through calcium chelation [107].

Other factors found to be important are: (a) a low loading of NaN_3 (ca. 20 wt%) gives improved rates and regioselectivity; (b) the coexistence of an optimum amount of water (ca. 20 wt%) is required to enhance the nucleophilicity of azide and to balance the degree of surface acidity; (c) choice of a low-polarity solvent. In practice, 0.195 g of NaN_3 is supported on CaY (0.78 g) and reacted with 1 mmol of epoxide. Although most of the supported NaN_3 is dispersed inside the small cavities of CaY, the two regioisomeric products have a similar size and therefore selectivity is closely related to the nature of the exchangeable cations rather than the geometric constraints imposed by the pore geometry.

$$ (5.31) $$

5.10.2 Glycoside formation

The traditional promoters for glycosylations involving 1-bromosugars are salts of Ag and Hg, which assist in the removal of the halide-leaving group. A major problem in the formation of the glycosidic bonds is that the configuration of the anomeric centre may become scrambled. Thus, glycosylation reactions often result in roughly equal amounts of two possible anomeric products, α and β. Insoluble silver silicates [108] and Ag zeolite A [109], however, induce selective β-D-mannopyranoside formation. In addition it has been demonstrated that β-bond formation is more favourable for galacto- and manno-pyranosyl glycons than for the corresponding glucopyranosyl derivatives [110].

Using cation exchanged zeolites [109, 111] it is now possible to improve the glycosylation process for a wide range of monosaccharides to give clean inversion or alternatively clean retention of stereochemistry, depending on the nature of the zeolite, glycon and aglycon.

Silver zeolite A is in general the most useful reagent (e.g. equation (5.32)) [109]. However, certain bromosugars tend to be unstable in the presence of silver-based promoters. Thus, Whitfield *et al.* have investigated 'softer' non-oxidizing metal cation forms (e.g. Tl, Co, Cd) [111]. Although Ag, Tl, Co and Cd exchanged zeolites behave similarly they show synthetically useful differences both in terms of the yields of glycosides and in terms of their anomeric ratios. In particular, Tl NaA has been shown to be a suitable alternative when the glycosyl bromide is unstable towards Ag^+ promoters (equation (5.33)). A further advantage carried by zeolite based promoters is their excellent drying capacity, which protects the product from unwanted hydrolytic side reactions.

$$(5.32)$$

$$(5.33)$$

(90%)

5.10.3 Formation of ethers from alkyl halides

KY promotes the reaction of alcohols with alkyl halides (equation (5.34)) [112]. The zeolite has an ideal mixture of acidic and basic sites with which to polarize the alkyl halide (e.g. benzyl chloride) and deprotonate the alcohol hydroxyl group, and the liberated HCl is trapped to give KCl and HY. Polar solvents (e.g. THF) bind strongly to the acid sites in KY and retard the ether-forming reaction. Thus, non-polar solvents such as hexane or chlorobenzene give the best yields. The geometry of the alcohol and the steric bulkiness around the hydroxyl group greatly affect the ease of benzylation within the KY cavities. In particular, the ease of benzylation is determined by whether or not the molecules of benzyl chloride and alcohol can be adequately arranged on the acid and base sites. Comparative experiments for a series of different primary alcohols indicate that the ease of benzylation over KY is 1-decanol > cyclohexylmethanol ≫ 1-adamantylmethanol, whilst in free solution (conventional method using NaH, THF) there is little difference.

$$R^1OH + R^2Cl \xrightarrow[\text{reflux}]{\text{KY, hexanes}} R^1OR^2 + [HCl] \qquad (5.34)$$

5.10.4 Alkylation of amines and amides

Amines and amides are successfully alkylated (equation (5.35)) to give the corresponding monoalkylation products selectively by the use of zeolite KY [113]. It has been suggested that the shape-selective nature of the zeolite lattice is responsible for the high monoalkylation selectivity. KY is much more selective for monoalkylation than conventional bases such as Na_2CO_3 or DBU.

$$(5.35)$$

5.11   REDUCTION REACTIONS

5.11.1   Reductions of carbonyl compounds

Meerwein–Ponndorf–Verley (MPV) reductions of carbonyl compounds using secondary alcohols such as 2-propanol (IPA) for hydrogen transfer and a heterogeneous catalyst have long been established [114]. Shabtai et al. found that simple aldehydes undergo clean and selective reduction using the system IPA/NaX at 180°C [115]. Conversely, cyclic ketones larger in ring size than cyclohexanone are reduced only very slowly. In contrast to the reagent IPA on alumina (see Chapter 3), IPA on NaX fails to reduce α,β-unsaturated aldehydes (e.g. 2-butenal, citral). Reaction of citronellal(13) with IPA over zeolite CsX (large cation) favours MPV reduction, giving citronellol (15) as the main product (92% selectivity) at 150°C (Scheme 7). On the other hand, reaction over NaX (small cation) allows conversion to isopulegol (14) via a cyclic transition state (intramolecular Prins addition, also promoted by microwave irradiation [142]). Thus, the pore system of CsX forces the citronellal into a stretched conformation within the zeolite cavity due to the smaller available pore diameter [115].

Risbood and Ruthven report [116] selective reduction of aldehydes using $LiBH_4$ on zeolite A or X. Several aldehydes could be reduced (in up to 70% yield) in competition with various ketones (90% recovery). However, it would appear that the

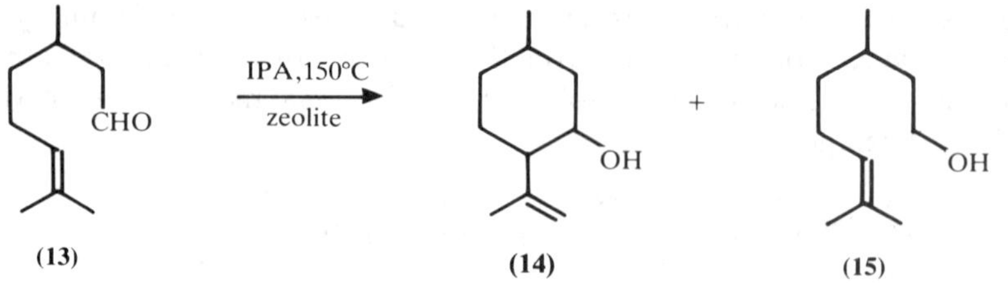

Scheme 7 — Reduction of citronellal with 2-propanol over zeolites.

bulk of the hydride is present on the external surface of the zeolite and the pore system is not important in discriminating between substrates.

5.11.2   Reductions of alkenes

Several groups have investigated metal-loaded zeolites as shape-selective catalysts for hydrogenation of alkenes [117–119]. Huang and Schwartz have developed a rhodium hydride catalyst [117] coordinatively bound to zeolite X. In a sequence of steps, soluble triallylrhodium is bonded to HX and reduced *in situ* to the dihydrido species (zeolite-O-RhH$_2$). A significant proportion of Rh-H sites are located on the external surface of the zeolite but by adding bulky phosphines (e.g. tri-*n*-butylphosphine), it is possible selectively to deactivate these external sites. (Smaller phosphines (e.g. trimethylphosphine) are able to penetrate the cavities and deactivate all Rh sites). The rate of hydrogenation (at 1 atm H$_2$, 20°C) of alkenes larger than cyclohexene in critical diameter is very small. Moreover, the shape-selective nature of the catalyst differentiates between a wide range of alkene structural types. The order of 'apparent' reactivity for a series of alkenes is given in Scheme 8.

Zero-valent Rh catalysts can also be generated by *in situ* reduction of Rh^{3+} exchanged zeolites. Thus Corbin and coworkers have demonstrated that Rh^{3+}X or Rh^{3+}ZSM-11 can be reduced to the corresponding Rh° zeolites and be used for selective hydrogenation of cyclopentene in the presence of 4-methylcyclohexene [118].

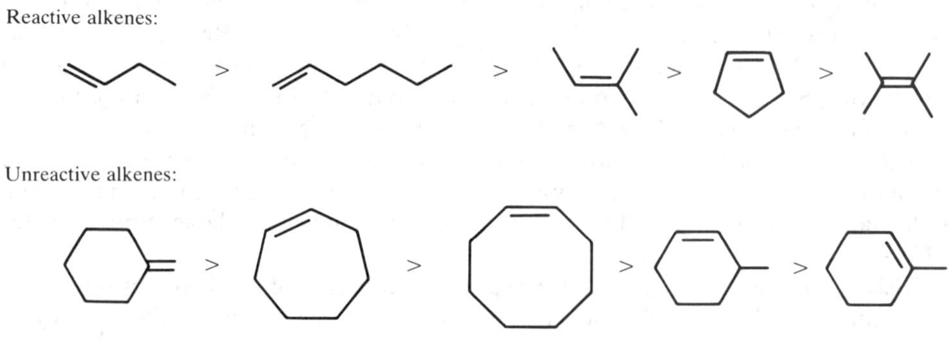

Reactive alkenes:

Unreactive alkenes:

Scheme 8 — Reactivities of alkenes towards hydrogenation over Rh hydride-zeolite X.

When some tri-*n*-butylphosphine is employed to block external sites, selectivity ratios in the region of 47/1 are achieved (3 atm H_2, 60°C). It is also found that the selectivity is affected by water content (2% w/w is optimum). In a similar study, 1-hexene was selectively hydrogenated (selectivity 90/1) in the presence of 4,4-dimethyl-1-hexene using the shape-selective Pd/Cs–ZSM-5 catalyst [120].

Takahashi and coworkers, have found a method [119] for preparing Rh° zeolite Y catalysts in which the Rh° is exclusively sited within the cavities, and addition of phosphines is not necessary to achieve shape selectivity. A Rh^{3+} ion-exchanged zeolite NaY (Rh content 1.5 wt%) is treated with wet carbon monoxide (70 atm, 130°C) to give a pink Rh carbonyl cluster which is then decarbonylated with H_2(1 atm, 120°C). Relative rates of hydrogenation (in hexane, 50°C, 1 atm H_2) decrease in the order; 1-hexene $\simeq$ cyclohexene > cyclooctene $\gg$ cyclododecene. Cyclohexene and cyclooctene are hydrogenated at approximately the same rate over Rh°/carbon, whereas over Rh°/NaY there is a five-fold rate difference. In competition experiments with equimolar amounts of cyclohexene and cyclododecene, 100% conversion of cyclohexene is achieved in 8 h with 0% conversion of cyclododecene (equation (5.36)). Similarly, ethyl acrylate is hydrogenated quantitatively (1.5 h) in the presence of cyclohexyl acrylate (only 2% conversion after 1.5 h).

$$\text{(5.36)}$$

5.12 OXIDATION REACTIONS

5.12.1 Oxidation of alcohols

Pyridinium chlorochromate (PCC) [121] and pyridinium dichromate (PDC) oxidize a wide range of alcohols. Unfortunately, some difficulty lies in the working up of the reaction mixtures. Antonakis and Herscioci have found that PCC and PDC in the presence of a variety of molecular sieves (reactivity 5A < 4A < 3A) perform efficient oxidations of nucleosides at room temperature and in good yields (equation (5.37)) [122, 123]. This technique has several advantages over traditional methods, including mild reaction conditions and very simple workup. The utility of PCC/3A has been further demonstrated for oxidation of 2-nitroalkanols (equation (5.38)) [124] (very susceptible to retroaldol and β-elimination) and carbohydrates (equation (5.39)) [125].

The titanium zeolite TS-1 [126] is a potent catalyst for oxidation of alcohols with H_2O_2 (equations (5.40) and (5.41)) [147]. The alcohol starting material is mixed with an aqueous solution of hydrogen peroxide (and possibly a cosolvent) in the presence of the TS-1 catalyst and heated to between 20 and 100°C. A great advantage of TS-1 compared with the conventional oxidation catalysts, such as OsO_4 and V_2O_5, is that

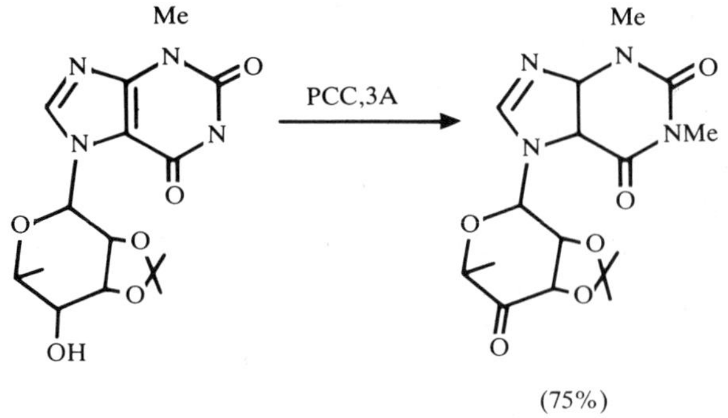

(5.37)

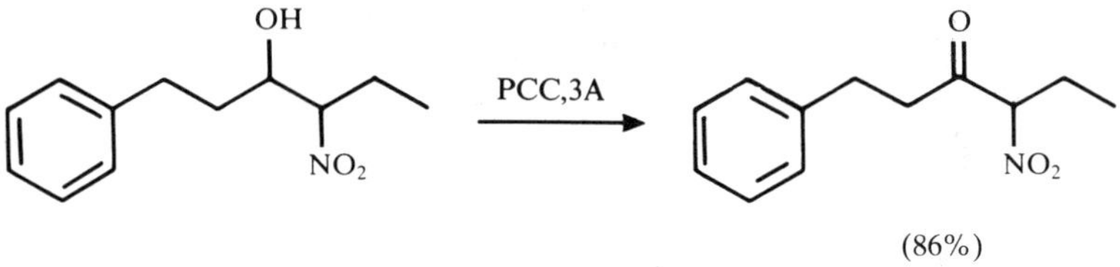

(75%)

(5.38)

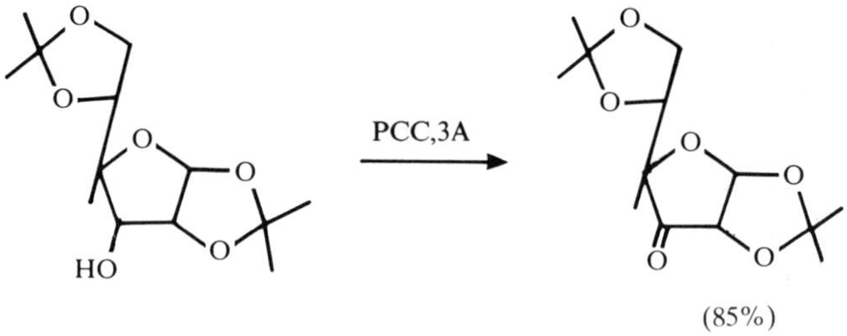

(86%)

(5.39)

(85%)

it is possible to use dilute aqueous peroxide solution in practically stoichiometric amounts.

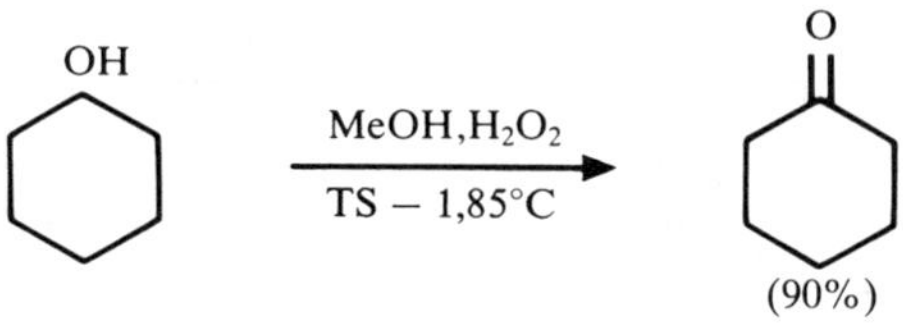

(5.40)

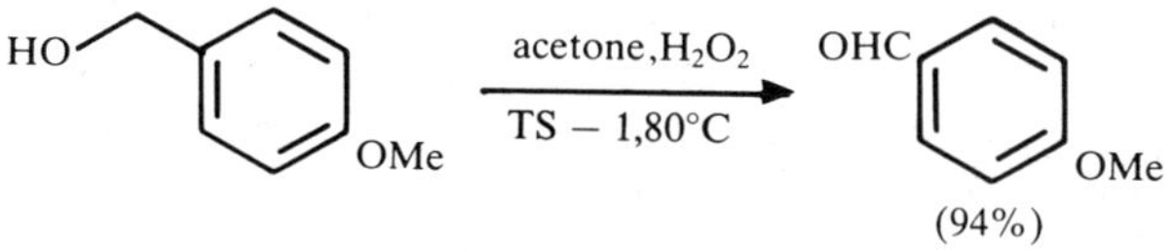

(90%)

(5.41)

(94%)

5.12.2 Oxidation of alkenes

The PDC/4A reagent can be used in the presence of iodine to effect conversion of alkenes into iodohydrins [128] which can subsequently undergo alumina promoted ring closure to the corresponding epoxides (equation 5.42)). The method appears limited to trisubstituted alkenes and sometimes gives the iodohydrin in a regio- and stereospecific manner (normally favouring trans diaxial relationship for OH and I groups).

$$(5.42)$$

Liquid phase epoxidation of alkenes or dienes with H_2O_2 in methanol or acetone is promoted by TS-1 between 0 and 80°C [129]. As an alternative to the chlorohydrin route, TS-1-catalysed oxidation of propene to propylene oxide is environmentally attractive and commercially viable. The chemistry is quite general for other alkenes, with selectivities of 75–96% (based on alkene) and usually quantitative conversion of H_2O_2. If the process is carried out in methanol, the main byproduct formed is a glycol monomethyl ether, and this product can be favoured by doing the reaction at 110°C (equation (5.43)). If the reaction is carried out in acetone, the main byproduct is a glycol ketal. When styrenes are epoxidized using the H_2O_2/TS-1 method (in acetone or methanol, 80°C) it is also possible to isomerize the styrene oxide to the corresponding arylacetaldehyde with selectivities higher than 90% at conversions between 90 and 100% (equation (5.44)) [130].

$$(5.43)$$

$$(5.44)$$

(90 – 100%)

5.12.3 Oxidation of phenols

Selective oxidation of 2,6-dialkylphenols by *tert*-butyl hydroperoxide in Co^{2+} zeolite X has been reported by Oudejans and van Bekkum [131]. Apparently, the *tert*-butyl hydroperoxide decomposes by a radical oxidation-reduction mechanism to give *tert*-butylperoxy radicals (generated during decomposition) that serve to oxidize 2,6-dialkylphenols to the corresponding 2,6-dialkyl-*p*-benzoquinones (equation 5.45)).

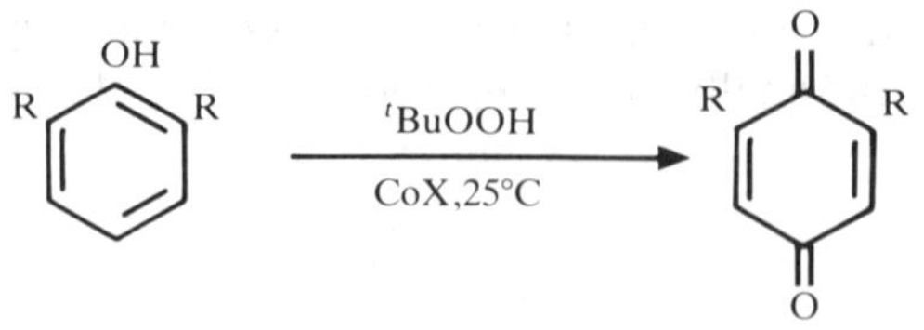

$$\text{R} \overset{\text{OH}}{\underset{}{\bigcirc}} \text{R} \xrightarrow[\text{CoX,25°C}]{^{t}\text{BuOOH}} \text{R} \overset{\text{O}}{\underset{\text{O}}{\bigcirc}} \text{R} \tag{5.45}$$

5.12.4 Oxidation of alkanes

Zeolite-entrapped metal phthalocyanine complexes are perhaps the most sophisticated oxidation catalysts to date [132–135]. Most porphyrin catalyst systems suffer from bimolecular catalyst destruction. Metal phthalocyanine complexes can be prepared within the cavities of zeolites X and Y. The procedure involves heating molten 1,2-dicyanobenzene (200°C/4 h) in the presence of a metal ion-exchanged zeolite, followed by extensive washing/Soxhlet extraction to remove any non-encapsulated phthalocyanine (Scheme 9). The resultant complexes are like a 'ship in a bottle', since once made inside the zeolite they cannot be removed without destroying the 'bottle'. Self-destruction of phthalocyanine molecules by bimolecular processes (analogous to porphyrins) is precluded in the zeolite since the catalyst species are held apart and cannot diffuse together. Herron and coworkers have studied the oxidation of hydrocarbons with iodosylbenzene as the oxidant and FePcY as catalyst [132–135]. Competitive oxidation of cyclohexane and cyclododecane or n-pentane and n-octane shows a preference for the smaller substrate. Moreover, reactions of methyl-cyclohexane and n-octane demonstrate catalyst selectivity for oxidation at the ends of the long molecular axis — owing to zeolite channel orientation. Oxidation of norbornane with the PhIO–FePcY system gives a change from the normal ratio of stereoisomers produced. Thus, the orientating effects of the zeolite channels allow one of the two diastereomeric hydrogen atoms to approach the FePc complex more closely and so be oxidized more readily.

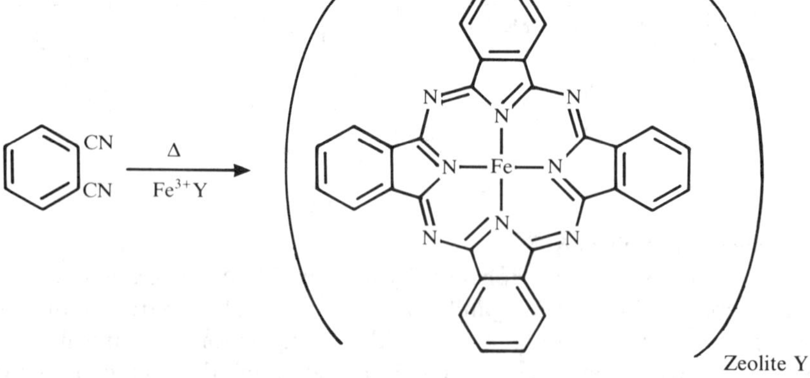

Scheme 9 — Formation of FePcY, a phthalocyanine complex within a zeolite cavity.

In competition experiments, the selectivity for cyclohexane over cyclododecane can be improved from 1.2/1 to 10/1 by adjusting pore openings with exchanged Rb or NH_4 cations. The turnover of the catalyst is limited to about 6, owing to pore blockage by iodoxybenzene ($PhIO_2$) formed by a side reaction.

Dehydrogenation of tetrahydronaphthalenes can be achieved [142] using CuO-on-NaY with microwave irradiation.

5.13 PHOTOCHLORINATION OF ALKANES

The photochlorination of n-alkanes adsorbed on pentasil zeolites proceeds with up to 20-fold greater selectivity for monochlorination of terminal methyl groups than comparable homogeneous reactions [136]. This chemistry provides a novel means of synthesizing terminally functionalized linear alkanes. In a typical reaction, a sample of n-alkane (e.g. dodecane) is first applied to a portion of predried (calcination at 500°C) ZSM-5 (Al/Si ratio 1/24) as a solution in n-pentane, and the solvent is evaporated to constant weight in a desiccator. The dried dodecane/ZSM-5 solid is then charged to a quartz reactor and purged with Cl_2 (0.7% in N_2) in the presence of UV light (low-pressure Hg lamp). On completion of reaction, the products are isolated by CH_2Cl_2 extraction and analysed by GC. This method gives remarkable selectivity for photochlorination of the terminal position, providing 1-chlorododecane as the main product. Expressing terminal selectivity as a proportion (weighted on the basis of available secondary $C-H$ bonds), selectivity ratios in the region of 7.8 are realized. Under normal liquid phase conditions, the terminal selectivity for photochlorination of dodecane is only 0.4.

The terminal selectivity is sensitive to percentage weight loading of substrate, Al/Si ratio of the ZSM-5, moisture content and extent of alkane conversion [136]. The technique can be extended to different alkane chain lengths and to 1-substituted dodecanes with equal success. It is proposed that the pentasil structure adsorbs n-alkanes into the channel system in a manner that protects the more reactive backbone CH_2 groups from attack by diffusing chlorine atoms, which instead encounter the methyl hydrogen atoms 'head on'.

REFERENCES

[1] P. B. Venuto and P. S. Landis, *Adv. Catalysis*, 1986, **18**, 259; K. G. Ione, *Izv. Sibir. Otdel. Akad. Nauk SSSR, Ser. Khim. Nauk*, 1990, 103.

[2] W. Hölderich, M. Hesse and F. Näumann, *Angew. Chem. Int. Ed. Engl.*, 1988, **27**, 226.

[3] P. Laszlo (ed.), *Preparative Chemistry Using supported Reagents*, Academic Press, New York, 1987.

[4] A. F. Cronstedt, *Akad. Handl. Stockholm*, 1756, **17**, 120.

[5] Y. Gao, R. M. Hanson, J. M. Klunder, S. Y. Ko, H. Masamune and K. B. Sharpless. *J. Am. Chem. Soc.*, 1987, **109**, 5765.

[6] H. van Bekkum and H. W. Kouwenhoven, in *Heterogeneous Catalysis and Fine Chemicals*, M. Guisnet et al. (Eds), *Stud. Surf. Sci. Catal.*, 1988, **41**, 45.

[7] R. M. Cursetji, A. N. Singh and A. V. Deo; *Adv. Catalysis*, 1985, **34**, 345.

[8] Y. Aoki, Jap. Pat. 61072727 (14/4/86).

[9] P. B. Weisz, *Pure Appl. Chem.*, 1980, **52**, 2091.

[10] J. H. Markgraf, E. W. Greeno, M. D. Miller and W. J. Zaks, *Tetrahedron Lett.*, 1983, **24**, 241.

[11] M. Nomura and F. Yoshihito, *Nippon Nogei, Kagaku Kaishi*, 1983, **57**, 139; *Chem. Abstr.* **99**: 22711d.

[12] Q. Wang and Z. Chen., *Kexue Tongbao*, 1984, **29**, 1130; *Chem. Abstr.*, **102**, 45280g.

[13] Y. Konai and M. Hino, *Kureha Kagaku Kogyo*, DE 3620512 (87-000925) (02.01.87); *J. Synth. Meth.*, 1988, **14**, 75989C.

[14] D. P. Roelofsen, E. R. J. Wils and H. Van Bekkum, *Recl. Trav. Chim. Pays Bas*, 1971, **90**, 1141.

[15] D. P. Roelofsen and H. van Bekkum, *Synthesis*, 1972, 419.

[16] Th. M. Wortel, W. H. Esser, G. van Minnen-Pathuis, R. Taol, D. P. Roelofsen and H. van Bekkum, *Recl. Trav. Chim. Pays Bas*, 1977, **96**, 44.

[17] D. P. Roelofsen and H. van Bekkum, *Recl. Trav. Chim. Pays Bas*, 1972, **91**, 605.

[18] J. Gasteiger and U. Strauss, *Chem. Ber*, 1981, **114**, 2336.

[19] M. G. Ryzhor, USSR Patent SU899535 (01745 J-E) (23.1.82); *J. Synth. Meth.*, 1983, **9**, 75500Y.

[20] D. Taub, Merck and Co. Inc. Patent ZA-8209141 (84-017692-B) (28.09.83); *J. Synth. Meth.*, 1984, **10**, 76082Z.

[21] D. L. Boger, J. S. Panek and M. M. Meier. *J. Org. Chem.*, 1982, **47**, 895.

[22] Y. Ono, *Heterocycles*, 1981, **16**, 1755.

[23] K. Hatada, M. Shimada, Y. Ono and T. Keii, *J. Catal.*, 1975, **37**, 166.

[24] F. Seel, M. Van Blon and A. Dessauer, *Z. Naturforsch.*, *1982*, **37**, 820.

[25] J. Gasteiger and C. Herzig, *Tetrahedron*, 1981, **37**, 2607.

[26] Z. Chen, Q. Li and R. Liang, *Huaxue Shijie*, 1988, **29**, 14; *Chem. Abstr.* **110**: 7971r.

[27] T. Takabatake and M. Hasegawa, *J. Heterocyclic Chem.*, 1987, **24**, 529.

[28] E. Santacesaria, G. Carra and F. Silva., *J. Catal.*, 1984, **85**, 519.

[29] Y. Zhang, L. Han and S. Cai, *Shiyou Huagoug*, 1986, **15**, 411. *Chem. Abs.*, 105: 190458n.

[30] D. P. Roelofsen, J. W. M. De Graaf, J. A. Hagendoorn, M. H. Verschoor and H. van Bekkum, *Recl. Trav. Chim. Pays Bas*, 1970, **89**, 193.

[31] J. C. Gilbert and T. A. Kelly, *J. Org. Chem.*, 1988, **54**, 449.

[32] R. M. Barrer and J. L. Lopez, *J. Appl. Chem. Biotechnol.*, 1973, **23**, 189.

[33] H. Yuki, K. Hatada, K. Nagata and K. Kajiyama, *Bull. Chem. Soc. Jpn*, 1969, **42**, 3546.

[34] A. R. Banks, R. F. Fibiger and T. Jones, University Patents US 4258204 (27213D-E) (24.03.81); *J. Synth. Meth.*, 1981, **7**, 76540W.

[35] L. M. Weinstock, US Pat. 4,053, 286 (1977); E. Paul, *Chem. Ind. (London)*, 1990, **10**, 320.

[36] Y. Chen, *Faming Zhuanli, Shenqing Gongkai Shuomingshu CN*, 1986, 85,100, 584; *Chem. Abstr* **106**: 83999p.

[37] Q. Wang, Y. Zhai, W. Zhang, P. Liu, J. Yao and Y. Meng, *Gaodeng. Xuexiao Huaxue Xuebae*, 1987, **8**, 438; *Chem. Abstr.*, **108**, 37777 h.

[38] J. D. Sherman and C. C. Chao, *Proceedings 7th Int. Zeol. Conf.*, Y. Murakami, A. Iijima and J. W. Ward (eds), Elsevier, New York, 1986, 1025.

[39] K. Miwa and K. Tada Jap. Pat. 61,183,248 (8.2.85); *Chem. Abstr.* **106**: 49764y.

[40] H. A. Zinnen and S. T. Franczyk, US 4,717,778 (5.1.88); *Chem. Abstr.*, **108**, 188905z.

[41] R. W. Neuzil and R. A. Jensen, *Development of the Sarex Process for Saccharide Separation*, 85th National Meeting of AIChE, Philadelphia, June 1978.

[42] R. W. Neuzil and J. W. Priegnitz, UK Patent 1,574,915 (May 1977); *Chem. Abstr.*, **87**, 184896x.

[43] Th. M. Wortel and H. van Bekkum, *Recl. Trav. Chim. Pays Bas*, 1978, **97**, 156.

[44] A. P. G. Kieboom, T. Spoormaker, A. Sinnema, J. van der Toorn, and H. van Bekkum, *Recl. Trav.Chim. Pays Bas*, 1975, **94**, 53.

[45] C. B. Ching and D. M. Ruthven, *Zeolites*, 1988, **8**, 68.

[46] N. J. Turro and P. Wan, *Tetrahedron Lett.*, 1984, **25**, 3655.

[47] N. J. Turro, C-C. Cheng and W. Mahler, *J. Am. Chem. Soc.*, 1984, **106**, 5022.

[48] N. J. Turro and P. Wan, *J. Am. Chem. Soc.*, 1985, **107**, 678.

[49] N. J. Turro, C-C. Cheng, X-G. Lei and E. M. Flanigen, *J. Am. Chem. Soc.*, 1985, **107**, 3739.

[50] D. R. Corbin, D. F. Eaton and V. Ramamurthy, *J. Org. Chem.*, 1988, **53**, 5384.

[51] D. Avnir, L. J. Johnston, P. de Mayo and S. K. Wang, *J. Chem. Soc., Chem. Commun.*, 1981, 958.

[52] L. J. Johnston, P. de Mayo and S. K. Wang, *J. Chem. Soc., Chem. Commun.*, 1982, 1106.

[53] H. L. Casol and J. C. Scaiano, *Can. J. Chem.*, 1984, **62**, 628; 1985, **63**, 1308.

[54] A. J. Schipperijn and J. Lukas, *Tetrahedron Lett.*, 1972, 231.

[55] A. J. Schipperijn and J. Lukas, *Recl. Trav. Chim. Pays Bas*, 1973, **92**, 572.

[56] H. Reimlinger, U. Kruërke and E. de Ruiter, *Chem. Ber.*, 1970, **103**, 2317, see also U.S. Pat. 3,444,253 (3.05.69); *Chem. Abstr.*, **71**, 38433x.

[57] I. E. Maxwell, R. S. Downing and S. A. J. van Langen, *Acta. Phys. Chim.*, 1978, **24**, 215.

[58] I. E. Maxwell, R. S. Downing, J. J. de Boer and S. A. van Langen, *J. Catal.*, 1980, **61**, 485,493.

[59] P. Pichat, J. C. Vedrine, P. C. Gallezat and B. Imalik, *J. Catal.*, 1974, **32**, 190.

[60] B. K. Nefedov and N. S. Seergeva, *Neftekhimiya*, 1977, **17**, 516; *Chem. Abstr.*, **87**, 167586d.

[61] R. M. Dessau, U.S. Pat. 4,384,153 (17.05.83); *Chem. Abstr.*, **99**, P70285q.

[62] J. Ipaktschi, *Z. Naturforsch.* 1986, **41b**, 496.

[63] R. A. Pabon, D. J. Bellville and N. L. Bauld, *J. Am. Chem. Soc.*, 1983, **105**, 5158.

[64] S. Ghosh and N. L. Bauld, *J. Catal.*, 1985, **95**, 300.

[65] W. A. Smit, S. L. Kineev, O. M. Nefedov and V. A. Tarasov, *Tetrahedron Lett.*, 1989, **30**, 4021.

[66] Y. BenTaarit, Y. Diab, B. Elleuch, M. Kerkani and M. Chihaoui, *J. Chem. Soc., Chem. Commun.*, 1986, 402.

[67] H. Bönnemann, *Angew. Chem. Int. Ed. Engl.*, 1978, **17**, 505.

[68] H. A. Bouncer and I. A. Cody, Eur. Pat. 160144 (06.11.85); *Chem. Abstr.*, **104**, P148457c.

[69] R. Q. Kluttz and L. H. Slough. U.S. Pat. 4395372 (26.07.83); *Chem. Abstr.*, **99**, P139469x.

[69a] R. A. Grey, U.S. Pat. 4731497 (15.03.88).

[70] A. Corma, H. Garcia and J. Primo, *J. Chem. Research(S)*, 1988, 40.

[71] D. E. Pat-3700917 (1988); *J. Synth. Meth.*, 1988, **14**, 77683D.

[72] M. N. Magerramov and A. G. Lyutfaliev, *Azerb. Khim. Zh.* 1985, 25; P. Y. Chen and M. C. Chen, *Stud. Surf. Sci. Catal.*, 1986, 739; *Chem. Abstr.*, **108**, 37375a.

[73] D. E. Pat. 2633458 (26.01.78); *Chem. Abstr.*, **88**, 137904j.

[74] B. Chiche, A. Finiels, C. Gauthier and P. Geneste, *J. Org. Chem.*, 1986, **51**, 2128.

[75] Y. Higuchi and T. Suzuki, EP 118851 (04.07.84), EP 154236 (19.09.85); *J. Synth. Meth.*, 1984, **10**, 77881Z.

[76] K. Smith, M. Butters and B. Nay, *Synthesis*, 1985, 1157.

[77] M. Butters, PhD Thesis, U.C. Swansea, 1986.

[78] T. M. Wortel, D. Oudijn, C. J. Vleugel, D. P. Roelofsen and H. van Bekkum, *J. Catal.*, 1979, **60**, 110.

[79] J. van Dijk, J. van Daalen and G. Paerels, *Recl. Trav. Chim. Pays Bas*, 1974, **93**, 72.

[80] F. de la Vega and Y. Sasson, *J. Chem. Soc., Chem. Commun.*, 1989, 653.

[81] S. Kai, J. Pat 60224644 (09.11.85); *Chem. Abstr.*, **104**, 148469w.

[82] M. Onaka and Y. Izumi, *Chem. Lett.*, 1984, 2007.

[83] L. C. Sams, T. A. Reames and M. A. Durrance, *J. Org. Chem.*, 1978, **43**, 2273.

[84] C. D. Chang and S. D. Hellring, U.S. Patent 4578521 (1986); see also U.S. Pat. 4,410,501 (1983).

[85] B. Notari in *Innovation in Zeolite Science*, P. J. Grobet *et al.* (eds), 1987, 413.

[86] *Chem. Abstracts*, **96**, 199290e (1982); see also Brit. Pat. 2116974 (29.09.85); *Chem. Abstr.*, **100**, 22409a.

[87] G. Perego, G. Bellusi, C. Corno, M. Taramasso, F. Buonomo and E. Esposito, *Proceedings 7th Int. Zeol. Conf.*, 1986, 129.

[88] K. Smith, K. Fry, M. Butters and B. Nay, *Tetrahedron. Lett.*, 1989, **30**, 5333; M. Butters, K. Smith, K. Fry and B. Nay, E. P. 0356091A2. (11.8.89).

[89] S. M. Nagi, E. A. Zubkov and V. B. Shubin, *Bull. Acad. Sci. USSR, Div. Chem. Sci.*, 1989, **38**, 1780.

[90] Th. M. Wortel, S. Rozendaal and H. van Bekkum, *Recl. Trav. Chim. Pays Bas*, 1979, **98**, 505.

[91] E. Narita, N. Horiguchi and T. Okabe, *Chem. Lett.*, 1985, **6**, 787.

[92] P. A. Risbood and D. M. Ruthven, *J. Am. Chem. Soc.*, 1978, **100**, 4919.

[93] R. M. Dessau, *J. Am. Chem. Soc.*, 1979, **101**, 1344.

[94] K. Smith, *Bull. Soc. Chim. Fr.*, 1989, **2**, 272; *Studies in Surface Science and Catalysis*, 1991, **59**, 55; K. Smith and K. B. Fry, *J.C.S., Chem. Commun.*, 1992, 187.

[95] EP 224116 (1987); *J. Synth. Meth.*, 1987, **13**, 77027C.

[96] M. Nomura and Y. Fujihara, *Kinki Daigaku, Kogakubu, Kenkyu, Hokoku*, 1985, **19**, 1.

[97] J. C. Oudejans, J. Kaminska, A. C. Kock-van Dalen and H. van Bekkum, *Recl. Trav. Chim. Pays Bas*, 1986, **105**, 421.

[98] M. Onaka, H. Kita and Y. Izumi, *Chem. Lett.* 1985, 1895.

[99] Y. Yashima, Y. Nahida, M. Ebisawa and N. Hara, *J. Catal.*, 1975, **36**, 320.

[100] L. Bonneviat, D. Oliver and M. Che, *J. Mol. Catal.*, 1983, **21**, 415.

[101] S. J. Miller, DE 3,235,280 (25.09.81); *Chem. Abstr.*, **98**, 215165s.

[102] D. L. Sikkenya, U.S. 4,503,282 (05.03.85); *Chem. Abstr.*, **103**, 6846x.

[103] L. R. M. Martens, W. Vermeiren, P. Grobet and P. Jacobs, *Preparation of Catalysts IV*, 1987, 531.

[104] B. Burczyk and F. Kukla, *Zesz. Probl. Postepow, Nauk Roln.*, 1981, **211**, 99; *Chem. Abstr.* **99**, 7201u; F. L. Johnston US 4438281 (20.03.84); *Chem. Abstr.*, **100**, 209154f.

[105] US Pat. 4,620,044 (1986); *Chem. Abs.* **106**: 84049r.

[106] M. Onaka, M. Kawai and Y. Izumi, *Chem. Lett.*, 1985, 779.

[107] M. Onaka, K. Sugita and Y. Izumi, *J. Org. Chem.*, 1989, **54**, 1116.

[108] H. Paulsen and R. Lebuhn, *Annalen*, 1983, 1047, and references therein.

[109] P. Garegg, C. Hendrichson, J. Norberg and P. Ossowski, *Carbohydr. Res.*, 1983, **119**, 95.

[110] C. A. van Boekel, T. Beetz, A. C. Kock-van Dalen and H. van Bekkum, *Recl. Trav. Chim. Pays Bas*, 1987, **106**, 596.

[111] D. M. Whitfield, R. N. Shah, J. P. Carver and J. Krepinsky, *Synth. Commun.*, 1985, **15**, 737.

[112] M. Onaka, M. Kawai and Y. Izumi, *Bull. Chem. Soc. Jpn*, 1986, **59**, 1761, and references therein.

[113] M. Onaka, A. Umezono, M. Kawai and Y. Izumi, *J. Chem. Soc., Chem. Commun.*, 1985, 1202, and references therein.

[114] R. A. W. Johnstone, A. H. Wilby and I. D. Entwistle, *Chem. Rev.*, 1985, **85**, 129.

[115] J. Shabtai, R. Lazar and E. Biron, *J. Mol. Catal*, 1984, **27**, 35.

[116] P. A. Risbood and D. M. Ruthven, *J. Org. Chem.*, 1979, **44**, 3969.

[117] T-N. Huang and H. Schwartz, *J. Am. Chem. Soc.*, 1982, **104**, 5244.

[118] D. R. Corbin, W. C. Seidel, L. Abrams, N. Herron, G. D. Stucky and C. A. Tolman, *Inorg. Chem.*, 1985, **24**, 1800.

[119] I. Yamaguchi, Y. Joh and S. Takahashi, *J. Chem. Soc., Chem. Commun.*, 1986, 1412.

[120] J.P. 59156437 (05.09.84); *Chem. Abstr.*, **102**, 28107p.

[121] E. J. Corey and J. W. Suggs, *Tetrahedron Lett.*, 1979, 399.

[122] J. Herscovici and K. Antonakis, *J. Chem. Soc., Chem. Commun.*, 1980, 561.

[123] J. Herscovici, M-J. Egron and K. Antonakis, *J. Chem. Soc., Perkin Trans. 1*, 1982, 1967.

[124] G. Rosini and R. Ballini, *Synthesis*, 1983, 543.

[125] S. Czernecki, G. Georgoulis, G. L. Stevens and K. Vijaykumaran, *Tetrahedron Lett.*, 1985, **26**, 1699.

[126] M. Taramasso, US Pat. 4410501 (1983); *Chem. Abstr.*, **95**, 206272k.

[127] A. Esprosito, C. Neri, and F. Buonomo, EP 0102655 A2 (28.07.83); *J. Synth. Meth.*, 1984, **10** 76299Z.

[128] R. Antonioletti, M. D'Aurian, A. DeMico, G. Piancatelli and A. Scettri, *Tetrahedron*, 1983, **39**, 1765.

[129] C. Neri, F. Buonomo and B. Anfossi, EP 0100118 Al (08.02.84) and EP0100119 Al (28.07.82); *J. Synth. Meth.*, 1984, 76275.

[130] C. Neri and F. Buonomo, Eur. Pat. Appl., 102, 997 (1984).

[131] J. C. Oudejans and H. van Bekkum, *J. Mol. Catal.*, 1981, **12**, 149.

[132] *Chem. Eng. News*, 1984, 17 September, 30.

[133] N. Herron, G. D. Stucky and C. A. Tolman, *J. Chem. Soc., Chem. Commun.*, 1986, 1521.

[134] N. Herron and C. A. Tolman, *J. Am. Chem. Soc.*, 1987, **109**, 2837.

[135] C. A. Tolman and N. Herron, *Symposium on Hydrocarbon Oxidation*, Division of A.C.S., 1987, p. 798.

[136] N. J. Turro, J. R. Fehlner, D. P. Hessler, K. M. Welsh, W. Ruderman, D. Firnberg and A. M. Braun, *J. Org. Chem.*, 1988, **53**, 3731.

[137] B. Chiche, A. Finiels, C. Gauthier and P. Geneste, *Appl. Catal.*, 1987, **30**, 365.

[138] K. G. Ione, *Izv. Akad. Nauk. SSSR, Ser. Khim.*, 1990, **3**, 103.

[139] G. Perot and M. Guisnet, *J. Mol. Catal.*, 1990, **61**, 173.

[140] M. J. Climent, H. Garcia, J. Primo and A. Corma, *Catalysis Lett.*, 1990, **4**, 85.

[141] Z. Chen, X. Wang, W. Lu and J. Yu, *Synlett*, 1991, **2**, 121.

[142] J. Ipaktschi and M. Brück, *Chem. Ber.*, 1990, 1591.

[143] J. N. Kim and E. K. Ryu, *Heterocycles*, 1990, **31**, 1693.

[144] A. R. A. S. Deshmukh, T. I. Reddy, B. M. Bhawal, V. P. Shivalkar and S. Rajappa, *J. Chem. Soc., Perkin. Trans. 1*, 1990, **4**, 1217.

[145] J. Haggin, *Chem. Eng. News*, 1990, 6 August, 29.

6

Reactions involving polymeric resins — a survey

J. M. Maud

6.1 INTRODUCTION

The support of organic substrates, reagents or catalysts on polymeric resins may offer great advantage in synthetic organic chemistry. The use of a polymeric support can facilitate reaction workup because a supported substrate can easily be removed from a soluble reagent, or vice versa, simply via filtration. On the other hand, supported, and possibly expensive, catalysts are easily removed from the reaction medium and, in principle at least, they may be recycled many times. This relative ease of workup may permit the use of an excess of reagent, in order to drive a reaction to or near to completion, a situation which is often avoided in solution becase of subsequent separation difficulties. Supported reagents and substrates are easily handled and, for example, can be used in flow reactors in automated processes. In addition, toxic and potentially explosive reagents can often be more safely handled when attached to a polymeric support. Finally, but importantly, many polymer-supported reactions offer greater selectivity than the corresponding solution reactions. Unfortunately, in the case of other reactions, the converse is true. Some reactions cannot successfully utilize a solid support. For many others, the reaction may be noticeably slower because of diffusional constraints. Nevertheless, even in these cases, the advantage of ease of workup may outweigh the disadvantage of a relatively slow reaction.

This chapter surveys organic reactions involving substrates, reagents or catalysts supported on organic polymeric resins. Given its size, the survey is representative rather than comprehensive or exhaustive. More extensive reviews [1–4] are available.

The survey divides reactions involving polymeric supports into two categories: (a) those in which an organic substrate is attached to the polymer, reacted with another organic fragment, or otherwise functionalized, and then cleaved from the resin; and (b) those in which a reagent or catalyst is supported on the polymer. It is recognized that the inclusion of a particular reaction within one of the classes may be arbitrary. Thus, in an effort to locate all polymer-supported phosphines within one section, the Wittig reaction is placed within the second category, with the polymer-supported phosphine being viewed as a 'coupling agent' rather than, following its conversion to an ylide, an organic substrate. This division, taken with inclusion of the prime example of the first category, solid-phase peptide synthesis, within a separate chapter of this monograph, means that the first category is rather shorter than the second. Within the second category, reactions are divided into classes which depend upon the type of reagent, e.g. acid, base, electrophile, nucleophile, etc., that is supported on the polymer.

The survey makes frequent reference to 'polymer supported' reagents, catalysts or substrates. Except where explicitly stated otherwise, the polymer referred to is a cross-linked polystyrene of some kind. The degrees of polymer cross-linking and functionalization are not usually indicated. Although these parameters are of great importance in terms of experimental results and their reproduction, very few reactions have been carried out with sufficient variation in one or both of them to permit an assessment of their significance. Some of the exceptions are discussed in Chapter 2.

In order to simplify chemical structures, equations and schemes, a shorthand notation is used to designate a polymeric support, with the most significant pendent group being displayed. Thus, depending on context, cross-linked chloromethylpolystyrene (1) is represented by one of the structures (2) or (3). Note that even the conventional figure (1) does not adequately describe the true structure, in that the cross-linking is not represented. In particular cases, a polymer-supported functional group may be covalently attached to two regions of the same polymeric network. Such situations are abbreviated by the structure (4).

(1)   (2)   (3)   (4)

6.2 POLYMER-SUPPORTED ORGANIC SUBSTRATES

6.2.1 Polymer-supported carbanions

The crossed-condensation between two different carbonyl compounds is sometimes complicated by the self-condensation of one of them. Attachment to a polymeric support can isolate any one molecule of this component and permit smooth reaction with the second component. Thus, potassium 3-phenylpropanoate reacts [5] with chloromethylated polystyrene to give the polymer-supported ester (5). Acylation with 4-nitrobenzoyl chloride, with trityllithium as base, followed by cleavage from the polymer and decarboxylation, gives a high yield of the ketone (6). Notwithstanding the use of the expensive base, the polymer-supported carbanion avoids (a) the use of the low temperatures normally required for the alkylation and acylation of esters, and

(b) the predominant formation of the self-condensation product (7). Acylation of (5) with acetyl chloride, and alkylation with benzyl and butyl bromides also occur efficiently [5].

(5)   (6)

(7)

6.2.2 Polymer-supported acylating agents

Polymer-supported trityllithium (8) serves [6] to generate the relevant carbanions from soluble ketones, nitriles and amides, which then react with a polymer-bound active ester (9). Thus, acetophenone gives a high yield of dibenzoylmethane, in contrast to the corresponding solution reaction where a yield of only ca. 50% is obtained.

(8)   (9)

The potential of macrocyclic lactone antibiotics [7–9] lends great importance to methods for the synthesis of medium-sized rings. Conventional methods employ high dilution or template syntheses to reduce bimolecular reactions which lead to polymers or oligomers. The former is tedious, and the latter is not completely general. In a series of model experiments [10], polystyrene-supported 12-hydroxydodecanoic thiol esters (10) gave 13% and 19% yields of the monolide (11) and the diolide (12) respectively when treated with mercuric trifluoroacetate. Yields for a high-dilution solution reaction under similar conditions were less than 2% in each case. Other polymer-supported methods for the synthesis of medium and large ring lactones have been reported [11–14].

(10)   (11)   (12)

6.2.3   Polymer-supported protecting groups

An extensive review [15] of the use of polymer-supported protecting groups is available. Applications include the synthesis of natural products such as alkaloids, β-lactam antibiotics and pheromones. Thus, for example, the reaction of octane-1,8-diol with polystyrene-supported trityl chloride gives the functionalized polymer (13) [16]. Subsequent conversion to the mesylate, reaction with lithium hexynide, and acidolytic cleavage from the resin gives the tetradecynol (14), which is then converted, using conventional solution techniques, to the pheromone (15), the sex attractant of the fall army worm moth.

$$ \boxed{P} \!\!-\!\! C_6H_4C(C_6H_5)_2\!\!-\!\!O(CH_2)_8OH \qquad\qquad HO(CH_2)_8C\!\equiv\!C(CH_2)_3CH_3 $$

(13)   (14)

$$ AcO(CH_2)_8 \diagdown\!\!=\!\!\diagup (CH_2)_3CH_3 $$

(15)

Polymeric supports in general, and polymeric protecting groups in particular, may be expensive to prepare. Thus, in many cases it is important that the support should be regenerable. An interesting recent development [1] is represented by the polymer-supported trialkylsilyl trifluoroacetate (16). Reaction with an alcohol affords the polymer-supported protected alcohol (17) and deprotection with trifluoroacetic acid regenerates the polymer (16). Polymer (16) is prepared by trifluoroacetic acid treatment of a 4-[3-(phenyldimethylsilyl)propyl]styrene–styrene–DVB copolymer.

$$ \boxed{P'}\!-(CH_2)_3 - \underset{\underset{CH_3}{|}}{\overset{\overset{CH_3}{|}}{Si}} - OCOCF_3 \qquad\qquad \boxed{P}\!-(CH_2)_3 - \underset{\underset{CH_3}{|}}{\overset{\overset{CH_3}{|}}{Si}} - OR $$

(16)   (17)

6.3   POLYMER-SUPPORTED REAGENTS AND CATALYSTS

6.3.1   Acids

Sulphonated cross-linked polystyrene resins, originally prepared as cation exchange resins, have found considerable application in the acid catalysis of organic reactions. Very nearly all those organic reactions that are catalysed by acid in solution have been performed with the aid of a corresponding cation exchange resin. Reaction types include acetal and ketal formation, ether and ester formation, epoxidation, dehydration, condensation and alkylation. An extensive review is available [17]. The commercially available Amberlyst-15, comprising a sulphonated macroreticular polystyrene resin, has proved to be particularly popular. An example of its use is the recently reported acid-catalysed conversion of the ketoacetonide (18) into the bicyclodioxolane **(19)** in a short synthesis of a bark beetle aggregation pheromone, (+)-*exo*-brevicomin **(20)**, from tartrate [18].

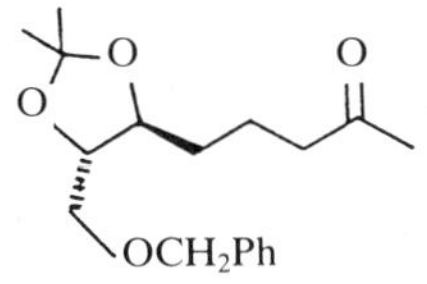

(18)

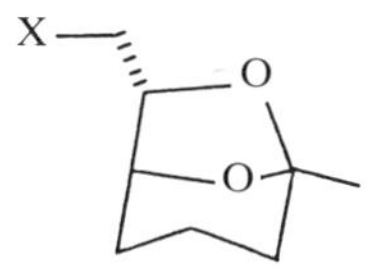

(19)   X = PhCH$_2$O

(20)   X = CH$_3$

Attempts [19] to obtain *trans-p*-mentha-1,8-dien-5-ol **(21)** via dehydration of diol **(22)** were unsuccessful, giving intractable mixtures. However, **(21)** was obtained via the action of Amberlyst-15 on **(23)**, the immediate precursor of **(22)** and itself obtainable from *trans*-verbenol in two steps.

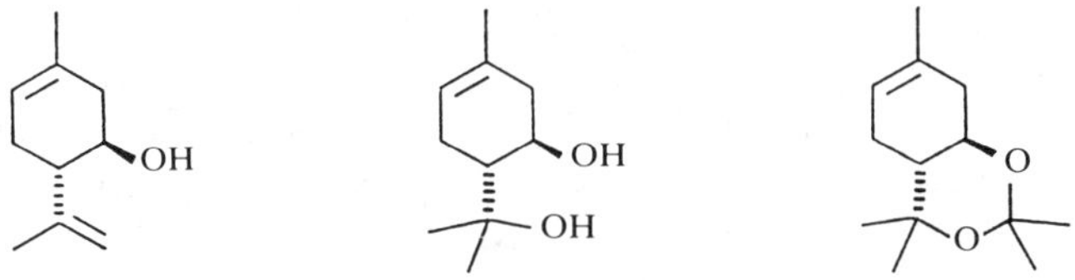

(21)   (22)   (23)

At an industrial level, methyl *t*-butyl ether, used as an 'anti-knock' petroleum additive as an alternative to tetraethyllead, is manufactured in many plants throughout the world via the reaction between methanol and isobutene, which is catalysed by a macroporous poly(styrenesulphonic acid) [20, 21]. Reaction [21] between phenol and acetone to give bisphenol-A **(24)**, a commercially important polycarbonate precursor, is also performed on an industrial scale with the use of an acidic resin.

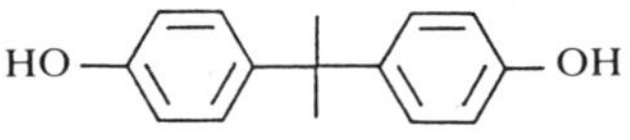

(24)

Nafion **(25)**, a commercially available perfluorinated polymer, serves as a solid superacid catalyst with a variety of synthetic applications. Examples include the Fries rearrangement [22], the rearrangement of α-ethynyl alcohols **(26)** to α,β-unsaturated ketones **(27)** (equation (6.1)) [23] the pinacol rearrangement [24], the synthesis of *gem*-diacetates [25], acetals and thioacetals [26], aromatic acylations and deacylations [27], hydrations of alkynes [28] and epoxides [29], and the facile preparation of methoxymethyl ethers from alcohols and dimethoxymethane [30]. This list is far from being exhaustive. An excellent review by Olah *et al.*, the principal architects of much of this work, is available [31].

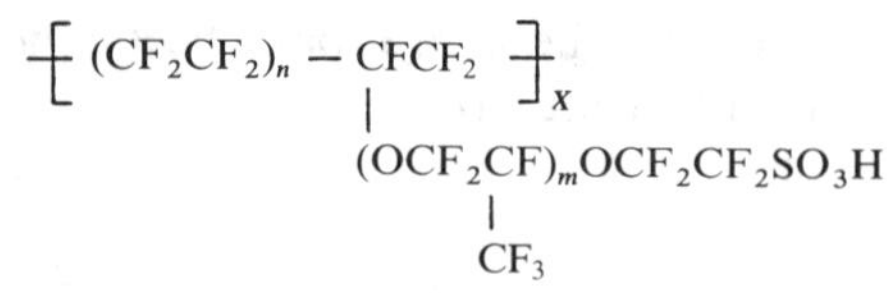

$$\left[(CF_2CF_2)_n - \underset{\underset{\underset{CF_3}{|}}{(OCF_2CF)_mOCF_2CF_2SO_3H}}{CFCF_2} \right]_x$$

(25)

(6.1)

(26)   (27)

6.3.2 Basic and nucleophilic catalysts

It was stated earlier that almost all of the reactions that occur under acid-catalysed conditions in solution have been carried out with use of an acidic ion-exchange resin. A similar statement can be applied to base-catalysed reactions. Amberlyst-A21, a weakly basic dimethylamino-functionalized polystyrene, exemplifies the use of basic polymeric resins. Thus, for example, A-21 catalyses the Michael addition of a wide variety of nitroalkanes (28) to acrylate ester (29), to give, following a Nef reaction, 4-oxoalkanoates (30) in good overall yield (equation (6.2)) [32]. The same catalyst effects the nitroaldol condensation between the nitroalkene (31) and acetaldehyde to give (32) as one step in the synthesis of the pheromone (33) (Scheme 1) [33].

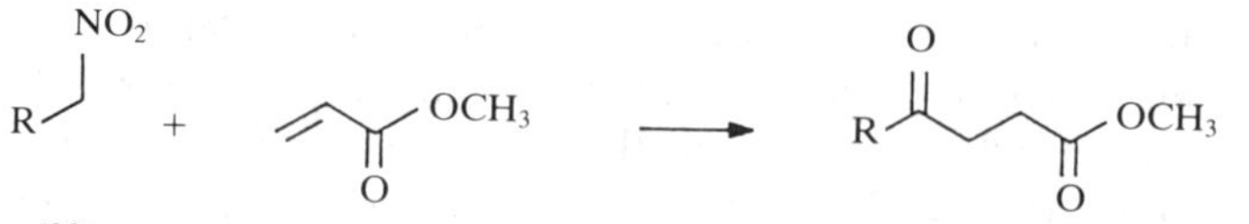

(6.2)

(28)   (29)   (30)

Commercially available poly(vinylpyridine)s (PVPs) (34) have been used in the acylation and trimethylsilylation [34] of amines and alcohols. Although the acid salts of the PVP are readily converted back into the free base form, this is hardly justified in view of their low price.

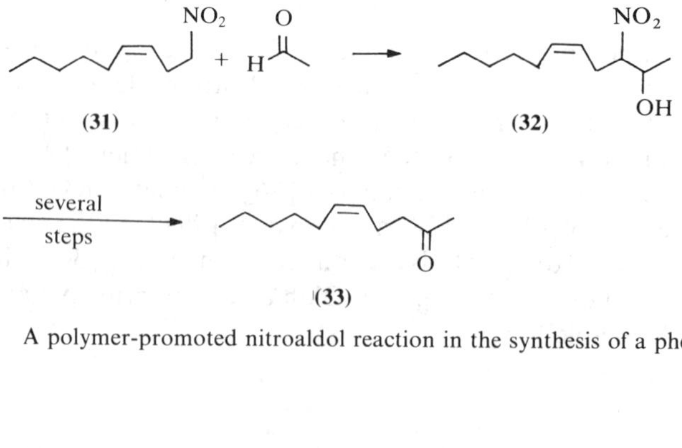

(31)   (32)

several
steps

(33)

Scheme 1. A polymer-promoted nitroaldol reaction in the synthesis of a pheromone.

(34)

N,N-Dialkylaminopyridines **(35)** are recognized as good nucleophilic catalysts [35–37]. Several polymeric analogues [38, 39] have also been shown to be effective catalysts in a wide variety of acyl-transfer reactions, and recent examples **(36, 37)** are claimed to offer distinct advantage in terms of activity and chemical stability [1]. Copolymers of 4-(diallyamino)pyridine have been shown to be effective catalysts for both the formation and the hydrolysis of esters [124]. The importance of the catalyst microenvironment within the polymeric matrix has been studied [1].

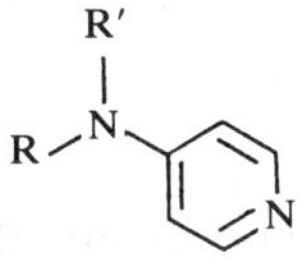

(35) 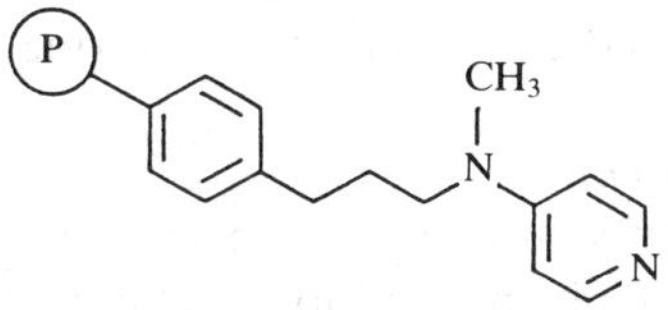   **(36)**

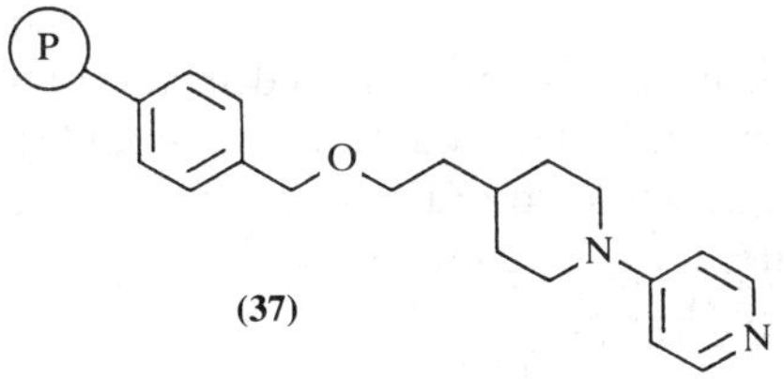

(37)

The nucleophilicity of pyridine bases may sometimes prove to be a disadvantage, and the highly hindered non-nucleophilic pyridine **(38)** has been shown [40] to be the base of choice in vinyl triflate syntheses using triflic anhydride. Polymeric analogue **(39)**, prepared via functionalization of chloromethylpolystyrene, has been shown to offer some advantage in this type of reaction [41].

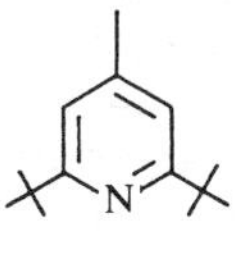

(38) 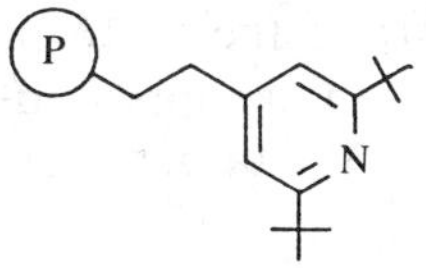   **(39)**

6.3.3 Electrophiles

Given the relative reactivity of alkyl-substituted benzene rings, the number of electrophilic reagents that have been successfully supported on polystyrene is small.

One notably successful reagent is perbromide supported on quaternary pyridinium
(40) [42, 43], alkylammonium (41) [44, 45], and phosphonium (42) [46, 47] polymers,
all of which are capable of effecting the addition of bromine to alkenes and the α-
bromination of ketones and aldehydes in high yields.

(40)   (41)

(42)

In addition to perbromide, Br_3^-, the mixed perhalide $BrCl_2^-$ has been supported on
A-26 [47]. The reagent effects the chlorobromination of alkenes. Similarly, polymer-
supported ICl_2^- serves as a chlorine source, effecting the α-chlorination of aldehydes
and ketones and the addition of chlorine to alkenes [48].

6.3.4   Nucleophiles

A wide variety of anions may be supported on quaternary ammonium functionalized
polymers via conventional ion-exchange techniques, and some commercial quaternary ammonium polymers such as Amberlyst A-26 are available in a range of different
anionic forms.

Amberlyst A-26 in its acetate form was used in the preparation of 2-amino-1,3-
diols (43) from the corresponding iodoaminoalcohols (44) [49]. Use of supported
acetate avoided an aqueous work up of the water-soluble product and hydroxyaziri-
dines (45) were not obtained as byproducts.

(43)   (44)   (45)

The latter could be obtained, however, by use of Amberlyst A-26 in its carbonate
form. The same authors used the same two forms of Amberlyst A-26, together with a
third, in a synthesis of the important β-blocker, propranolol [50]. The synthesis
included the alkylation of supported α-naphthoxide with the mesylate (46), followed
by hydrolysis to give the target propranolol (47) (equation (6.3)).

(46)   (47)

Fluoride, chloride, bromide and iodide have all been supported on Amberlyst A-26 and used in the preparation of the corresponding alkyl halides via exchange reactions with other alkyl halides [51]. The reaction of supported fluoride with mesylates and tosylates gives stereochemically clean S_N2 reactions, as shown by the preparation of epimerically pure 3-fluorosteroids and both enantiomers of ethyl 2-fluoropropanoate [52].

N,N-Dialkylaminopyridinium salts **(48, 49)** supported on polystyrene catalyse the reaction between potassium fluoride and 4-chloronitrobenzene to give 4-fluoronitrobenzene in high yield [53]. The polymeric catalyst can be reused at least eight times, offering advantage over the expensive and only moderately active unsupported catalyst **(50)**.

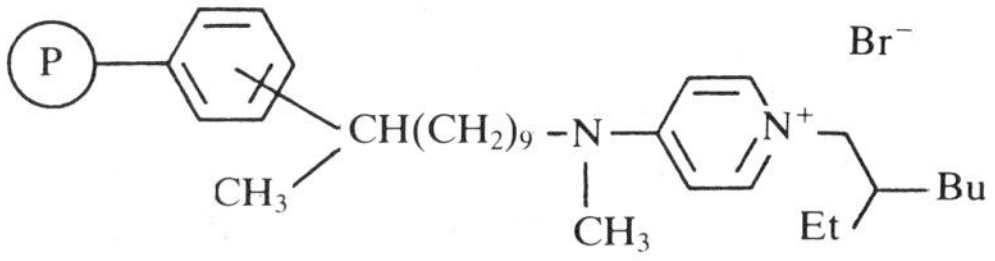

(48)

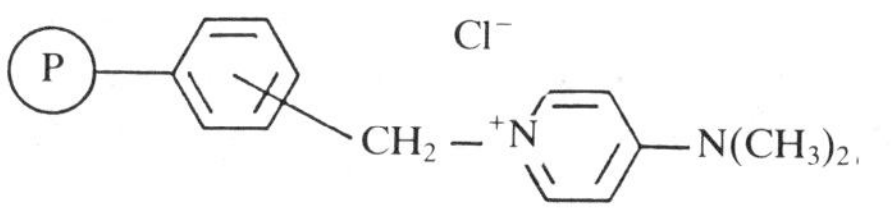

(49) 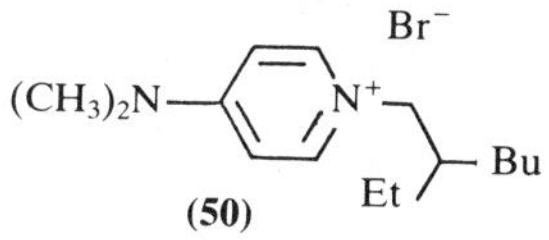   **(50)**

Alkyl and acyl cyanides may be obtained from the corresponding halides in the presence of Amberlyst A-26 in its cyanide form in a procedure that avoids protic conditions [54]. The same polymer catalyses the benzoin condensation [55].

Alkyl halides may be converted into nitroalkanes using the polymer-supported nitrite [56]. Similarly, supported thiocyanate converts alkyl halides into the corresponding alkyl thiocyanates [54, 57].

6.3.5 Oxidizing agents

Various Cr(VI) compounds, such as chromate, dichromate and more complex forms, have been supported on poly-4-vinylpyridinium and trimethylammoniomethyl functionalized polystyrene [58–61]. Such reagents oxidize primary and secondary alcohols cleanly to aldehydes and ketones respectively, although reaction may be slow and require excess reagent for completion. Polymer-supported hydrogen chromate oxidizes benzylic and allylic halides to give aldehydes [61].

Moffatt oxidation of primary and secondary alcohols to the corresponding carbonyl compounds can be achieved using dimethylsulphoxide and the polymeric carbodiimide **(51)** [62]. In a similar vein, iodooctane is converted into octanal using a polymer-bound sulphoxide [63].

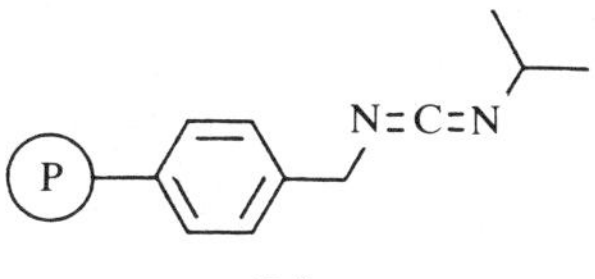

(51)

Hydrogen peroxide oxidizes commercially available cross-linked polymethacrylic acid to give a polymer-supported peracid which successfully epoxidizes a variety of alkenes [64]. Sulphides and sulphoxides are oxidized to sulphones, and ketones to esters and lactones [65]. Unfortunately, the use of polymeric peracid reagents of this type is limited by their tendency to detonate.

Chloromethylated polystyrene has been converted into a polymer-supported perbenzoic acid [66], which offers greater stability than the peralkanoic acids. Epoxidations of alkenes and oxidations of sulphur compounds parallel the corresponding solution reactions.

The iodate and periodate forms of the commercially available Amberlyst A-26 resin can be prepared via standard ion-exchange procedures [67]. 1,2-Diols are cleaved, quinols are oxidized to quinones, and sulphides and phosphines are converted to the corresponding oxides.

Osmium tetroxide supported on poly-4-vinylpyridine (52) or on 1,4-diazabicyclo[2.2.2]octane (DABCO)-functionalized polystyrene (53) has been used as a catalyst in: (a) the *cis*-hydroxylation of alkenes, with hydrogen peroxide, *tert*-butyl hydroperoxide and trimethylamine *N*-oxide as secondary oxidants; and (b) the oxidative cleavage of alkenes in the presence of periodate [68, 69]. Polymers (52) and (53) are comparable in the former case, with *tert*-butyl peroxide being the oxidant of choice. However, in the case of oxidative cleavage, PVP offers distinct advantage over the DABCO support.

(52)   (53)

Chloromethylated polystyrene reacts with 3,6-dibromocarbazole and phenothiazine to give polymers (54) and (55), which can be oxidized to the respective radical

(54)   (55)

cations with antimony pentafluoride [70]. The oxidized form of polymer (55) serves as a one-electron oxidant in the cleavage of a dinuclear di-iron complex.

6.3.6 Reducing agents

Ionic reagents such as borohydride and cyanoborohydride may be supported on the wide variety of ion-exchange resins that are commercially available. Thus borohydride supported on quaternary-ammonium-modified polymers reduces aldehydes, although the reagent is less active than sodium borohydride [71]. Polymer-supported borohydride also reduces aryl azides to arylamines, and arenesulphonyl azides to arenesulphonamides [125]. The relative ease of product isolation, by filtration and evaporation, is noted as an important feature of the use of the polymer-supported reagent. Similarly, supported cyanoborohydride effects expected reactions such as reductive amination and reduction of pyridinium cations, although again the reactions are slower than the corresponding reactions in solution, with higher temperatures being required to obtain good yields [72].

Sodium borohydride reacts with poly(4-vinylpyridine) as its hydrochloride salt to give the polymeric pyridine–borane (56), which reduces a variety of aromatic aldehydes and ketones to the corresponding alcohols [73]. Optimum results were obtained with use of a reticulated support, with a hot solution of the substrate being passed down a column of the supported reducing agent. In a similar vein, polymer-supported sulphide–borane (57) reduces acetophenone in quantitative yield and offers ease of handling and lack of odour relative to dimethyl sulphide–borane itself [74].

(56)   (57)

Polymer-supported organotin dihydrides (58) reduce alkyl iodides and bromides to the corresponding alkanes [75]. Aromatic aldehydes and ketones are also reduced by the reagent, although long reaction times are necessary. However, good selectivity in the reduction of dicarbonyl compounds is obtained, with the preferential formation of hydroxycarbonyl compounds rather than diols.

(58)

(59)   (60)

The synthesis and application of soluble polyethylene- **(59)** and polystyrene-bound organotin compounds **(60)** have been described recently [76]. The polymers catalyse the reduction of alkyl halides using a suspension of sodium borohydride in toluene, in the presence of a crown ether as phase transfer catalyst. The polymeric organotin catalysts are insoluble in the solvent at normal temperatures and reaction is carried out under reflux. On cooling, at the end of the reaction, the polymers precipitate from solution and can be recovered quantitatively. A more recent development involves the use of a polymeric catalyst incorporating both the tin dichloride and a covalently attached crown ether [121].

$$(CH_3)_3SiO \left[\begin{array}{c} H \\ | \\ Si - O \\ | \\ CH_3 \end{array} \right]_n Si(CH_3)_3$$

(61)

Poly(methylhydrosiloxane) **(61)** is a versatile, low-cost hydride-transfer reagent, with reactions being catalysed by bis(dibutylacetoxytin) oxide or Pd(0) compounds [77–79]. Thus, aldehydes and ketones are reduced in the presence of bis(dibutylacetoxytin) oxide, hydrogenation of alkenes occurs in the presence of palladium on carbon, and the highly chemoselective reductive cleavage of allylic acetates is catalysed by Pd(PPh$_3$)$_4$. Although a soluble polymeric reagent, the siloxane co-products are insoluble and easily removed by filtration.

Dimethylviologen, also widely known as paraquat in its role as a herbicide, can act as an electron transfer catalyst in a wide variety of situations. When covalently attached to an insoluble polymer resin, such as polystyrene or poly(4-vinylpyridine), this 4,4'-bipyridinium moiety can be used to catalyse a variety of reductions. Thus, polymer **(62)** catalyses the dithionite-mediated reduction of azobenzene to diphenylhydrazine [80], and both p-benzoquinone and ethyl phenylglyoxalate are reduced by zinc in the presence of polymer **(63)** [81]. In both cases, reaction in the absence of polymer-supported viologen is either exceedingly slow at room temperature or does not occur at all. More recently, methyl phenylglyoxalate has been reduced by 1-benzyl-1,4-dihydronicotinamide, again with viologen polymer **(63)** as catalyst [120].

(62)

(63)

6.3.7   Coupling agents

6.3.7.1   Phosphines
A wide variety of synthetic transformations, most notably the Wittig reaction, involve phosphines as reagents. Triphenylphospine is commonly used, appearing as the phosphine oxide co-product at the end of the reaction. The solubility properties of triphenylphosphine oxide are such that removal of this co-product often presents serious problems. In this respect, at least, the use of polymer-supported phosphines offers particular advantage.

The use of polymer-bound Wittig reagents has been reviewed [82], and only representative examples are included here. Polymer-bound methyl- and benzyl-phosphonium salts, from phosphine-modified polymer, react with ketones in the presence of base to give alkenes in yields comparable to the analogous solution reactions. With polymeric methylenetriphenylphosphorane, yields greater than 90% were obtained, even in the cases of bulky ketones such as nonadecan-10-one and cholest-4-en-3-one [83]. The phosphine oxide co-product remains polymer bound, and may be recycled. Polymer-supported phosphonates have been used in the Wadsworth–Horner–Emmons variation of the Wittig alkene synthesis [84]. Thus, passage of a dilute solution of the diethyl cyanomethylphosphonate through a column of Amberlyst-A26, in its hydroxide form, gave an ionically bound carbanion (64) which reacts with aldehydes and ketones to give α, β-unsaturated nitriles in high yields [84]. The corresponding ester (65) reacted only with aldehydes.

$$\boxed{P}\!\!-\!CH_2\overset{+}{N}(CH_3)_3 \quad (EtO)_2\overset{-}{P}\overset{\cdot}{C}HX$$
$$\qquad\qquad\qquad\qquad\qquad\quad \overset{\|}{O}$$

(64) X = CN
(65) X = COOEt

The reagent derived from triphenylphosphine and carbon tetrachloride may be used to effect a variety of useful synthetic transformations, usually under exceptionally mild conditions. In all cases, difficulties arising from the removal of the phosphine oxide co-product may arise. The use of polymer-supported triphenylphosphine obviates these difficulties. Thus, the reagent derived from polymeric phosphine and carbon tetrachloride has been used: (a) to effect the condensation between carboxylic acids and amines to give amides [85]; (b) in the conversion of alcohols into the corresponding chlorides [86]; and (c) as a dehydrating agent for the conversion of amides and oximes to nitriles [87].

6.3.7.2   Sulphides
The dimethyl(polystyryl)sulphonium salt (66) reacts with benzaldehyde, acetophenone and benzophenone under phase transfer conditions to give the corresponding epoxides (67) in virtually quantitative yields (equation (6.4)) [88]. Analogous reactions in solution give lower yields of ketone derived epoxides.

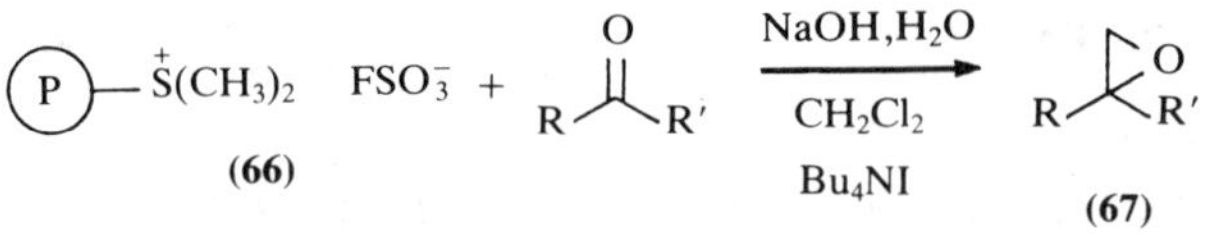

$$(6.4)$$

6.3.7.3 Carbodiimides

The use of N,N'-dialkylcarbodiimides for effecting condensation reactions is sometimes complicated by the difficulty of separating the desired product from the N,N'-dialkylurea co-product. Use of the polystyrene-supported carbodiimide (51) avoids this complication [62].

6.3.7.4 Azodicarboxylates

Polystyrene-bound azodicarboxylate (68), in conjunction with triphenylphosphine, serves to activate a wide variety of oxygen functions in the Mitsunobu reaction [89]. Thus, the hydroxyacid (69) was cyclized to the corresponding macrocyclic lactone (70) in 42% yield. The polymer shows no tendency to detonate and can be recycled up to five times without loss of activity.

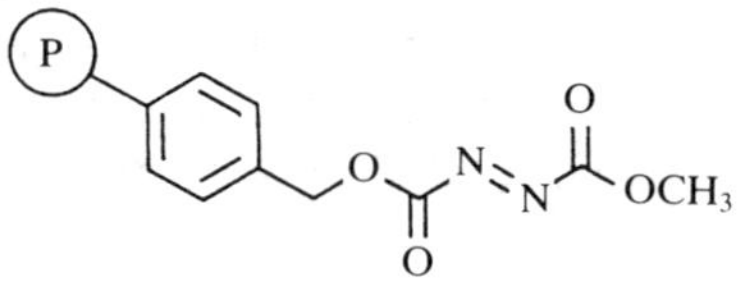

(68)

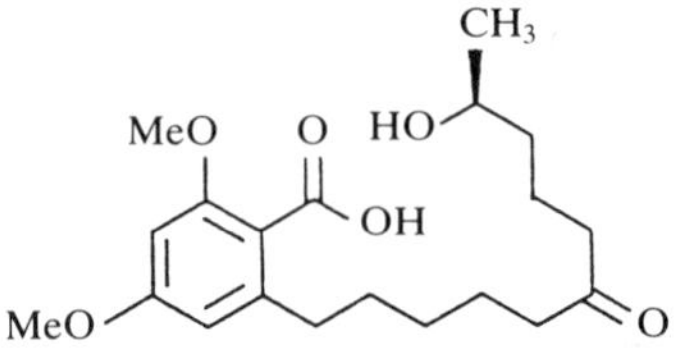

(69)

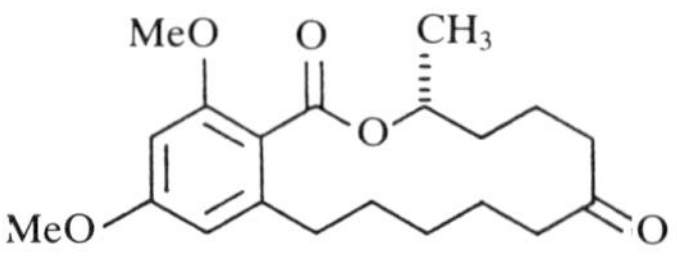

(70)

6.3.8 Catalysts

Specific examples of several types of catalyst (e.g. acid, base, nucleophile) have been described in earlier sections of this chapter. This section discusses some examples of polymer-supported catalysts that do not conveniently fit into those sections.

6.3.8.1 Phase transfer catalysts

Reactions involving phase transfer catalysis offer a number of advantages relative to conventional techniques. These include the use of inexpensive reagents, e.g. aqueous alkali metal hydroxides instead of metal alkoxides under anhydrous and/or aprotic conditions and, in principle at least, the relative ease of workup. Nevertheless, the phase transfer catalyst must be removed at the end of the reaction, and this sometimes involves difficult separations and the loss of expensive material such as crown ethers. It is not surprising, therefore, that polymer-supported catalysts have found application. Examples [90–94] include polymer-supported quaternary ammonium and phosphonium salts [90, 91], the poly(ethylene glycol) ether **(71)** and the crown ethers **(72)** and **(73)** [93, 94]. Reactions effected with the use of polymer-supported catalysts include a wide variety of nucleophilic substitutions, α-alkylation of nitriles, and cyclopropane syntheses via dichlorocarbene generated from chloroform and aqueous sodium hydroxide. A variety of polyamides including dimethylacrylamide–styrene copolymers [122] and amide functionalized polystyrenes [123] have also been used as phase transfer catalysts.

P—CH$_2$(OCH$_2$CH$_2$)$_8$OCH$_3$

(71)

(72)

(73)

6.3.8.2 *Transition metal catalysts*

Examples of polymer-supported transition metal compounds have been included in earlier sections of this chapter, under the discussion of oxidizing agents, for example. This section is concerned with particular examples of the wide range of transition metal catalysts which have been supported on polymers. More extensive reviews are available [95–97].

Palladium catalysts form the basis of a wide variety of industrial processes, particularly those involving the oxidation of hydrocarbons. Examples include the Wacker synthesis of acetaldehyde from oxygen and ethylene, and the reaction between ethylene and acetic acid in the presence of Pd(OAc)$_2$ to give vinyl acetate. Polystyrene-supported Pd(II) salts appear to offer advantage in the latter reaction. The former is particularly well catalysed by Pd(II) supported on a sulphonated quinone polymer [98]. No Cu(II) co-catalyst is required, and the quinone unit appears to play an important role in the catalytic mechanism, with Pd(II) salts supported on a conventional ion-exchange resin being significantly less active.

Low-valent Pd complexes play an important role in a wide variety of synthetic transformations on a laboratory scale. Examples include use of a polymer-supported palladium(0) phosphine complex in the cyclization of the vinylepoxide **(74)** to give the macrocyclic lactone **(75)** in high yield [13]. High concentrations of substrate can be used, in contrast to the 'high dilution' conditions normally required for cyclizations of this type in solution.

(74)   (75)

Copper(II) salts catalyse the oxidative polymerization of 2,6-dimethylphenol **(76)** to give the commercially important poly-2,6-dimethyl-1,4-phenylene oxide **(77)** [99]. Water is produced as a co-product and may reduce the lifetime of the copper catalyst. The incorporation of the Cu(II) salts into a hydrophobic polymer seems to offer some protection against this sort of degradation and 4-vinylpyridine copolymers [100,101], imidazole-based polymers [102] and polymers incorporating 4-aminopyridine [103,104] have found particular use.

(76)   (77)

Copper-loaded polymers **(78)**, designed to catalyse the hydrolysis of organophosphorus-based 'nerve gases', have been shown to promote six different synthetically useful reactions [105], including a Diels–Alder cycloaddition, an epoxide ring opening and the hydrolysis of an aryl iodide. In general, the copper-loaded polymers offer higher yields and shorter reaction times than the analogous copper salts in solution.

Partial replacement of the acidic protons in Nafion **(25)** with transition metal ions offers new opportunities for catalysis. There exists the possibility of interaction between acidic protons and the transition metal centre. In addition, the superior physical properties and general chemical inertness of the perfluorinated support offer the possibility of using reaction conditions too hostile for conventional polymer supports such as polystyrene. Rhodium(III) exchanged Nafion catalyses the conversion of toluene into toluic acid with extremely high (94%) *para*-selectivity [106].

R = CH$_3$, CH$_3$(CH$_2$)$_{13}$

(78)

Cerium exchanged Nafion catalyses the selective oxidation of secondary alcohols to ketones in the presence of primary alcohols with sodium bromate as primary oxidant [107].

The epoxidation of alkenes represents an important synthetic transformation both at an industrial level and on a laboratory scale, and polymer-supported transition metal catalysts have found some use in this area. Recent results using molybdenum supported on the polystyrene polymers (**79**) and (**80**) via the pyridine-based ligands appear particularly encouraging [108]. The polymer-supported molybdenum catalysts compare favourably with comparable catalysts in solution for the *t*-butyl hydroperoxide-mediated epoxidation of cyclohexene. In addition, the catalysts appear to be particularly stable, with only a small loss of activity after first re-use and with subsequent stability during nine further cycles. Iron(III) porphyrins supported on ion-exchange resins have been used as catalysts for the alkene epoxidation using iodosylbenzene in methanol [126,127].

(79)   (80)

6.3.9   Chiral reagents and auxiliaries

A recent critical overview of chiral polymer catalysts in preparative organic chemistry provides an entry into the literature [109]. Two examples only are described here.

Polymer-bound *N*-alkylnorephedrines (**81**, R = Me, Et, Pr, Bu) catalyse the enantioselective addition of dialkylzincs to both aromatic and aliphatic aldehydes [110]. Although the product alcohols are obtained with high enantioselectivities, there appears to be considerable sensitivity towards the exact nature of the supported catalyst. Thus, the catalyst derived from *N*-methylnorephedrine, i.e. ephedrine itself, afforded good enantiometric efficiencies with *aromatic* aldehydes whereas that derived from *N*-ethylnorephedrine gave good results in additions to *aliphatic* aldehydes. Polymer-bound ephedrine has also been used as a chiral auxiliary in the lithium aluminium hydride reduction of acetophenone. This work is discussed in more detail in Chapter 2, where the influence of the degree of functionalization is discussed.

(81)   (82)

Polystyrene-supported chiral aminoalcohol (82) reacts with two equivalents of borane to give a polymer-supported complex which effects the reduction of aromatic ketones with near quantitative optical yields [111, 112]. The same complex reduces oxime ethers to amines, again with virtually quantitative optical efficiency. The products do not bind to the polymer, thus offering the possibility of using a flow system, in which the chiral auxiliary could be used in truly catalytic rather than stoichiometric amounts.

6.3.10   Polymer-modified electrodes

Oxidations and reductions involve respectively the removal from or addition to a substrate of one or more electrons, and conventional oxidizing and reducing agents, either in solution or on a solid support, are normally used to achieve these ends. However, in certain cases, electrochemical oxidation or reduction may prove to be advantageous, with an electrode in an electrochemical cell acting as an electron sink or source. Modification of the electrode with an electroactive polymeric coating may serve a number of purposes. It may serve to catalyse a particular reaction, thus bringing the working potential of the electrode into a range where other, and undesired, reactions do not take place. Alternatively, it may provide some degree of selectivity between closely related substrates. Although most studies of modified electrodes have been carried out from a physicochemical point of view [113–115], some direct approaches to their application in synthetic organic chemistry have been made.

For example, electrochemical reduction of 4-acetylpyridine and ethyl phenylglyoxalate, at a graphite electrode modified with (S)-phenylalanine methyl ester, yields products in which one enantiomer predominates (ee $\leqslant$ 15%) [116]. Rather better results are obtained in the electrochemical oxidation of isopropyl phenyl sulphide to the corresponding sulphoxide using a poly-(S)-valine-polypyrrole-modified platinum electrode [117]. An enantiomeric excess of 73% has been obtained in this case. Poly(4-nitrostyrene) coated on a platinum electrode serves to mediate the reduction of stilbene dibromide to stilbene itself [118]. 1,2-Dibromides are also reduced at electrodes modified with cobalt(II) or copper(II) porphyrins [119].

6.4   CONCLUSIONS

A great variety of synthetic transformations may be carried out with the aid of polymer-supported substrates, reagents or catalysts. In many cases the polymer-supported reaction offers real advantage over a corresponding solution reaction. The field is an active one, with many important recent developments, and many more expected in the future.

REFERENCES

[1] J. M. J. Fréchet, G. D. Darling, S. Itsuno, P.-Z. Lu, M. V. de Meftahi and W. A. Rolls Jr., *Pure Appl. Chem.*, 1988, **60**, 353.

[2] *Preparative Chemistry Using Supported Reagents* (Ed. P. Laszlo), Academic Press, London, 1987.

[3] *Polymeric Reagents and Catalysts* (ed. W. T. Ford), *A.C.S Symp. Ser.*, No. 308, American Chemical Society, Washington D.C., 1986.

[4] *Polymer-supported Reactions in Organic Synthesis* (Eds P. Hodge and D. C. Sherrington), Wiley, New York, 1980.

[5] Y. H. Chang and W. T. Ford, *J. Org. Chem.*, 1981, **46**, 5364.

[6] B. J. Cohen, M. A. Kraus and A. Patchornik, *J. Am. Chem. Soc.*, 1981, **103**, 7620.

[7] K. C. Nicolau, *Tetrahedron*, 1977, **33**, 683.

[8] S. Masamune, G. S. Bates and J. W. Corcoran, *Angew. Chem., Int. Ed. Engl.*, 1977, **16**, 585.

[9] T. G. Back, *Tetrahedron.*, 1977, **33**, 3041.

[10] S. Mohanraj and W. T. Ford, *J. Org. Chem.*, 1985, **50**, 1616.

[11] L. T. Scott and J. O. Naples, *Synthesis*, 1976, 738.

[12] S. L. Regen and Y. Kimura, *J. Am. Chem. Soc.*, 1982, **104**, 2064.

[13] B. M. Trost and R. W. Warner, *J. Am. Chem. Soc.*, 1983, **105**, 5940.

[14] R. A. Amos, R. W. Embildge and N. Havens, *J. Org. Chem.*, 1983, **48**, 3598.

[15] J. M. J. Fréchet, in Ref. [4], p. 293.

[16] C. C. Leznoff, T. M. Fyles and J. Weatherstone, *Can. J. Chem.*, 1977, **55**, 1143.

[17] D. C. Sherrington, in Ref. [4], p. 157.

[18] B. Giese and R. Rupaner, *Synthesis*, 1988, 219.

[19] M. Bulliard, G. Balme and J. Gore, *Synthesis*, 1988, 972.

[20] F. Ancillotti, M. M. Mauri and E. Pescarollo, *J. Catal.*, 1977, **46**, 49.

[21] H. Widdecke, *Br. Polym. J.*, 1984, **16**, 188.

[22] G. A. Olah, M. Arvanaghi and V. V. Krishnamurthy, *J. Org. Chem.*, 1983, **48**, 3359.

[23] G. A. Olah and A. P. Fung, *Synthesis*, 1981, 473.

[24] G. A. Olah and D. Meidar, *Synthesis*, 1978, 358.

[25] G. A. Olah and A. K. Mehrotra, *Synthesis*, 1982, 962.

[26] G. A. Olah, S. C. Marang, D. Meidar and G. F. Salem, *Synthesis*, 1981, 282.

[27] G. A. Olah, K. Laali and A. K. Mehrotra, *J. Org. Chem.*, 1983, **48**, 3360.

[28] G. A. Olah and D. Meidar, *Synthesis*, 1978, 671.

[29] G. A. Olah, A. P. Fung and D. Meidar, *Synthesis*, 1981, 280.

[30] G. A. Olah, A. Husain, B. G. B. Gupta and S. C. Narang, *Synthesis*, 1981, 471.

[31] G. A. Olah, P. S. Iyer and G. K. S. Prakash, *Synthesis*, 1986, 513.

[32] R. Ballini, M. Petrini and G. Rosini, *Synthesis*, 1987, 711.

[33] G. Rosini, R. Ballini and M. Petrini *Synthesis*, 1986, 46.

[34] J. M. J. Fréchet and M. V. de Meftahi, *Br. Polym. J.*, 1984, **16**, 179.

[35] L. M. Litvinenko and A. I. Kirichenko, *Dokl. Akad. Nauk SSSR, Ser. Kim.*, 1967, **176**, 97; *Chem. Abstr.*, 1968, **68**, 68325.

[36] W. Steglich and G. Höfle, *Angew. Chem., Int. Ed. Engl.*, 1969, **8**, 981.

[37] E. F. V. Scriven, *Chem. Soc. Rev.*, 1983, 129.

[38] S. Shinkai, H. Tsuji, Y. Hara and O. Manabe, *Bull. Chem. Soc. Jpn*, 1981, **54**, 631.

[39] M. Tomoi, Y. Akada and H. Kakiuchi, *Makromol. Chem., Rapid Commun.*, 1982, **3**, 527.

[40] P. J. Stang and W. Treptow, *Synthesis*, 1980, 283.

[41] M. E. Wright and S. R. Pulley, *J. Org. Chem.*, 1987, **52**, 5036.

[42] J. M. J. Fréchet, M. J. Farral and L. J. Nuyens, *J. Macromol. Sci., Chem.*, 1977, **A11**, 507.

[43] Y. J. Zupan and B. Sket, *J. Chem. Soc., Perkin Trans. 1*, 1982, 2059.

[44] S. Cacchi, L. Caglioti and E. Cernia, *Synthesis*, 1979, 64.

[45] A. Bongini, G. Cainelli, M. Contento and F. Manescalchi, *Synthesis*, 1980, 143.

[46] A. Akelah, M. Hassenain and I. Abidel-Galil, *Polym. Prepr., Am. Chem. Soc., Div. Polym. Chem., 1983*, **24**, 467.

[47] A. Akelah, M. Hassenain and I. Abidel-Galil, *Eur. Polym. J.*, 1984, **20**, 221.

[48] A. Bongini, G. Cainelli, M. Contento and F. Manescalchi, *J. Chem. Soc., Chem. Commun.*, 1980, 1278.

[49] G. Cardillo, M. Orena, G. Porzi and S. Sandri, *J. Chem. Soc., Chem. Commun.*, 1982, 1309.

[50] G. Cardillo, M. Orena and S. Sandri, *J. Org. Chem.*, 1986, **51**, 713.

[51] G. Cainelli and F. Manescalchi, *Synthesis*, 1976, 472.

[52] S. Colonna, A. Re, G. Gelbard and E. Cesarotti, *J. Chem. Soc., Perkin Trans. 1*, 1979, 2248.

[53] Y. Yoshida and Y. Kimura, *Tetrahedron Lett.*, 1989, **51**, 7199.

[54] C. R. Harrison and P. Hodge, *Synthesis*, 1980, 299.

[55] J. Castells and E. Dunach, *Chem. Lett.*, 1984, 1859.

[56] G. Gelbard and S. Colonna, *Synthesis*, 1977, 113.

[57] G. Cainelli, F. Manescalchi and M. Panuzio, *Synthesis*, 1979, 141.

[58] J. M. J. Fréchet, J. Warnock and M. J. Farral, *J. Org. Chem.*, 1978, **43**, 2618.

[59] J. M. J. Fréchet, P. Darling and M. J. Farral, *Polym. Prepr., Am. Chem. Soc., Div. Polym. Chem., 1980*, **21**, 272.

[60] T. Brunelet, C. Jouitteau and G. Gelbard, *J. Org. Chem.*, 1986, **51**, 4016.

[61] G. Cainelli, G. Cardillo, M. Orena and M. Panuzio, *J. Am. Chem. Soc.*, 1976, **98**, 6737.

[62] N. M. Weinshenker and C.-M. Shen, *Tetrahedron Lett.*, 1972, 3285.

[63] J. L. Foureys, T. Hamaide, E. Yaacoub and P. Le Perchec, *Eur. Polym. J.*, 1985, **21**, 221.

[64] T. Takagi, *Polym. Lett., 1967*, **5**, 1031.

[65] T. Takagi, *J. Appl. Polym. Sci., 1975*, **19**, 1649.

[66] C. R. Harrison and P. Hodge, *J. Chem. Soc., Perkin Trans. 1*, 1976, 605; *ibid.* 2252.

[67] C. R. Harrison and P. Hodge, *J. Chem. Soc., Perkin Trans. 1*, 1982, 509.

[68] G. Cainelli, M. Contento, F. Manescalchi and L. Plessi, *Synthesis*, 1989, 45.

[69] G. Cainelli, M. Contento, F. Manescalchi and L. Plessi, *Synthesis*, 1989, 47.

[70] M. E. Wright and M.-J. Jin, *J. Org. Chem.*, 1989, **54**, 965.

[71] H. W. Gibson and F. C. Bailey, *J. Chem. Soc., Chem. Commun.*, 1977, 815.

[72] R. O. Hutchings, N. R. Natale and I. M. Taffler, *J. Chem. Soc., Chem. Commun.*, 1978, 1088.

[73] M. L. Hallensleben, *J. Polym. Sci., Polym. Symp.*, 1974, **47**, 1.

[74] G. A. Crosby, USP 3 928 293 (1975); *Chem. Abstr.*, 1976, **84**, 106499u.

[75] N. M. Weinshenker, G. A. Crosby and J. Y. Wong, *J. Org. Chem.*, 1975, **40**, 1966.

[76] D. E. Bergbreiter and S. A. Walker, *J. Org. Chem.*, 1989, **54**, 5138.

[77] J. Lipowitz and S. A. Bowman, *J. Org. Chem.*, 1973, **38**, 162.

[78] E. Keinan and N. Greenspoon, *Isr. J. Chem.*, 1984, **24**, 82.

[79] E. Keinan and N. Greenspoon, *J. Org. Chem.*, 1983, **48**, 3545.

[80] Y. Saotome, T. Endo and M. Okawara, *Macromolecules*, 1983, **16**, 881.

[81] M. Okawara, T. Hirose and N. Kamiya, *J. Polym. Sci., Polym. Chem. Ed., 1979,* **17**, 927.

[82] W. T. Ford, in Ref. [3], p. 155.

[83] M. Bernard and W. T. Ford, *J. Org. Chem.*, 1983, **48**, 326.

[84] G. Cainelli, M. Contento, F. Manescalchi and R. Regnoli, *J. Chem. Soc., Perkin Trans., 1*, 1980, 2516.

[85] P. Hodge and G. Richardson, *J. Chem. Soc., Chem. Commun.*, 1975, 622.

[86] C. R. Harrison and P. Hodge, *J. Chem. Soc., Chem. Commun.*, 1978, 813.

[87] C. R. Harrison, P. Hodge and W. J. Rogers, *Synthesis*, 1977, 41.

[88] M. J. Farrall, T. Durst and J. M. J. Fréchet, *Tetrahedron Lett.*, 1979, 203.

[89] L. D. Arnold, H. I. Assil and J. C. Vederas, *J. Am. Chem. Soc.*, 1989, **111**, 3973.

[90] S. L. Regen, *Angew. Chem., Int. Ed. Engl.*, 1979, **18**, 421.

[91] E. Chiellini, R. Solaro and S. D'Antone, *Makromol. Chem. Suppl., 1981*, **5**, 82.

[92] W. T. Ford and M. Tomoi, *Adv. Polym. Sci.*, 1984, **55**, 49.

[93] E. Blasius, K. P. Janzen, H. Klotz and A. Toussaint, *Makromol. Chem., 1982*, **183**, 1401.

[94] K. H. Pannell and A. J. Mayr, *J. Chem. Soc., Perkin Trans. 1*, 1982, 2153.

[95] C. U. Pittman Jr., in *Comprehensive Organometallic Chemistry* (Eds G. Wilkinson, F. G. A. Stone and E. W. Abel), Pergamon, Oxford, 1983, vol. 8, p. 553.

[96] N. L. Holy, in *Homogeneous Catalysis by Metal Phosphine Catalysts* (Ed. L. H. Pignolet), Plenum, New York, 1983, p. 443.

[97] D. C. Bailey and S. H. Langer, *Chem. Rev.*, 1981, **81**, 109.

[98] H. Arai and M. Yashiro, *J. Mol. Catal.*, 1977/78, **3**, 427.

[99] A. S. Hay, *Adv. Polym. Sci., 1967*, **4**, 496.

[100] E. Tsuchida and H. Nishide, *Adv. Polym. Sci.*, 1977, **24**, 1.

[101] J. P. Verlaan, J. P. C. Bootsma, C. E. Koning and G. Challa, *J. Mol. Catal.*, 1983, **18**, 159.

[102] J. P. Verlaan, R. Zwiers and G. Challa, *J. Mol. Catal.*, 1983, **19**, 223.

[103] C. E. Koning, J. J. W. Eshius, F. J. Viersen and G. Challa, *Reactive Polym.*, 1986, **4**, 293.

[104] C. E. Koning, G. Challa, F. B. Hulsbergen and J. Reedijk, *J. Mol. Catal.*, 1986, **34**, 355.

[105] F. M. Menger and T. Tsuno, *J. Am. Chem. Soc.*, 1989, **111**, 4903.

[106] F. J. Waller, *Br. Polym. J.*, 1984, **16**, 239.

[107] S. Kanemoto, H. Saimoto, K. Oshima and H. Nozaki, *Tetrahedron Lett.*, 1984, **25**, 3317.

[108] D. C. Sherrington, *Pure Appl. Chem., 1988*, **60**, 401.

[109] M. Aglietto, E. Chiellini, S. D'Antone, G. Ruggeri and R. Solaro, *Pure Appl. Chem.*, *1988*, **60**, 415.

[110] K. Soai, S. Niwa and M. Watanabe, *J. Chem. Soc.*, *Perkin Trans. 1*, 1989, 109.

[111] S. Itsuno, M. Nakano, K. Miyazaki, H. Masuda, K. Ito, A. Hirao and S. Nakahama, *J. Chem. Soc.*, *Perkin Trans. 1*, 1985, 2039.

[112] S. Itsuno, M. Nakano, K. Ito, A. Hirao, M. Owa, N. Kanda and S. Nakahama, *J. Chem. Soc.*, *Perkin Trans. 1*, 1985, 2615.

[113] L. R. Faulkner, *Chem. Eng. News*, 1984, **62**, 28.

[114] R. W. Murray, *Acc. Chem. Res.*, 1980, **13**, 135.

[115] M. Fujihira, in *Topics in Organic Electrochemistry* (Eds A. J. Fry and W. E. Britton), Plenum, New York, 1986, p. 255.

[116] B. F. Watkins, J. R. Behling, E. Kariv and L. L. Miller, *J. Am. Chem. Soc.*, 1975, **97**, 3549.

[117] T. Komori and T. Nonaka, *J. Am. Chem. Soc.*, 1984, **106**, 2656.

[118] J. B. Kerr, L. L. Miller and M. R. Van De Mark, *J. Am. Chem. Soc.*, 1980, **102**, 3383.

[119] R. D. Rocklin and R. W. Murray, *J. Phys. Chem.*, 1981, **85**, 2104.

[120] T. Endo, T. Takada, A. Kameyama and M. Okawara, *J. Polym. Sci.*, *Polym. Chem. Ed.*, *1991*, **29**, 135.

[121] J. R. Blanton and J. M. Salley, *J. Org. Chem.*, 1991, **56**, 490.

[122] S. Kondo, N. Nakashima, H. Hado and K. Tsuda, *J. Polym. Sci.*, *Polym. Chem. Ed.*, 1990, **28**, 2229.

[123] S. Kondo, Y. Inagaki, H. Yasui, M. Iwasaki and K. Tsuda, *J. Polym. Sci.*, *Polym. Chem. Ed.*, 1991, **29**, 243.

[124] G. Cei and L. J. Mathias, *Macromolecules*, 1990, **23**, 4127.

[125] G. W. Kabalka and P. P. Wadgaonkar, *Synth. Commun.*, 1990, **20**, 293.

[126] D. R. Leanord and J. R. Lindsay Smith, *J. Chem. Soc.*, *Perkin Trans. 2*, 1990, 1917.

[127] D. R. Leanord and J. R. Lindsay Smith, *J. Chem. Soc.*, *Perkin Trans. 2*, 1991, 25.

Part III

Solids in biological chemistry and molecular biology

7

Solid phase peptide synthesis

J. S. Davies

7.1 INTRODUCTION

In his Nobel Prize lecture [1], Bruce Merrifield stated that the impetus that had turned his thoughts to developing solid-supported synthesis of peptides came from a realization that such syntheses in the solution phase 'were difficult and time consuming'. He also alluded to four main advantages of the solid-supported technique: (a) it simplifies and accelerates the multi-step synthesis, since only a single reaction vessel is used, which avoids the attendant losses involved in the repeated transfer of materials; (b) it avoids large losses due to isolation and purification of intermediates; (c) excess reagents can be used to give high yields and force reactions to completion; and (d) there is increased solvation and decreased aggregation of the intermediate products. By the start of the 1990s, Merrifield's plans have been fully justified in that there are now a plethora of commercially available [2] peptide synthesizers on the market and solid phase techniques have been accepted as the method of choice in most of the syntheses currently being attempted. The technique fills the vacuum left by the dwindling numbers of trained personnel with sufficient expertise to attempt the more demanding solution phase approach [3]. Solid phase synthesis has probably reduced the time for making oligopeptides from months to a matter of weeks.

Merrifield's initial approaches have since been modified and developed by a number of practitioners, and the aim of this chapter is to summarize and highlight, from the very large amount of painstaking and pioneering work by a number of experts, the essentials of contemporary thought and methodology in this field.

The practical aspects of protocol and methodology have been the subject of two books [3,4], which provide the technical details essential for proceeding with a synthesis. Recent reviews by Kent [5] and by Barany *et al.* [6] are also essential background reading to currently recognized acceptable methodology.

During the development of the solid phase technology, great progress was continually being made in the chemistry of protection and deprotection of functional groups at both the amino-terminal and the carboxyl-terminal ends of amino acids and of the side-chains of trifunctional amino acids. Much of this work has been reviewed authoritatively [7], especially by practitioners of the solution phase approach [8]. Side-by-side with developments in the chemistry, the availability of high-performance liquid chromatography must rank as a most significant contributory factor to the development of the solid phase approach, both in its analytical and in its purification role.

7.2 SYNTHETIC STRATEGY FOR PEPTIDE SYNTHESIS

7.2.1 General aspects

Table 7.1 lists the twenty or so amino acids that are the genetically encoded building blocks necessary in the construction of a peptide. The challenge of any synthetic strategy is to couple together the individual amino acid residues in the correct sequence while preserving the chiral integrity (the L-form in the genetically encoded forms) of the individual residues. Scheme 1 summarizes the strategy required, which involves the following steps:

(a) carboxyl-protection at the *C*-terminal residue;
(b) amino group protection at the *N*-terminal residue;
(c) side-chain protection of trifunctional amino acids;
(d) use of a coupling agent or equivalent means of forming a peptide bond.

To achieve chain elongation of the peptide it is necessary to choose *C*-terminal and side-chain protecting groups compatible with the need to remove the amino group protection for further coupling with other residues. The *C*-terminal and side-chain protecting groups need therefore to be more *permanent* than the removable *N*-terminal protecting group, which is usually designated a *temporary* protecting group.

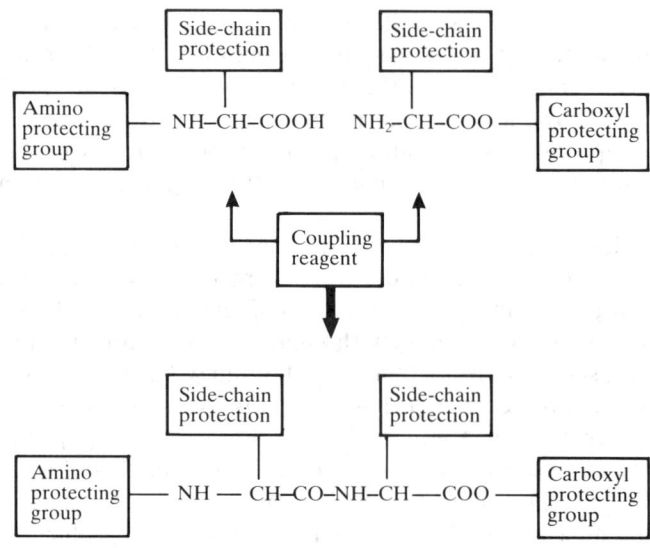

Scheme 1 — The essential strategy of peptide synthesis.

Table 7.1 — The common coded amino acids found in natural peptides

TYPE A (no functional group in the side chain)

$NH_3^+CH_2COO^-$ (Gly) [G][a] $NH_3^+CHCOO^-$ (Ala) [A] $NH_3^+CHCOO^-$ (Val) [V]
                                          | |
                                         CH_3 $CH(CH_3)_2$

$NH_3^+CHCOO^-$ (Leu) [L] $NH_3^+CHCOO^-$ (Ile) [I] $NH_3^+CHCOO^-$ (Phe) [F]
       | | |
$CH_2CH(CH_3)_2$ $CH(CH_3)CH_2CH_3$ CH_2-⬡

$H_2N^+-CHCOO^-$ (Pro) P

TYPE B (amide side chains)

$NH_3^+CHCOO^-$ (Asn) [N] $NH_3^+CHCOO^-$ (Gln) [Q]
       | |
CH_2CONH_2 $CH_2CH_2CONH_2$

TYPE C (sulfur-containing side chains)

$NH_3^+CHCOO^-$ (Cys-Cys) $NH_3^+CHCOO^-$ (Cys) [C] $NH_3^+CHCOO^-$ (Met) [M]
       | | |
$CH_2S-S-CH_2$ CH_2SH $CH_2CH_2SCH_3$
       |
$NH_3^+CHCOO^-$

TYPE D (hydroxyl groups in the side chains)

$NH_3^+CHCOO^-$ (Ser) [S] $NH_3^+CHCOO^-$ (Thr) [T] $NH_3^+CHCOO^-$ (Tyr) [Y]
       | | |
CH_2OH $CH(OH)CH_3$ CH_2-⬡$-OH$

TYPE E (basic side chains)

$NH_3^+CHCOO^-$ (Arg) [R] $NH_3^+CHCOO^-$ (His) [H] $NH_3^+CHCOO^-$ (Lys) [K]
       | | |
$(CH_2)_3NH(C=NH)NH_2$ CH_2-⬠NH $(CH_2)_4NH_2$
                                      N=

$NH_3^+CHCOO^-$ (Trp) [W]
       |
CH_2-

TYPE F (acidic side chains)

$NH_3^+CHCOO^-$ (Asp) [D] $NH_3^+CHCOO^-$ (Glu) [E]
       | |
CH_2COOH CH_2CH_2COOH

[a] Three-letter abbreviations for amino acids are in parentheses; one-letter abbreviations are in square brackets.

The ideal in any strategy would be to have readily removable temporary protecting groups and a set of side-chain protectors and a C-terminal anchor that are stable to all conditions except the final deprotection to release the free peptide.

Experience has shown [5] that stepwise solid phase synthesis starting from the C-terminal residue has advantages over other options such as fragment condensation, and the currently accepted strategy is the linking of the C-terminal residue to the 'solid' support and the addition of the other residues in stepwise fashion as depicted in Scheme 2.

$$AA_1 - \textcircled{P}$$

$$AA_2 - AA_1 - \textcircled{P}$$

$$AA_3 - AA_2 - AA_1 - \textcircled{P}$$

$$AA_4 - AA_3 - AA_2 - AA_1 - \textcircled{P}$$

$$AA_5 - AA_4 - AA_3 - AA_2 - AA_1 - \textcircled{P}$$

Scheme 2 — Stepwise attachment of amino acid residues in solid phase peptide synthesis; AA represents an amino acid residue and $\textcircled{P}$ represents a polymeric support.

One main advantage of the stepwise approach is that control of the chiral integrity of amino acid derivatives during 'activation' is possible through the judicious choice of amino-protecting groups (usually urethane derivatives). No such protection is afforded if a dipeptide or a larger peptide is activated during the course of segment condensation, unless the C-terminal residue is glycine or proline. In addition, the solid phase approach removes complications arising from the unpredictability of the solubility of protected peptide fragments in the solution phase.

7.2.2 Overcoming the racemization of residues

Racemization during the formation of a peptide bond has been an area of immense mechanistic investigation [9]. While a number of mechanisms can be postulated for removal of an α-proton, the most likely general mechanism under the activation conditions of routine peptide synthesis is the result of the intermediate formation of the oxazolone (**1**), which can enolize readily owing to its pseudo-aromatic character (Scheme 3).

When R^1 (Scheme 3) is a chiral amino acid residue, racemization of (**1**) is a fairly facile reaction under the basic conditions of peptide coupling, but when $R^1 = R^3O$ (which represents an amino acid protected as its urethane derivative — see section 7.2.5) the 'aromaticity' of the anion of (**1**) is modified by the lone pair of electrons exocyclic to the ring system [10]. The chiral stability of proline in contrast to other N-alkylated amino acids has been explained [11] by the constraints of the pyrrolidine ring at the oxazolone intermediate stage.

In routine solid phase synthesis the extensive use of urethane protecting groups for amino-terminal protection has greatly minimized the chances of racemization, and

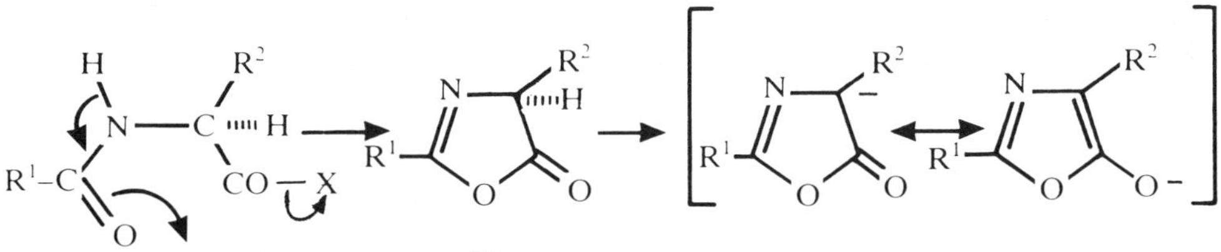

Scheme 3 — Oxazolone formation and deprotonation during activation of peptides.

only in certain instances does this problem need control. Significant cases are the result of imidazole heteroatoms in histidine undergoing neighbouring group participation [12] and the need for relatively vigorous conditions for linking the first amino acid residue on to the linker group attached to the polymer support [13].

7.2.3   *C*-Terminal protection and the 'linker group' in solid phase synthesis

Carboxyl group protection initially evolved from the use of alkyl, *t*-butyl and benzyl esters in the solution phase to use of the benzyl link to the 'Merrifield' polymer [14] as depicted in Scheme 4, with the first amino acid being linked to the resin via its caesium salt.

$$Me_3COCONHCH(R^1)COO^- Cs^+ + ClCH_2C_6H_4 - Polymer$$

$$\downarrow$$

$$Me_3COCONHCH(R^1)COOCH_2C_6H_4 - Polymer$$

Scheme 4 — C-terminal linkage of an amino acid residue to the 'Merrifield' polymer.

However, the need to add more stability to the link led to further modification to include the 'Pam' (*p*-hydroxymethylphenylacetamidomethyl) linker (**2**) [15]. This has given the opportunity of making sure that the resin is free of chloromethyl groups, which can cause problems.

Final release of the synthesized peptide from the resin depends on the stability of the link. In the examples already mentioned, the main reagent of choice has been anhydrous hydrogen fluoride, which is a strong acid and known to give rise to side reactions such as the alkylation of Tyr, Trp, Met and Cys side chains, and catalysis of the cyclization of aspartic acid residues to form aspartimide, an intermediate in the $\alpha \rightarrow \beta$-aspartyl rearrangement. A scavenger molecule to pick up the released carbocations from treatment with HF is required, and the general conditions used to be 90% HF + 10% anisole at 0°C for 1 h. However, further research by the Merrifield group has shown that a two-stage protocol utilizing first low (25% HF/65% DMS) and then high (90% HF) concentrations of HF in dimethylsulphide (DMS) [16] or, alternatively, trifluoromethanesulphonic acid in trifluoroacetic acid [17], can be advantageous.

HF and trifluoromethanesulphonic acid are very toxic and need special expertise in handling. These disadvantages have inspired a further search for ideal linkers.

$$HOCH_2 \text{—}\langle\text{benzene ring}\rangle\text{—} CH_2CONH\text{—}CH_2\text{—}\langle\text{benzene ring}\rangle\text{—}\textcircled{P}$$

(2)

Useful examples are shown as (3)–(6). Linker (3), developed by Wang, allows the *C*-terminal residue to be removed in trifluoroacetic acid [18]. Structures such as (4), based on the 4,4′-dimethoxybenzhydryl-protecting group [19], not only provide easy removal in trifluoroacetic acid, but also allow the *C*-terminal residue to be released as an amide, which is a structural feature often found in peptide hormones such as LHRH, secretin and oxytocin. Trifluoroacetic acid (1%) is all that is needed to remove a residue from (5) [20], which often allows selective cleavage of the peptide from the resin without concomitant loss of side-chain protecting groups, advantageous if further *C*-terminal activation is required. A very recent development involves (6) [21], which results in attachment of the residue as an allylic ester anchor which is stable to the usual acidic and basic protocol of solid phase synthesis. Fully protected synthetic peptides can be released from this Hycram [21] support by activation with tetrak-is(triphenylphosphine)palladium in 50% DMSO, with DMF as a typical solvent, in the absence of oxygen. Linker technology is therefore well developed, and these examples are only representative of others listed in Table 2 of ref. [6].

$$HOCH_2\text{—}\langle\text{benzene ring}\rangle\text{—}OCH_2CH_2CONHCH_2\text{—}\textcircled{P}$$

(3)

$$NH_2\text{—}CH\text{—}\langle\text{benzene ring}\rangle\text{—}O(CH_2)_3CONHCH(CH_3)CONHCH_2\text{—}\textcircled{P}$$

OCH$_3$

(4)

$$HO\text{—}CH_2\text{—}\langle\text{benzene ring}\rangle\text{—}OCH_2\text{—}\textcircled{P}$$

CH$_3$O

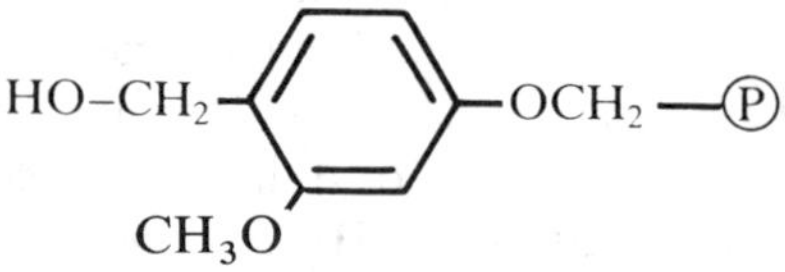

(5)

$$HOCH_2CH\text{=}CHCONH(CH_2)_2CONH\text{—}CH_2\text{—}\textcircled{P}$$

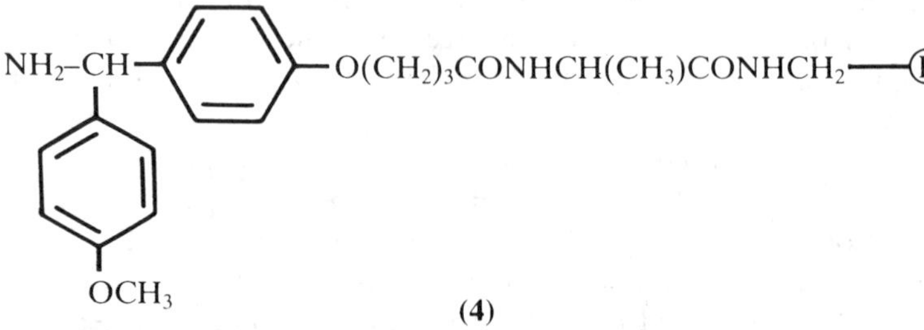

(6)

7.2.4   The solid support

7.2.4.1   Cross-linked polystyrene

Although a number of supports have been explored, popularity and universal acceptance have been achieved by two types of polymer: the original cross-linked styrene–divinylbenzene polymer of Merrifield [14], and the later development of polyacrylamide resins by Sheppard and coworkers [22]. The Merrifield resin is prepared by suspension copolymerization of styrene and 1% of divinylbenzene as cross-linkage (see Chapter 2). The beads obtained from the polymer are spherical (approx 50 μm) in diameter in the absence of solvent, but in the presence of organic solvents such as dichloromethane (and in solvents even up to the polarity of dimethylformamide [23]) they swell to five or six times their original volume and to even greater volumes when the growing peptide chain is also linked. The description of the swollen support which has evolved is that of a polystyrene-type matrix linked to the peptide which, due to high solvation, freely allows the chemical reagents to diffuse through it. This state of affairs assists the maximization of coupling yield and efficiency in the washing stages of the synthetic cycles. The peptide chain and the polymer support become completely mixed and seem to exert solvating effects on one another. The physical mobilities of the growing peptide chains within the polymer are on a par with peptides in solution as deduced by ^{1}H and ^{13}CNMR measurements [24]. The overall dimensions of the polymer beads change with increasing length of the peptide on the polymer carrier [25]. The swollen peptide-resin is approximately 80–90% solvent by volume, with of the order of 10^{12} growing peptide chains per polymer bead [26].

Functionalization of the beads to give the appropriate point of attachment to the resin has involved a number of methods [3], but for the polystyrene-type matrix, chloromethylation using Friedel–Crafts conditions, aminomethylation and the introduction of benzhydrylamino groups represent the most popular routines. The future demands of adapting the solid phase technique to commercial production of peptides on a large scale has justified work on trying to increase the loading ratio of peptide to polymer support [27]. In this context, ultra-high load solid (or gel) phase synthesis can be carried out on cross-linked poly[N-[2-(4-hydroxyphenyl)ethyl]acrylamide.

The routine protocol adopted for the polystyrene matrix support involves a batchwise approach where the polymeric resin is reacted, coupled and washed within a single reaction vessel. However, continuous flow of the solutions through a column of resin has advantages that have not been so successfully harnessed for polystyrene-based resin when compared to its polyacrylamide counterpart. However, there are recent reports of multiple continuous flow techniques being performed under low-pressure conditions using only the styrene–1% divinylbenzene copolymer [28]. The flow through the reactors is maintained by a moderate inert gas overpressure of about 20–50 kPa, which prevents a flow rate greater than is tolerated by the interstitial cavities among the soft gelatinous resin beads.

7.2.4.2   Polyacrylamides

In order to satisfy the need for full solvation of peptide and polymer in the more polar solvents such as dimethylformamide, Sheppard and coworkers have successfully

$$\begin{array}{ccccc}
\text{CONMe}_2 & & \text{CONMe}_2 & & \text{CONMe}_2 \\
| & & | & & | \\
-\text{CH}-\text{CH}_2-\text{CH}-\text{CH}_2 & \text{CHCH}_2 & \text{CHCH}_2 & \text{CHCH}_2- \\
| & & | & & \\
\text{CONMe}_2 & & \text{CONH}-\text{CH}_2 & &
\end{array}$$

$$\begin{array}{ccc}
\text{CONMe}_2 & & \text{CONH}-\text{CH}_2 \\
| & & | \\
-\text{CH}-\text{CH}_2-\text{CH}-\text{CH}_2\text{CHCH}_2\text{CHCH}_2\text{CHCH}_2- \\
| & & | & & | \\
\text{CONMe}_2 & & \text{CONMe}_2 & & \text{COR.}
\end{array}$$

(7)

developed a solid support (7), based on poly(dimethylacrylamide) insolubilized by cross-linking with bis(acryloyl)ethylenediamine [22].

Since the development of the original polymer, other variations such as a polyacryloylpyrrolidine [29] and a polyacryloylmorpholine [30] have been developed. Details of the syntheses of these polymers have been summarized recently [4]. A polymer support (7, R = NMeCH$_2$CO$_2$Me) can be made by suspension polymerization from a mixture of N,N-dimethylacrylamide (CH$_2$ = CHCONMe$_2$) and acryloyl-sarcosine methyl ester (CH$_2$ = CHCONMeCH$_2$CO$_2$Me). The average polymer bead size from this process varies between 50 and 100 μm and is marketed commercially (distributed by Mutagen) as Pepsyn. The R group (7) can be further modified into an aminopolymer with R = NMeCH$_2$CONH(CH$_2$)$_2$NH$_2$. Such resins show good swelling characteristics, increasing to about 10 times the original volume in water, dimethylformanide, dichloromethane, pyridine or methanol (5 times in dioxan).

The general advantages of continuous-flow methodology in terms of efficiency in time, savings on solvents, fast reaction rates and the possibility of concurrent synthesis of a number of sequences have led to investigation of such an approach with polyacrylamide beads. As with the polystyrene support, the polymer beads on their own tend to be too easily deformed under pressure to be applied to continuous-flow methodology. The problems have been overcome, however, by polymerizing the gel resin within a rigid macroporous framework, custom designed to maintain good solvent flow and rapid diffusion of reactants in and out of the supported gel. Macroporous Kieselguhr (commercially available from Sterling Organics as 'Macrosorb SPR') has the rigidity, mechanical strength and large pore size to make it a successful carrier for gel resins, and it can be efficiently packed into glass columns [31]. Resins with capacities of 0.25 meq g^{-1} and 0.1 meq g^{-1} are available on the Macrosorb matrix. Recently, other supported gel resins have been reported [32] where the polydimethylacrylamide has been supported and bonded to macroporous polystyrene (Polyhipe) or has been supported on polyoxyethylene/polystyrene (Rapp resin).

Both the beaded polyacrylamide gel and the Kieselguhr-supported methyl ester resins need to be converted into an aminomethyl resin for coupling to linker molecules. The conversion is usually carried out using redistilled ethylenediamine, and the derived resin has to be linked immediately to, for example, an internal reference amino acid (such as norleucine) via the usual activation procedures or a linkage agent, that can stabilize the aminopolymer.

7.2.5   N-Terminal protection — a temporary requirement

Since the peptide chain 'grows' from the C-terminal end, protection of the amino group needs to be only transient. Stability of the groups is paramount during the activation and coupling stages, but facile removal under the mildest of conditions is also a required property. While the development of N-protecting groups in solution involved a wide selection of possibilities [33], the solid phase approach has relied heavily on the properties of urethane-protecting groups. In the protocol based on polystyrene, the acid labile *t*-butyloxycarbonyl (Boc) group (**8**) [24] has proved popular and can be removed with trifluoroacetic acid (TFA) (20–50%) in dichloromethane or 4 M hydrogen chloride in dioxan. Sufficient sensitivity to acid is also shown by the 3,5-dimethoxyphenyl-2-propyl-2-oxycarbonyl (Ddz) group (**9**), and this has good potential for large-scale synthesis [35]. The Ddz group is cleaved by 1–5% trifluoroacetic acid in dichloromethane. Boc derivatives of amino acids have become generally available from a number of commercial sources.

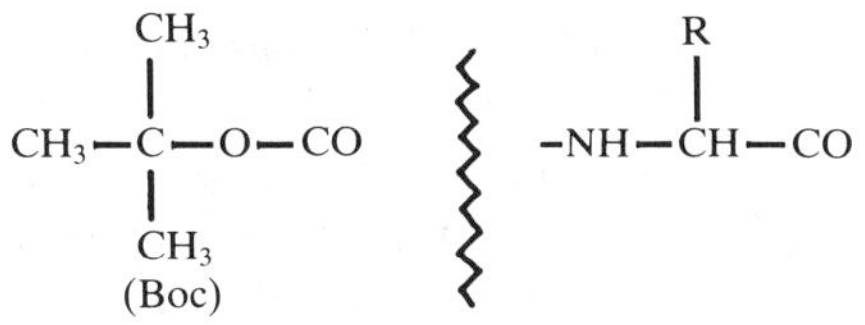

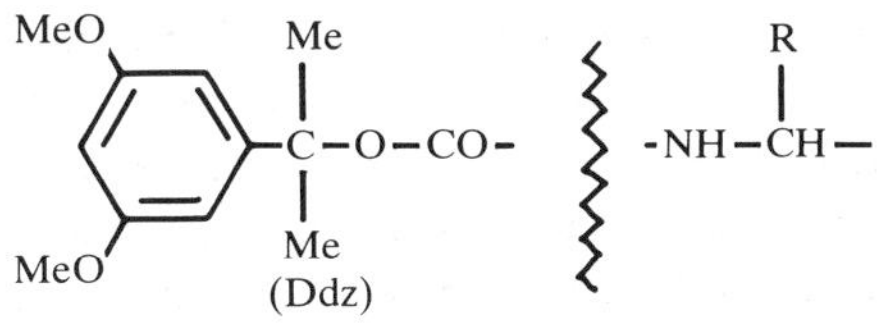

Research on polyacrylamide solid supports provided the impetus for an alternative to the use of acid-labile groups, and use of the 9-fluorenylmethoxycarbonyl (Fmoc) group (Scheme 5) [36] has received universal acceptance [37]. The Fmoc group is readily removed via β-elimination initiated by a secondary organic base such as piperidine and allows the use of acid-sensitive protecting groups for side chains.

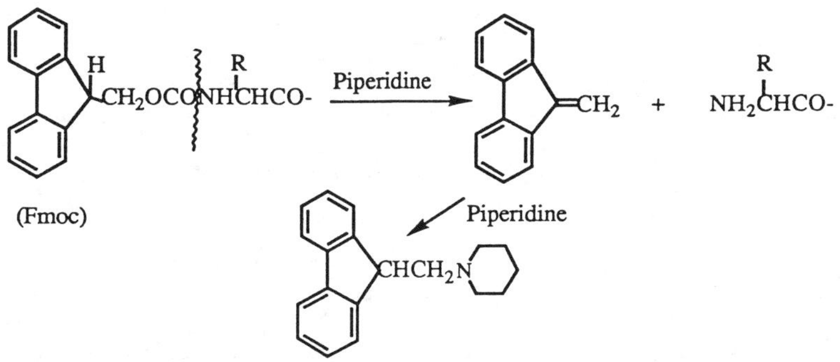

Scheme 5 — Removal of Fmoc protecting groups.

The popularity of the Fmoc-polyacrylamide strategy has also meant that Fmoc derivatives of protected amino acids have become commercially available [4].

7.2.6 Side-chain protecting groups

As can be seen in Table 7.1, only approximately half of the common natural amino acids are trifunctional and require extra protection of their side chains. A good choice of protecting group can both protect and enhance the solubility of the amino acid derivatives. The choice of derivative is also governed by whether the Fmoc-polyacrylamide protocol or the Boc-polystyrene protocol is being used. In a detailed discussion of the potential of each amino acid residue to give rise to side reactions, it has been noted that only alanine and leucine are generally free from problems in solution, with even the glycine residue a potential source of trouble [38]. However, in general, the bifunctional residues (those containing only aliphatic side chains and phenylalanine) have not proved troublesome in solid phase work as long as it is recognized that residues such as valine and isoleucine have highly hindered side chains and require longer coupling times.

Of the trifunctional residues, the side-chain amide groups of asparagine and glutamine, the methionine side chain, and in the Fmoc strategy the tryptophan side chain can be left unprotected. All other residues require appropriate protection. Some care must be taken, however, to prevent dehydration of the side-chain amide group during the activation of asparagine and glutamine with dicyclohexcarbodiimide. In the case of lysine (amino-protection required) and tyrosine (phenolic hydroxyl group), functional group protection must be compatible with the protocol adopted, yet O-benzyltyrosine can undergo rearrangement to 3-benzyltyrosine [39]. The remaining amino acids are more troublesome in that side reactions, poor solubility and difficulty of complete removal of the protecting groups can cause problems.

Nominally, in chemical terms, a simple ester might be an adequate protection for the carboxyl groups of aspartic and glutamic acids. However, these two residues are prone under most peptide synthesis conditions to undergo side-chain interaction with the peptide chain amide groups as depicted in Schemes 6 and 7.

Some relief from the problems associated with these rearrangements can be gained through the use of sterically hindered esters (e.g. t-butyl [40], cyclohexyl [41]) and for the Merrifield Boc strategy, benzyl esters.

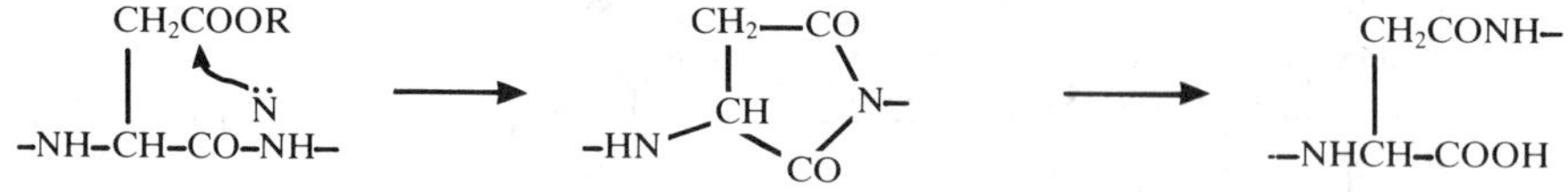

Scheme 6 — The $\alpha \rightarrow \beta$-aspartyl rearrangement.

Scheme 7 — Formation of a pyroglutamyl derivative.

The hydroxyl groups of serine and threonine are either protected as benzyl (Merrifield approach) or *t*-butyl ethers (Fmoc strategy). There is a possibility of α,β-elimination under acid-catalysed conditions during prolonged exposure (equation 7.1) [42].

$$R-CH{\overset{\frown}{-}}O-CH_2Ph \qquad \longrightarrow \qquad R-\underset{\parallel}{C}-H \qquad + PhCH_2OH \qquad (7.1)$$

The protection of the imidazole ring in histidine has been the subject of intensive study. In the Merrifield protocol, protecting groups (**10**) [43], (**11**) [44] and (**12**) [45] have been used while the Fmoc procedure favours (**13**) [46], (**14**) [4], (**15**) [4] and (**16**) [12]. It is recognized that maximal protection occurs when the protecting group is attached to the distant N atom.

(**10**)	$R = CH_3$ $\left\langle\!\!\left\langle\ \right\rangle\!\!\right\rangle$—$SO_2-$
(**11**)	R = 2,4-dinitrophenyl (structure (b))
(**12**)	$R = PhCH_2OCO$
(**13**)	$R = Ph_3C$ (structure (b))
(**14**)	R = Boc
(**15**)	R = Fmoc
(**16**)	$R = CH_2OBu^t$ (structure (a))

(a)   (b)

Arginine has a highly basic guanidino group (pK_a 12.5), which is not readily acylated under the usual coupling conditions, as it is in its protonated form. Yet protection is highly desirable for improved solubility. Tosyl (Tos) [47] and mesitylene-2-sulphonyl (Mts) groups [48] have been the most popular in the Merrifield protocol while the Mtr group (**17**) [49] is relatively successful in the Fmoc protocol. The main problem with all these arylsulphonyl compounds tends to be difficulty in their complete removal under acid treatment. Some good physical–organic rationalizations underpin the recent introduction of the Pmc group (**18**) [50], which can be cleaved in trifluoroacetic acid in less than 20 min.

(**17**)   (**18**)

Substituted benzyl derivatives such as 4-methylbenzyl [51] and 4-methoxybenzyl [52] are useful for cysteine protection during the usual trifluoroacetic acid protocol in the Merrifield approach. Final cleavage is possible during hydrofluoric acid treatment. t-Butyl [53], acetamidomethyl [54] and trityl [55] groups represent well tried protection in the Fmoc approach. The trityl group is removed during treatment with TFA in the last deprotection step. t-Butyl and acetamidomethyl groups can be advantageously cleaved with mercury salts.

7.2.7 'Activation' for the coupling step

In general, for amide bond formation to occur, the carboxyl entity being condensed with the amino terminal group of the growing peptide chains needs the right level of 'activation'. Traditional activated derivatives such as acid chlorides and anhydrides suffer from the disadvantages of side reactions (such as racemization) and insufficient 'shelf-life' for storage on automated instruments. The mixed anhydride approach [56] has found success in the solution phase and symmetrical anhydrides, freshly prepared before use, can be acceptable solid phase reagents, but in the latter context half a mole equivalent of anhydride is 'lost' at each stage.

Active esters [57] have found more extensive use in solid phase work, since they can be pre-synthesized and stored and afford an acceptable level of activation for peptide bond formation. The tried work-horses of the solid phase technique are pentafluorophenylesters [58] and 1-oxodihydrobenzotriazinyl(Dhbt) esters [59], which have better activation characteristics than p-nitrophenyl or pentachlorophenyl esters.

N,N'-Dicyclohexylcarbodiimide (DCC) has come to be regarded as the universal activation agent of solid phase work [60]. The mechanism of activation involves the O-isoacylurea (**19**) as a key activated intermediate capable of reacting with several types of nucleophile (Scheme 8).

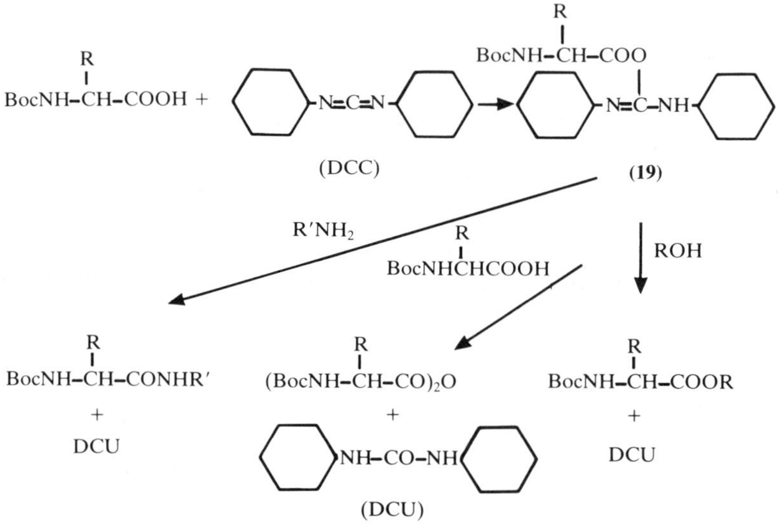

Scheme 8 — Activation of N-protected amino acids by dicyclohexylcarbodiimide.

Two aspects of this chemistry deserve caution: (a) the ready rearrangement of (**19**) to the corresponding *N*-acylurea; and (b) the product dicyclohexylurea (DCU) is fairly insoluble. These problems are less severe in solid phase work because the byproducts tend to be washed out in the subsequent washing cycles, but acylurea formation can also be eliminated by the addition of catalytic amounts of hydroxy components such as hydroxybenzotriazole (**20**) [59b]. This additive also reduces dehydration of carboxamide side chains [61]. There is a current trend to minimize all possible side rections by pre-forming and purifying the activated derivatives outside the reaction vessel [4]. A more soluble urea is formed as a product if diisopropylcarbodiimide [62] is used for the coupling.

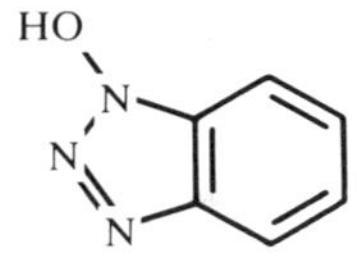

(**20**)

The benzotriazol-1-yloxytris(dimethylamino)phosphonium hexafluorophosphate (BOP) reagent [63] (**21**) has been shown to be a useful reagent for *in situ* formation of benzotriazolyl esters of protected amino acids, which have good characteristics and few side reactions, although the tris(dimethylamino)phosphonium hexafluorophosphate released has carcinogenic properties. The alternative competitor is benzotriazol-1-yloxy-1,1,3,3-tetramethyluronium tetrafluoroborate (TBTU) (**22**) [64], which could be a very useful reagent in larger-scale synthesis [35b].

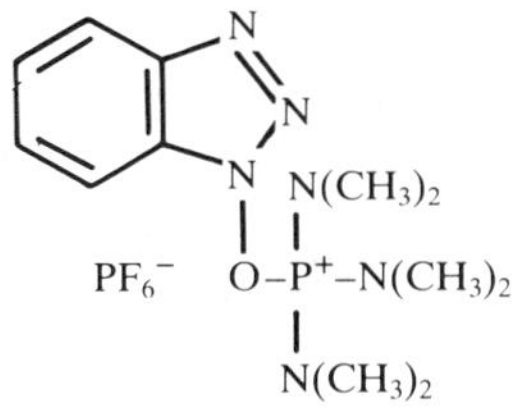

Benzotriazol-l-yloxytris(dimethylamino)
phosphonium hexafluorophosphate (BOP)

(**21**)

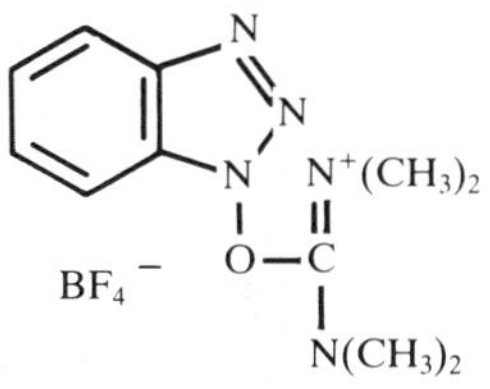

Benzotriazol-l-yloxy-1,1,3,3-
tetramethyluronium tetrafluoroborate
(TBTU)

(**22**)

7.3 THE AUTOMATED SOLID PHASE PROTOCOL

The 'Merrifield' approach from its infancy [65] has placed great emphasis on the advantages of routine automation of the technique, but it has taken the best part of two decades to optimize all of the complicated series of chemical reactions to give good quality oligopeptides. Even now, at the start of a new decade, there are still points of chemical detail that are often sequence-dependent and need careful assessment and judgement. Postassembly-deprotection, the peptide sequences often demand further purification using h.p.l.c. techniques.

7.3.1   Instrument format

The solid phase protocol can be said to have developed along the two traditions of the *t*-Boc/Fmoc-amino-protected-polystyrene resin approach [66] and the Fmoc-amino-protected-polyacrylamide resin approach [4]. Both protocols can utilize a single reaction vessel being 'serviced' by a predefined sequence of amino acid derivatives, coupling agents and wash cycles under either manual or computer control. A schematic for an optimized automated synthesizer using the single-reaction-vessel approach is represented in Fig. 7.1. A number of the features highlighted earlier in this chapter are incorporated, e.g. a facility for making symmetric anhydrides or active esters prior to entry to the reaction chamber, immediately before reaction. Although the coupling time has to rely on preset conditions, synthetic conversion and efficiencies can be assessed by release of a few milligrams of resin for testing by the ninhydrin reaction [67]. Quantitative Edman sequencing techniques can also be used to assess the accumulative efficiency of the deprotection steps and the coupling yields [68].

Information about troublesome coupling steps is increasing [69] and some rationalizations have been made: (a) there is a sequence-dependent tendency for aggregation, often due to high resin loadings, which can be overcome by using aprotic solvents such as DMF; (b) a persistent sequence-dependent 0.5–2% lowering of coupling efficiency due to a number of reasons which can be overcome by treatment of unreacted amino groups to give terminated species — the so-called 'capping' procedure [70], and (c) activated β-branched amino acids, e.g. Ile, Thr and Val give poor coupling yields.

The Fmoc-amino-protected-polyacrylamide routine is lending itself to more rapid development as a continuous-flow automatic procedure [4,72,73]. Fig. 7.2 summarizes the circulation loops required for a machine [71] based on Fmoc-amino-acid pentafluorophenyl esters or other pre-activated derivatives. Valves 5, 6, 2b and 3 define a route for introduction of the activated derivative into the column of Kieselguhr-supported-polyacrylamide resin. The availability of a recirculation route out of valve 3 through a UV monitoring cell gives this type of technology the advantage that each coupling and deprotection step can be monitored in real time. The UV-absorbing characteristics of the Fmoc group obviously is a benefit for monitoring the progress of the reaction, and the profiles obtained at each coupling step give a visual check on progress.

Fmoc-amino-acid-Dhbt esters have provided another possibility for continuously monitoring the progress of the coupling stages. During coupling using these esters a bright yellow colouration appears on the resin and then disappears when acylation is complete, as indicated by a separate ninhydrin test [59]. The transient colouration has been explained as being due to the species (**23**), which is only present when there are free amino groups on the resin. Prototype photometric cells to analyse this colour development and an automatic feedback to the control mechanism to indicate completion of coupling have been assessed. On a similar vein, counterion distribution

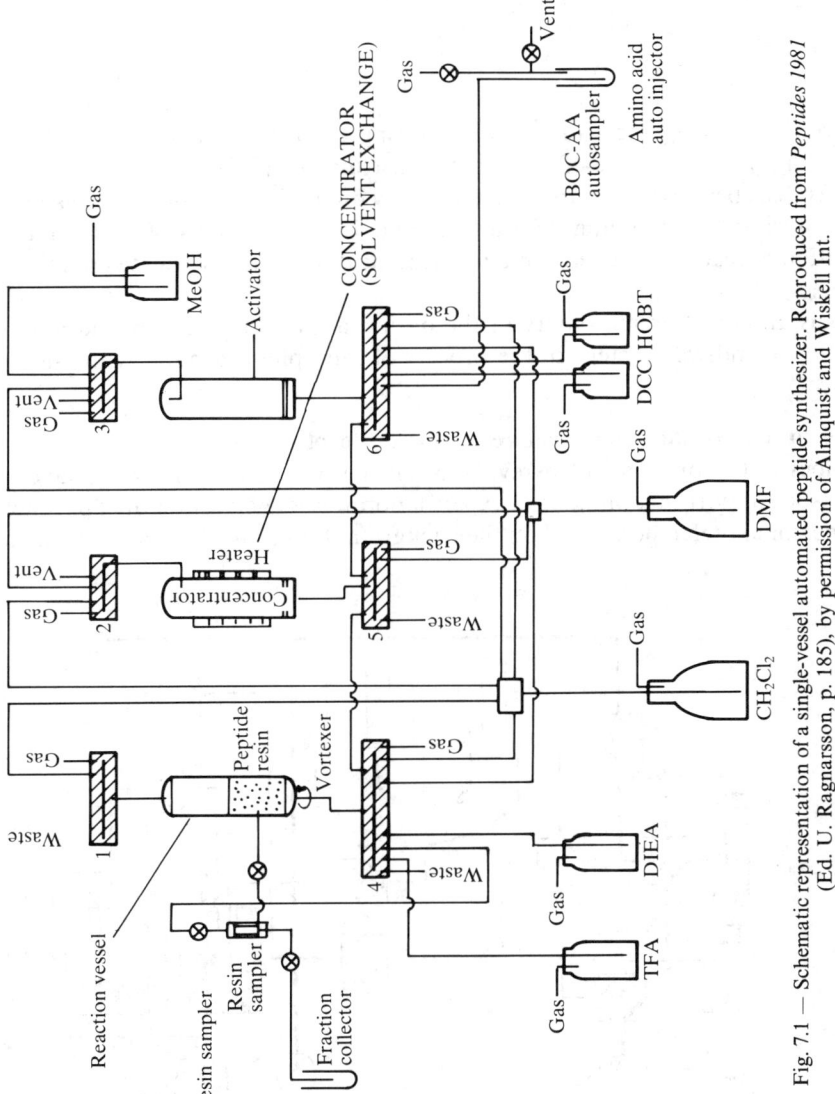

Fig. 7.1 — Schematic representation of a single-vessel automated peptide synthesizer. Reproduced from *Peptides 1981* (Ed. U. Ragnarsson, p. 185), by permission of Almquist and Wiskell Int.

$$ResinNH_3^+ \ O^-$$

(23)

monitoring, which operates in real time during the acylation procedure, has been developed [73a] for a range of commercial instruments [73b].

It has also been possible to devise [74] a conductivity sensor that responds to changes during the coupling of Fmoc-amino-acid pentafluorophenyl esters. The information is relayed to a real-time full graphical display to indicate completion of coupling.

This is an era of great activity in the optimization of conditions and automatic control of the individual steps in the process of solid phase peptide synthesis.

7.3.2 Summary of the stages required in the protocol

Space does not allow a detailed review of all the procedures that may have found enthusiastic support in a number of expert laboratories. Very often the approach has been a personal preference, based on the background training of the personnel, and up

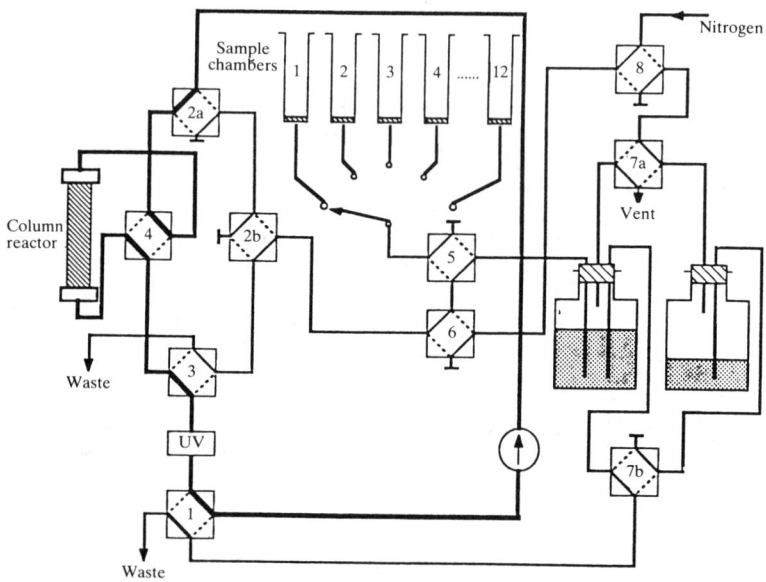

Fig. 7.2 — Schematic representation of a continuous-flow automated peptide synthesizer. Reproduced from *Solid Phase Peptide Synthesis* (Eds E. Atherton and R. C. Sheppard, p. 100), by permission of Oxford University Press.

to now few comparisons have been made of the various protocols, e.g. in the synthesis of a particular well defined peptide. This aspect of comparable assessment is currently becoming more feasible in the area of simultaneous solid phase synthesis [75].

In this chapter the main chemical protocols representing the *t*-Boc-amino-acid-polystyrene resin approach and the Fmoc-amino-acid-polyacrylamide resin strategy are summarized. In the former, the Boc temporary protecting group is removed in each cycle by treatment with TFA, so choice of side-chain protection is usually based on groups that will survive this acid treatment to the end of the assembly. On the other hand, the Fmoc group is removed under base-catalysis, so side chains are protected by groups later removable under mildly acidic conditions.

Figs. 7.3 and 7.4 represent 'idealized' composite summaries based on the protocols of Merrifield and coworkers [3,6] and Sheppard and coworkers [4] respectively, for the synthesis of pentapeptide (**24**). Amino acids have been chosen to represent typical side-chain-protecting choices for a hypothetical sequence.

7.3.3   Purification of the resultant peptides and their characterization

The more recent optimized conditions for peptide synthesis on the solid phase give rise to products that on h.p.l.c. analysis would, on average, be 60–80% pure. The difficulty arises if a significant amount of the impurity might be 'deletion' sequences due to incomplete coupling at any stage. These sequences could well be chromatographically very similar in properties. Traditional chromatographic techniques of ion-exchange and gel filtration can often be used with great benefit [4], but the availability of good quality peptides from the solid phase approach owes a great deal to the development of reversed-phase h.p.l.c. [76].

Use of a reversed-phase silica support (C_8 or C_{18} derivatized) has been shown to be generally successful with solvent systems such as a gradient of acetonitrile — 0.1% aqueous TFA, or a gradient of acetonitrile — 0.01 M aq ammonium acetate, pH 4.5. For 50–100 mg of peptide to be purified it is possible to adapt analytical h.p.l.c. techniques to take a higher loading, and preparative versions of such h.p.l.c. systems are also commercially available. Use of a cheaper reversed-phase support with larger particle size (15–20 μm) at low pressure in glass columns [77] is also useful to purify up to 0.3 g of peptide. Endcapped reversed-phase silica in glass columns (52 × 480 mm) have been reported to be capable of purifying gramme quantities of peptide [78].

However, use of h.p.l.c. alone as a check on purity should not be accepted as the only criterion. A timely reminder [79] of the care needed in accepting sharp single peaks in h.p.l.c. has come from a fast atom bombardment (FAB) MS and plasma desorption MS study of a single peak from a somatostatin solid phase synthesis, which picked up a number of failure sequences. So, in addition to the h.p.l.c., total amino acid analysis [80], and an Edman sequence check [81], together with FAB MS [82] and plasma desorption MS [83] would be an ideal total check on purity. An estimation of purity should not be left to FAB MS alone since, in a mixture of peptides, the rate of appearance of different peptides at the surface of the matrix solvent could mask the presence of impurities [84].

Stage 1.   Choose a resin and linker

e.g. Pam linker on to styrene – 1% divinylbenzene copolymer

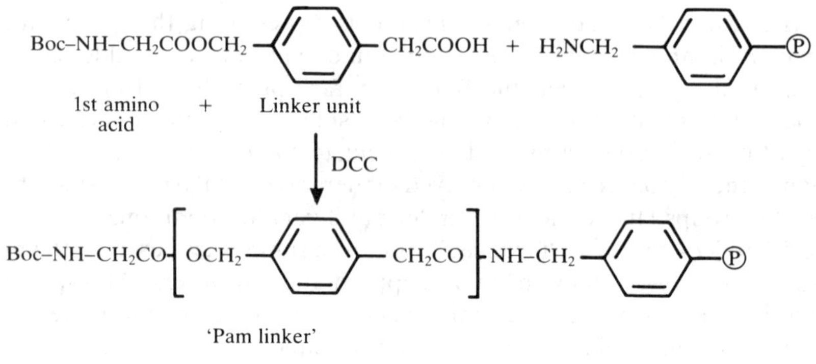

$$\text{Boc–NH–CH}_2\text{COOCH}_2\text{—}\bigcirc\text{—CH}_2\text{COOH} + \text{H}_2\text{NCH}_2\text{—}\bigcirc\text{—}\textcircled{P}$$

1st amino   +   Linker unit
acid

DCC ↓

$$\text{Boc–NH–CH}_2\text{CO}\!\left[\text{OCH}_2\text{—}\bigcirc\text{—CH}_2\text{CO}\right]\!\text{NH–CH}_2\text{—}\bigcirc\text{—}\textcircled{P}$$

'Pam linker'

Stage 2.   Deprotection cycle:

(i) CH_2Cl_2 wash (3 times);
(ii) $TFA/CH_2Cl_2(1/1)$ (30 min)
(iii) CH_2Cl_2 wash (4 times);
(iv) diisopropylethylamine/$CH_2Cl_2(1/19)$;
(v) CH_2Cl_2 wash (6 times).

Stage 3.   Addition of next amino acid derivative

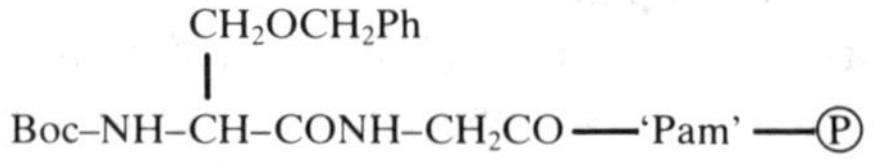

$$\begin{array}{c}\text{CH}_2\text{OCH}_2\text{Ph}\\ |\\ (\text{Boc–NHCH–CO})_2\text{O}\end{array}$$

$$\begin{array}{c}\text{CH}_2\text{OCH}_2\text{Ph}\\ |\\ \text{Boc–NH–CH–CONH–CH}_2\text{CO —'Pam' —}\textcircled{P}\end{array}$$

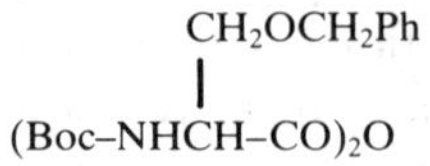

Stage 4.   Deprotection as in stage 2

Stage 5.   Next aminoacid derivative

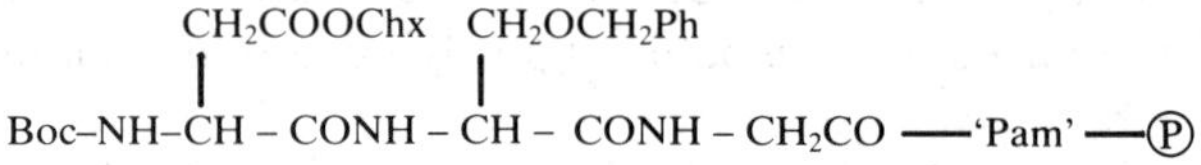

$$\begin{array}{c}\text{CH}_2\text{COOCyclohexyl (Chx)}\\ |\\ (\text{Boc–NH–CH–CO})_2\text{O}\end{array}$$

$$\begin{array}{c}\text{CH}_2\text{COOChx}\quad\text{CH}_2\text{OCH}_2\text{Ph}\\ |\qquad\qquad |\\ \text{Boc–NH–CH – CONH – CH – CONH – CH}_2\text{CO —'Pam' —}\textcircled{P}\end{array}$$

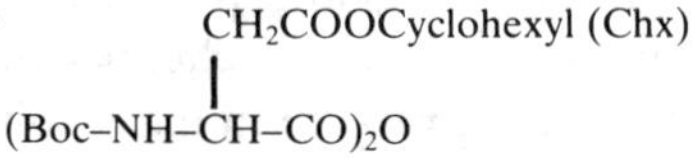

Stage 6.   Deprotection as in stage 2

Stage 7.   Next amino acid derivative

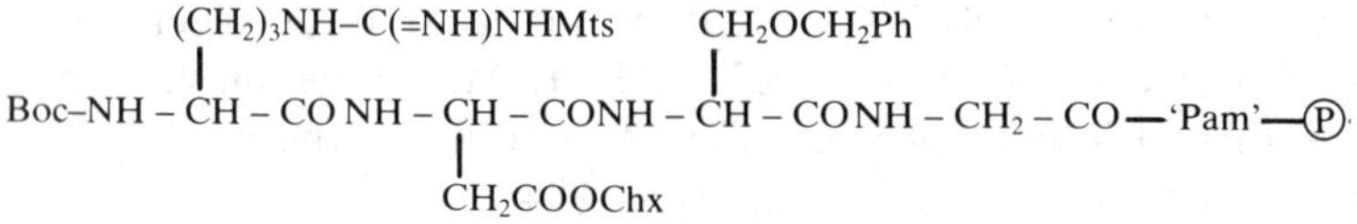

$$\begin{array}{c}(\text{CH}_2)_3\text{NH–C(=NH)–NH–Mts}\\ |\\ (\text{Boc–NH–CH–CO})_2\text{O}\end{array}$$

$$\begin{array}{c}(\text{CH}_2)_3\text{NH–C(=NH)NHMts}\qquad\text{CH}_2\text{OCH}_2\text{Ph}\\ |\qquad\qquad\qquad\qquad\quad |\\ \text{Boc–NH – CH – CO NH – CH – CONH – CH – CO NH – CH}_2\text{ – CO —'Pam'—}\textcircled{P}\\ |\\ \text{CH}_2\text{COOChx}\end{array}$$

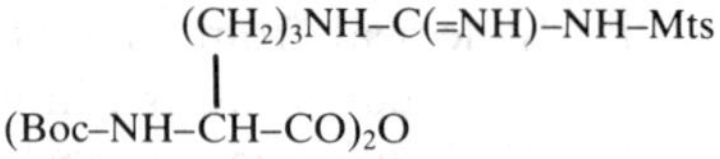

Stage 8. Deprotection as in stage 2

Stage 9. Next amino acid derivative

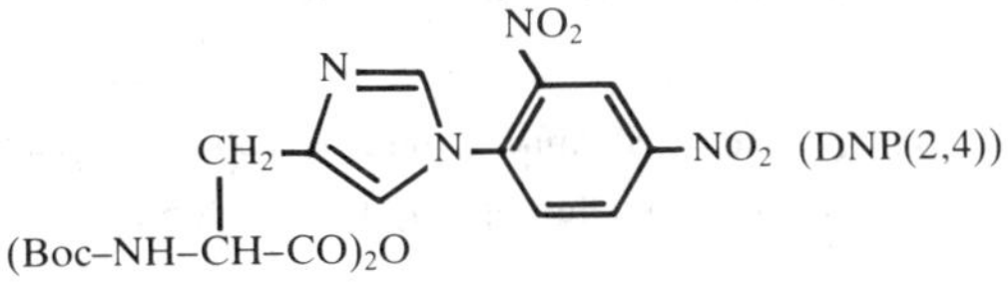

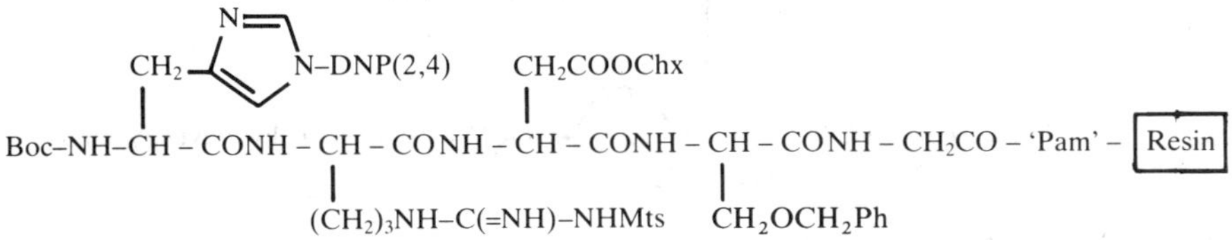

$$CH_2 \overset{N=}{\underset{|}{\smile}} N \overset{NO_2}{\underset{}{\bigcirc}} NO_2 \quad (DNP(2,4))$$

$(Boc–NH–CH–CO)_2O$

$$CH_2 \overset{N=}{\underset{|}{\smile}} N–DNP(2,4) \quad CH_2COOChx$$

$Boc–NH–CH – CONH – CH – CONH – CH – CONH – CH – CONH – CH_2CO – 'Pam' – \boxed{Resin}$

$(CH_2)_3NH–C(=NH)–NHMts \quad CH_2OCH_2Ph$

Stage 10. Final deprotection:
  (i) thiolysis for DNP derivative;
 (ii) TFA;
(iii) 25% HF in Me_2S + cation scavengers;
(iv) 90% HF in Me_2S + carbocation scavengers.

Workup should give (**24**), which would be purified by h.p.l.c. or other chromatographic techniques.

H – His – Arg – Asp – Ser – Gly – OH

(**24**)

Fig. 7.3 — Stages in the synthesis of a peptide by the Merrifield method.

7.4 ALTERNATIVE APPROACHES TO SOLID PHASE SYNTHESIS — SIMULTANEOUS (OR PARALLEL) MULTIPLE PEPTIDE SYNTHESIS (SMPS)

7.4.1 General introduction
The recognition that certain features in some biological systems, e.g. immunological responses, can uniquely select a particular sequence of amino acid residues and carry out its own 'biological selection' has spawned a strong desire to increase the synthetic output several fold in order to have as many different sequences as possible within a short period of time. Steamlining of the conventional solid phase approach to this challenge was initiated by Geysen and coworkers [85], who recognized that many of the protocols were routine and could be carried out at the same time on parallel solid support assemblies. There is a obviously a sacrifice in scale and rigour, but the reality that one individual can synthesize over 100 15-residue sequences in two weeks or so is a definite bonus, which could indirectly feed back useful information to conventional methodology. The SMPS approach provides identical conditions for a number of different sequences and can be used to assess and monitor techniques [86].

Stage 1.   Choose resin and linker and add 1st residue

e.g. Kieselguhr-polydimethylacrylamide resin.

Fmoc–NHCH$_2$COOPfp + HO–CH$_2$⟨benzene ring⟩–CH$_2$CO–Nle–Polydimethylacrylamide
(Pfp = pentafluorophenyl)   (Nle = Internal reference)

4-dimethylaminopyridine

Fmoc–NH–CH$_2$COO–CH$_2$⟨benzene ring⟩–OCH$_2$CO–Nle–Polydimethylacrylamide

Stage 2.   Deprotection cycle

(i)   DMF (5 times);
(ii)   20% piperidine in DMF (10 min);
(iii) DMF (10 times).

CH$_2$OBu^t
|
Stage 3.   Addition of next amino acid derivative   (Fmoc–NH–CH–COOPfp)

CH$_2$OBu^t
|
Fmoc–NH–CH–CONH–CH$_2$CO–Linker–Nle–Ⓟ

Stage 4.   Deprotection as in stage 2

CH$_2$COOBu^t
|
Stage 5.   Next amino acid derivative^a   (FmocNH–CH–COOPfp)

CH$_2$COOBu^t CH$_2$OBu^t
| |
FmocNH – CH – CONH – CH – CONH – CH$_2$CO Linker – Nle – Ⓟ

Stage 6.   Deprotection as in stage 2

(CH$_2$)$_3$NH – C(=NH) – NH – SO$_2$–⟨Pmc ring structure with Me groups⟩ (Pmc)
|
*Stage 7.   Next amino acid
derivative^a*   (FmocNH – CH – COOPfp)

(CH$_2$)$_3$NH–C(=NH)–NHPmc
|
FmocNH – CH – CONH – CH – CONH – CH – CONH–CH$_2$CO–Linker–Nle–Ⓟ
| |
CH$_2$COOBu^t CH$_2$OBu^t

Stage 8.   *Deprotection as in stage 2*

```
                                    ┌──N–Boc
                          CH₂──⟨    ⟩
                           │         N
```

Stage 9.   *Next amino acid derivative*   (Fmoc–NH–CH–COOPfp)

```
            ┌──N–Boc
   CH₂──⟨    ⟩                      CH₂COOBuᵗ
    │        N                       │
 FmocNH–CHCONH– CH– CONH –CH –CONH – CH – CONH–CH₂CO–Link–Nle–Ⓟ
                  │                  │
          (CH₂)₃NH–C(=NH)NHPmc   CH₂OBuᵗ
```

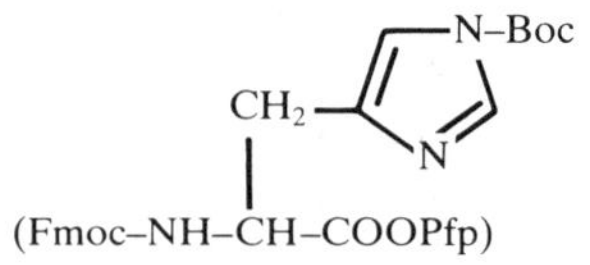

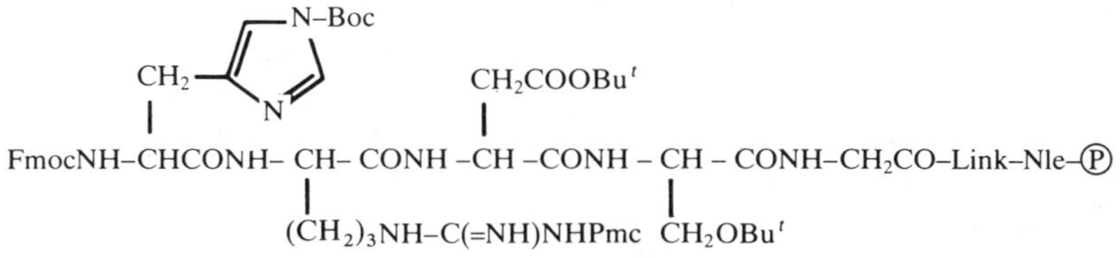

Stage 10.   *Final deprotection:*

(i) 20% piperidine in DMF to remove Fmoc group;
(ii) washed peptidic resin treated with TFA/anisole/ethandithiol (150:5:1.5) for 1.5 h;
(iii) purification by chromatography.

   Work up should give

H–His–Arg–Asp–Ser–Gly–OH

(24)

---

a These steps could include addition of 1-hydroxybenzotriazole.

Fig. 7.4 — Stages in the synthesis of pentapeptide (**24**) by the Sheppard method.

7.4.2   Peptide growth on the tips of polyethylene rods

Geysen chose polyethylene rods (4 mm diameter × 40 mm length), 'capped' with polyacrylic acid for support of the growing peptide chains. The rods were assembled into a polyethylene holder with the format and spacing of a microtitre plate. The reactions at the tip of the rods were carried out in a teflon tray with a matrix of wells to match the rod spacing. The link between polyacrylate and the peptide was made via the N^{ε}-amino group of lysine, which was coupled using Boc-L-Lys-methyl ester, and the loading was determined as 0.15–0.2 nmol/mm². Deprotection of the Boc–Lys group then released the amino group for subsequent assembly of Boc-protected amino acids using DCC/hydroxybenzotriazole (HOBt) in DMF as the coupling agent. Side chains were protected as follows: O-benzylation for Thr, Ser, Asp, Glu and Tyr; benzyloxycarbonyl for Lys; Tos for Arg; $4\text{-}CH_3C_6H_4CH_2$ for Cys; and 1-benzyloxycarbonylamino-2,2,2-trifluoroethyl for His. Removal of the peptides from the rods was achieved using boron tris(fluoroacetate) in TFA for 90 min. Using this technique, 208 possible overlapping hexapeptide sequences covering the 213 amino acid sequence of the immunologically important coat-protein of the foot-and-mouth virus type O_1 were synthesized without the need for expensvie equipment.

7.4.3   Use of compartmentalized resins

Another variation on this theme utilizes individually compartmentalized resins [87]. Batches of resin (25 mg to many grammes) are accommodated in perforated polypropylene bags (any solid phase resin for peptide synthesis can be used) (Fig. 7.5). For all the usual washings, deprotections, etc. the resin bags are all contained in a common reservoir. Then for individual coupling of protected amino acids, the resin packets are separated and added to individual solutions of activated protected amino acids, stirred and shaken. At the end of the coupling step, the packets are returned to the common reservoir. This SMPS approach has been found to work well with all solid phase chemistries, including Fmoc protection.

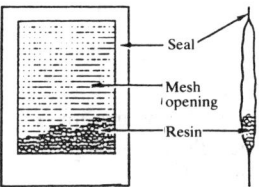

Fig. 7.5

7.4.4   Use of sheets as supports for peptide synthesis

Preliminary reports have appeared [88] that polyethylene sheets on to which are grafted high m.w. non-cross-linked polystyrene chains can also support synthesis of peptides, using either the single peptide approach or the SMPS approach. The synthesis can be undertaken on strips of the polystyrene-grafted polyethylene sheet which are aminomethylated to give a final substitution of $1.0 \, \text{mmol NH}_2 \, \text{g}^{-1}$ sheet. Addition of the first protected amino acid residue follows introduction of the Pam linker.

Functionalized cellulose paper [89] has also been researched as a possible support for SMPS, the initial attachment on to the paper being made via an Fmoc-amino-acid chloride [90]. Loading on the support comes to 1 to $2 \, \mu\text{mol cm}^{-2}$ and the cellulose ester bond between the paper and the first residue is stable under the reaction conditions of solid phase synthesis. Amino acids are coupled batchwise as their pre-activated HOBt esters. A block of teflon containing 96 wells [91], each equipped with filtration facilities, has been shown to be effective in carrying out 400 couplings per day using Fmoc amino-acid-DHBt esters attached to Macrosorb-SPR resin through an acid-labile linker, 4-hydroxymethylphenoxyacetic acid.

The large throughput of assemblies constructed via the SMPS approach provides a wealth of information for comparing different synthetic procedures under controlled and identical conditions. Polypropylene column reactors of 1 ml volume have been filled individually, or linked in series or in parallel to facilitate common treatments to study coupling efficiencies of carbodiimides and the BOP reagent [75]. In this way, the use of the BOP reagent in the presence of HOBt has been found to give more effective coupling than preformed anhydrides.

7.5 SELECTED EXAMPLES OF OLIGOPEPTIDE SYNTHESIS

Methodology in solid phase peptide synthesis has reached an era where quality oligopeptides with up to 50 residues can be routinely assembled, provided good chemical husbandry and analytical control are adhered to. A general review such as this cannot adequately represent the intrigues and experiences in the vast development work that has been published. However, reference to some recent accounts (see Table 7.2) of the assembly of representative peptides via solid phase synthesis is instructive. The annual chapter on peptide synthesis in Specialist Periodical Reports is a useful source of reference [92].

Table 7.2 — Oligopeptides synthesized by solid phase methods

Peptide	No. of residues	Technique[a]	Reference
Melanine-concentrating hormone (MCH)	17	B(CF)	[93]
Little gastrin	17	B(CF)	[94]
Interleukin-3	140	A	[95]
β-Lipotropin	89(60,29)[b]	A	[96]
Enkephalin analogues	5	C	[97]
Deaminoxytocin analogues	8	A, C	[98]
LHRH, FSH-releasing hormone (GnRH)	10	B, C	[99]
Conotoxin G1	13	A, B	[54], [100]
Somatostatin and analogues	14	A	[101]
Apamin	18	A	[102]
Thymosin α_1	28	A	[103]
Glucagon and analogues	29	A	[104]
β-Endorphin and analogues	31	B, A	[105]
Bombesin	14	B(CF), A	[106]
^{15}N-Gramicidins A, B and C	15	C	[107]
Conotoxin GI analogues	12/13	A	[108]
Trypsin inhibitor	28	A	[109]
Atral natriuretic peptides	23/25	A	[110]
Melittin	26	A	[111]
Myeloma immunoglobulin M603 heavy chain fragments	52	A	[112]
Human transforming growth factor α	50	B	[113]
Human apolipoprotein C-II	79	A	[114]
Thymopoietin (32–36)	5	B	[115]
Bovine apocytochrome-C fragments	23,37,44[b]	A	[116]
Dynorphin analogues	13	A	[117]
Substance P analogues	11	A	[118]
Ubiquitin	76(35,12,28)[b]	C, A	[119]
HIV-1 protease	99	A	[120]
Cholecystokinin-33	33	C	[121]
Microprotein EETI II	28	A	[122]
Endothelin	21	A	[123]

[a] *Technique A.*   Merrifield's Boc-polystyrene–1% divinylbenzene copolymer with a variety of linker units.
   Technique B.   Fmoc-polyacrylamide polymer–various linker groups (CF = continuous-flow techniques).
[a] *Technique C.*   Fmoc-polystyrene resin–various linkers.
[b] Additional numbers represent fragment sequences.

REFERENCES

[1] R. B. Merrifield, *Angew. Chem. Int. Edn. Engl.*, 1985, **24**, 799.
[2] For example, Instruments by Applied Biosystems, Pharmacia-LKB Biochrom, Milligen Biosearch, and Biotech Instruments Ltd.
[3] J. M. Stewart and J. D. Young, in *Solid Phase Peptide Synthesis*, 2nd edn., Pierce Chemical Co. Rockford, Illinois, 1984.
[4] E. Atherton and R.C. Sheppard, in *Solid Phase Peptide Synthesis: a Practical Approach*, IRL Press at Oxford University Press, 1989.
[5] S. B. H. Kent, *Ann. Rev. Biochem.*, 1988, **57**, 957.
[6] G. Barany, N. Kneib-Cordonier and D. G. Mullen, *Int. J. Pept. Protein Res.*, 1987, **30**, 705; G. Barany and R. B. Merrifield, in *The Peptides* (Eds E. Gross and J. Meienhofer), Academic Press, New York, 1979, Vol. 2, p. 1.
[7] E. Gross and J. Meienhofer (Eds) *The Peptides: Analysis, Synthesis, Biology*, Academic Press, New York, Vols 1–3, 1979–81.
[8] E. Wunsch, 'Synthese von Peptiden', in *Houben-Weyl's Methoden der Organischen Chemie* (Ed. E. Müller), Thieme, Stuttgart, 1974, Vol. 15, Pts I and II; M. Bodanszky, *Principles of Peptide Synthesis*, Springer-Verlag, Heidelberg, 1984; M. Bodanszky and A. Bodanszky, *The Practice of Peptide Synthesis*, Springer-Verlag, Heidelberg, 1984.
[9] J. Kovács, in *The Peptides: Analysis, Synthesis, Biology* (Eds E. Gross and J. Meienhofer), Academic Press, New York, 1980, Vol. 2A, Chapter 8, p. 486; D. S. Kemp, *ibid*, 1979, Vol. 1, Chapter 7, p. 317.
[10] N. L. Benoiton and F. M. F. Chen, *Can. J. Chem.*, 1981, **59**, 384; J. H. Jones and M. J. Witty, *J. Chem. Soc., Perkin Trans. 1*, 1979, 3203.
[11] J. S. Davies and A. K. A. Mohammed, *J. Chem. Soc., Perkin Trans. 1*, 1981, 2982.
[12] R. Colombo, F. Colombo and J. H. Jones, *J. Chem. Soc., Chem. Commun.*, 1984, 292.
[13] E. Atherton, N. L. Benoiton, E. Brown, R. C. Sheppard and B. J. Williams, *J. Chem. Soc., Chem. Commun.*, 1981, 336; T. Bolte, D. Yu, H-T. Stüwe and W. A. König, *Ang. Chem. Int. Edn. Engl.*, 1987, **26**, 331.
[14] R. B. Merrifield, *J. Am. Chem. Soc.*, 1963, **85**, 2149.
[15] J. P. Tam, S. B. Kent, M. Engelhard and R. B. Merrifield, *Synthesis*, 1979, 955.
[16] J. P. Tam, W. F. Heath and R. B. Merrifield, *J. Am. Chem. Soc.*, 1983, **105**, 6442.
[17] J. P. Tam, W. F. Heath and R. B. Merrifield, *J. Am. Chem. Soc.*, 1986, **108**, 5242.
[18] S. S. Wang, *J. Am. Chem. Soc.*, 1973, **95**, 1328.
[19] G. Breipohl, J. Knolle and W. Stüber, *Tetrahedron Lett.*, 1987, **28**, 5651; G. Breiphol, J. Knolle and W. Stüber, *Int. J. Pept. Protein Res.*, 1989, **34**, 262.
[20] R. C. Sheppard and B. J. Williams, *J. Chem. Soc., Chem. Commun.*, 1982, 587; *Int. J. Pept. Protein Res.*, 1982, **20**, 451; E. Atherton, M. Caviezel, H. Fox, D. Harkiss, H. Over and R. C. Sheppard, *J. Chem. Soc., Perkin Trans. 1*, 1983, 65; M. Mergler, R. Tanner, J. Costeli and P. Grogg, *Tetrahedron Lett.*, 1988, **29**, 4009.
[21] H. Kunz and B. Dombo, *Angew. Chem. Int. Edn. Engl.*, 1988, **27**, 711; G. Becker, H. Hguyen-Trong, C. Birr, B. Dombo and H. Kunz, in *Peptides 1988* (Eds G.

Jung and E. Bayer), De Gruyter, Berlin, 1989, p. 157; H. Kunz in *Innovation and Perspectives in Solid Phase Synthesis* (Ed. R. Epton), SPCC (UK) Ltd., Birmingham, 1990, p. 371; Hycram is a registered trade mark of Orpogen GmbH, Heidelberg, West Germany.

[22] E. Atherton, D. L. J. Clive and R. C. Sheppard, *J. Am. Chem. Soc.*, 1975, **97**, 6585; R. Arshady, E. Atherton, D. L. J. Clive and R. C. Sheppard, *J. Chem. Soc., Perkin Trans. 1*, 1981, 529.

[23] R. B. Merrifield and S. B. H. Kent, in *Peptides 1980* (Ed. K. Brunfeldt), Scriptor, Copenhagen, 1981, p. 328.

[24] K. K. Bhargava, V. K. Sarin, N. L. Trang, A. Cerami and R. B. Merrifield, *J. Am. Chem. Soc.*, 1983, **105**, 3247; D. H. Live and S. B. H. Kent, in *Peptides: Structure and Function* (Eds V. Hruby and D. Rich), Pierce Chemical Co., Rockford, Illinois, 1983, p. 65.

[25] V. K. Sarin, S. B. H. Kent and R. B. Merrifield, *J. Am. Chem. Soc.*, 1980, **102**, 5463; E. Girault, J. Rizo and E. Pedroso, *Tetrahedron*, 1984, **40**, 4141.

[26] R. B. Merrifield, in *Biopolymer Topics*, No. 3, Beckman Spinco Division, Palo Alto.

[27] R. Epton, D. A. Wellings and A. Williams, *Reactive Polymers*, 1987, **6**, 143.

[28] V. Krchnák, J. Vágner and O. Mach, *Int. J. Pept. Protein Res.*, 1989, **33**, 209.

[29] C. W. Smith, G. L. Stahl and R. Walter, *Int. J. Pept. Protein Res.*, 1979, **13**, 109.

[30] R. Epton, P. Goddard, G. Marr, J. V. McLaren and G. J. Morgan, *Polymer.*, 1979, **20**, 1444.

[31] E. Atherton, E. Brown, R. C. Sheppard and A. Rosevear, *J. Chem. Soc., Chem. Commun.*, 1981, 1151; A. Dryland and R. C. Sheppard, *J. Chem. Soc., Perkin Trans. 1*, 1986, 125.

[32] See page 45 of Ref. [4]; E. Bayer and W. Rapp, *Chem. Abstr.*, 1987, **106**, 50859q (Ger. Patent).

[33] R. Geiger and W. König, in *The Peptides* (Eds E. Gross and J. Meienhofer), Academic Press, New York, Vol. 3, Chapter 1, 1981, p. 1.

[34] L. A. Carpino, *J. Am. Chem. Soc.*, 1957, **79**, 4427; F. C. McKay, N. F. Albertson and A. C. McGregor, *ibid.*, 1957, **79**, 6180.

[35] (a) C. Birr, in *Aspects of the Merrifield Peptide Synthesis*, Springer, Heidelberg, 1978; (b) in Proceedings of an International Symposium *Innovation and Perspectives in Solid Phase Synthesis* (Ed. R. Epton), SPCC (UK) Ltd., Birmingham, 1990, p. 155; *Biochem. Soc. Trans.*, 1990., **18**, 1313.

[36] L. A. Carpino and G. Y. Han, *J. Org. Chem.*, 1972, **37**, 3404; C-D. Chang and J. Meienhofer, *Int. J. Pept. Protein Res.*, 1978, **11**, 246.

[37] E. Atherton, C. J. Logan and R. C. Sheppard, *J. Chem. Soc., Perkin Trans. 1*, 1981, 538.

[38] M. Bodanszky and J. Martinez, *Synthesis*, 1981, 333.

[39] P. Sorup, H. Braae, P. Villemoes and T. Christensen, *Acta Chem. Scand.*, 1979, **33B**, 653.

[40] R. W. Roeske, *J. Am. Chem. Soc.*, 1963, **28**, 1251.

[41] J. P. Tam, T.-W. Wong, M. W. Riemer, F.-S. Tjoeng and R. B. Merrifield, *Tetrahedron Lett.*, 1979, 4033.

[42] S-S. Wang, I. D. Kulesha and D. R. Winter, *J. Am. Chem. Soc.*, 1979, **101**, 253.

[43] T. Fujii and S. Sakakibara, *Bull. Chem. Soc. Japan*, 1974, **47**, 3146; J. M. van der Eijk, R. J. Nolte and J. W. Zwikker, *J. Org. Chem.*, 1980, **45**, 547.

[44] S. Shaltiel and M. Fridkin, *Biochemistry*, 1970, **9**, 5122; F. Chillemi and R. B. Merrifield, *ibid.*, 1969, **8**, 4344.

[45] J. Blake and C. H. Li, *Int. J. Pept. Protein Res.*, 1978, **11**, 315.

[46] P. Sieber and B. Riniker, *Tetrahedron Lett.*, 1987, **28**, 48.

[47] J. Ramachandran and C. H. Li, *J. Org. Chem.*, 1962, **27**, 4006.

[48] H. Yajima, ,M. Takeyama, J. Kanaki and K. Mitani, *J. Chem. Soc.*, *Chem. Commun.*, 1978, 482.

[49] E. Atherton, R. C., Sheppard and J. D. Wade, *J. Chem. Soc.*, *Chem. Commun.*, 1983, 1060; N. Fujii, A. Otaka, O. Ikemura, K. Akaji, Y. Funakoshi, Y. Hayashi, Y. Kuroda and H. Yajima, *J. Chem. Soc.*, *Chem. Commun.*, 1987, 274.

[50] R. Ramage and J. Green, *Tetrahedron Lett.*, 1987, **28**, 2287.

[51] B. W. Erickson and R. B. Merrifield, *J. Am. Chem. Soc.*, 1973, **95**, 3750; 3757.

[52] S. Akabori, S. Sakakibara, Y. Shimonishi and Y. Nobuhara, *Bull. Chem. Soc. Japan*, 1964, **37**, 433.

[53] O. Nishimura, C. Kitada and M. Fujino, *Chem. Pharm. Bull.*, 1978, **26**, 1576; J. J. Pastuszak and A. Chimiak, *J. Org. Chem.*, 1981, **46**, 1868; T. Johnson, R. C. Sheppard and R. Valerio, in *Peptides 1990* (Eds E. Giralt and D. Andreu), ESCOM Science Publishers, Leiden, 1991, p. 34.

[54] E. Atherton, R. C. Sheppard and P. Ward, *J. Chem. Soc.*, *Perkin Trans. 1.*, 1985, 2065; E. Atherton, M. Pinori and R. C. Sheppard, *ibid.*, 1985, 2057; D. F. Veber, J. D. Milkowski, S. L. Varga, R. G. Denkewalter and R. Hirschmann, *J. Am. Chem. Soc.*, 1977, **94**, 5456.

[55] S. Sakakibara, Y. Shimonishi, Y. Kishida, M. Okada and H. Sugihara, *Bull. Chem. Soc. Japan*, 1967, **40**, 2164; V. Caciagli, A. S. Verdini, S. Silvestri and A. Pessi, in *Peptides 1990* (Eds E. Giralt and D. Andreu), ESCOM Science Publishers, Leiden, 1991, p. 37.

[56] E. P. Heimer, C. D. Chang, T. Lambros and J. Meienhofer, *Int. J. Pept. Protein Res.*, 1981, **18**, 237.

[57] M. Bodanszky, in *The Peptides* (Eds E. Gross and J. Meienhofer) Academic Press, New York, Vol. 1, 1979, p. 105.

[58] L. Kisfaludy and I. Schon, *Synthesis*, 1983, 325; *ibid.*, 1986, 303.

[59] (a) E. Atherton, J. Holder, M. Meldal, R. C. Sheppard and R. M. Valerio, *J. Chem. Soc.*, *Perkin Trans. 1*, 1988, 2887; (b) W. König and R. Geiger, *Chem. Ber.*, 1970, **103**, 2024; (c) L. R. Cameron, J. L. Holder, M. Meldal and R. C. Sheppard, *J. Chem. Soc.*, *Perkin Trans. 1*, 1988, 2895.

[60] J. C. Sheehan and G. P. Hess, *J. Am. Chem. Soc.*, 1985, **77**, 1067.

[61] J. Blake and C. H. Li, *Int. J. Pept. Protein Res.*, 1975, **7**, 495; S. Mojsov, A. R. Mitchell and R. B. Merrifield, *J. Org. Chem.*, 1980, **45**, 555.

[62] D. Sarantakis, J. Teichman, E. L. Lien and R. L. Fenickel, *Biochem. Biophys. Res. Commun.*, 1976, **73**, 336.

[63] B. Castro, J. R. Domoy, G. Evin and C. Selve, *Tetrahedron Lett.*, 1975, **14**, 1219; P. Rivaille, J. P. Gautron, B. Castro and G. Milhauld, *Tetrahedron*, 1980, **36**, 3413.

[64] R. Knorr, A. Trzeciak, W. Bannwarth and D. Gillesen, *Tetrahedron Lett.*, 1989, **30**, 1927; in *Peptides 1990* (Eds E. Giralt and D. Andreu), ESCOM Science Publishers, Leiden, 1991, p. 62.

[65] R. B. Merrifield, *J. Am. Chem. Soc.*, 1963, **85**, 2149.

[66] S. B. H. Kent, L. E. Hood, H. Beilan, S. Meister and T. Geiser, in *Peptides 1984* (Ed. U. Ragnarsson), Almquist and Wiskell Int., Stockholm, 1984, p. 185.

[67] E. Kaiser, R. L. Colescott, C. D. Bossinger and P. I. Cook, *Analytical Biochem.*, 1970, **34**, 595; V. K. Sarin, S. B. H. Kent, J. P. Tam and R. B. Merrifield, *Anal. Biochem.*, 1981, **117**, 147.

[68] G. W. Tregear, *Biochemistry*, 1977, **16**, 2817.

[69] S. M. Meister and S. B. H. Kent, in *Peptides: Structure and Function* (Eds V. J. Hruby and D. H. Rich), Pierce Chemical Co., Rockford, Illinois, 1984, p. 103; E. Atherton, V. Wolley and R. C. Sheppard, *J. Chem. Soc., Chem. Commun.*, 1980, 970.

[70] G. Barany and R. B. Merrifield, in *The Peptides* (Eds E. Gross and J. Meienhofer), Academic Press, New York, 1979, Vol. 2, p. 159.

[71] A. Dryland and R. C. Sheppard, *Tetrahedron*, 1988, **44**, 843.

[72] Two representative commercial examples of the continuous-flow machines are Novabiochem's NovaSyn Crystal and the Milligen model 9050.

[73] R. C. Sheppard, *Chem. in Britain*, 1988, 557.

[73a] S. C. Young, P. D. White, J. W. Davies, D. E. I. A. Owen, S. A. Salisbury and E. J. Tremeer, *Biochem. Soc. Trans.*, 1990, **18**, 1311.

[73b] NovaSyn Gem and NovaSyn Crystal.

[74] J. E. Fox, *Biochem. Soc. Trans.*, 1990, **18**, 1308; J. E. Fox, R. Newton, P. Heegard and C. Shafer-Nielsen, in *Innovations and Perspectives in Solid Phase Peptide Synthesis* (Ed. R. Epton), SPCC (UK) Ltd., Birmingham, 1990, p. 141; information booklet supplied by Biotech Instruments Ltd., for its BT 7600 automatic peptide synthesizer.

[75] D. Hudson, *J. Org. Chem.*, 1988, **53**, 617.

[76] J. E. Rivier, *J. Chromatography* 1980, **202**, 211; P. Tempst, M. W. Hunkapiller and L. E. Hood, *Anal. Biochem.*, 1984, **137**, 188; *CRC Handbook of Hplc for Separation of Amino Acids, Peptides and Proteins*, (Ed. W. S. Hancock), Chemical Rubber Press, Boca Raton, Florida, 1984.

[77] Ref. [4], p. 157.

[78] A. Lifferth, G. Becker and C. Birr, in *Peptides 1988* (Eds E. Bayer and G. Jung), Walter de Gruyter, Berlin, 1989, p. 103.

[79] E. Bayer, W. Rapp and L. Zhang, in Ref. [78], p. 390.

[80] D. H. Spackman, W. H. Stein and S. Moore, *Anal. Chem.*, 1958, **30**, 1190; I. Betnér and P. Földi, *LC-GC International*, 1989, **2**, 44.

[81] R. M. Hewick, M. W. Hunkapiller, L. E. Hood and W. J. Dreyer, *J. Biol. Chem.*, 1981, **256**, 7990.

[82] M. Barber, R. S. Bordoli, R. D. Sedgwick and A. N. Tyler, *J. Chem. Soc., Chem. Commun.*, 1981, 325; D. H. Williams, C. Bradley, G. Bojesen, S., Santikarn and L. C. Taylor, *J. Am. Chem. Soc.*, 1981, **103**, 5700; K. Eckhart, H. Schwartz, K. B. Tomer and M. L. Gross, *J. Am. Chem. Soc.*, 1985, **107**, 6765.

[83] B. Sundqvist and R. D. Macfarlane, *Mass Spectrom. Rev.*, 1985, **4**, 421; R. J.

Cotter, *Anal. Chem.*, 1988, **60**, 781A; B. T. Chait, B. F. Gisin and F. H. Field, *J. Am. Chem. Soc.*, 1982, **104**, 5157; G. Lindeberg, A. Engstrom, A. G. Craig and H. Bennich, in Ref. [78], p. 121.

[84] S. Naylor, A. F. Findeis, B. W. Gibson and D. H. Williams, *J. Am. Chem. Soc.*, 1986, **108**, 6358.

[85] H. M. Geysen, R. H. Meloen and S. J. Barteling, *Proc. Natl. Acad. Sci.*, U.S.A., 1984, **81**, 3998; H. M. Geysen, S. J. Rodda, T. J. Mason, G. Tribbick and P. G. Schoofs, *J. Immunol. Methods*, 1987, **102**, 259.

[86] D. Hudson, in Ref. [78], p. 211.

[87] R. A. Houghten, *Proc. Natl. Acad. Sci.*, *U.S.A.*, 1985, **82**, 5131; R. A. Houghten and N. Lynam, in *Peptides 1988* (Eds E. Bayer and G. Jung), Walter de Gruyter, Berlin, 1989, p. 214; R. A. Houghten, J. H. Cuervo, J. M. Ostresh, M. K. Bray and N. D. Frizzell, in *Peptides: Proceedings of the 10th American Peptide Symposium* (Ed. G. R. Marshall), Escom. Sci., Netherlands, 1988, p. 166.

[88] R. H. Berg, K. Almdal, W. B. Pedersen, A. Holm, J. P. Tam and R. B. Merrifield, in *Peptides 1988* (Eds E. Bayer and G. Jung), Walter de Gruyter, Berlin, 1989, p. 198.

[89] J. Eichler, M. Beyermann, M. Biernert and M. Lebl, Ref. [88], p. 205.

[90] L. A. Carpino, B. J. Cohen, K. E. Stephens, S. Y. Sadat-Aalee, J. H. Tien and D. C. Langridge, *J. Org. Chem.*, 1986, **51**, 3732.

[91] A. Holm and M. Meldal, Ref. [88], p. 208.

[92] *Amino-acids and Peptides* (Ed. J. H. Jones), Specialist Periodical Reports, Royal Society of Chemistry, Vols 18–22, 1986–1990, Chapter 2.

[93] A. N. Eberle, E. Atherton, A. Dryland and R. C. Sheppard, *J. Chem. Soc., Perkin Trans. 1*, 1986, 361.

[94] J. P. Tam and R. B. Merrifield, *Int. J. Pept. Protein Res.*, 1985, **26**, 262; E. Brown, R. C. Sheppard and B. J. Williams, *J. Chem. Soc., Perkin Trans. 1*, 1983, 1161.

[95] I. Clark-Lewis, R. Aebersold, H. Ziltener, J. W. Schrader, L. K. Hood, and S. B. H. Kent, *Science*, 1986, **231**, 134.

[96] J. Blake and C. H. Li, *Proc. Natl. Acad. Sci.*, *U.S.A.*, 1983, **80**, 1556.

[97] P. W. Schiller, T. M. Nguyen and J. Mirre, *Int. J. Pept. Protein Res.*, 1985, **25**, 171; W. Stuber, J. Knolle and G. Breipohl, *Int. J. Pept. Protein Res.*, 1989, **34**, 215.

[98] M. Lebl and V. J. Hruby, *Tetrahedron Lett.*, 1984, **25**, 2067.

[99] M. J. Karten and J. Rivier, *Endocrine Reviews*, 1986, **7**, 44; G. Breipohl, J. Knolle and W. Stuber, *Int. J. Pept. Protein Res.*, 1989, **34**, 262; E. Pedroso, A. Grandas, M. A. Saralegui, E. Giralt, C. Granier and J. Van Rietschoten, *Tetrahedron*, 1982, **38**, 1183.

[100] W. R. Gray, F. A. Luque, R. Galyean, E. Atherton, R. C. Sheppard, B. L. Stone, A. Reyes, J. Alford, M. McIntosh, B. M. Olivera, L. J. Cruz and J. Rivier, *Biochemistry*, 1984, **23**, 2796.

[101] R. F. Nutt, D. F. Veber, P. E. Curley, R. Saperstein and R. Hirschmann, *Int. J. Pept, Protein Res.*, 1983, **21**, 66, and refs cited therein.

[102] E. Giralt, F. Albericio, E. Pedroso, C. Granier and J. Van Rietschoten, *Tetrahedron*, 1982, **38**, 1193; W. L. Cosand and R. B. Merrifield, *Proc. Natl. Acad. Sci.*, U.S.A., 1977, **74**, 2771.

[103] S. S. Wang, R. Makofske, A. Bach and R. B. Merrifield, *Int. J. Pept. Protein Res.*, 1980, **15**, 1; T. W. Wong and R. B. Merrifield, *Biochemistry*, 1980, **19**, 3233; S. S. Wang, S. T. Chen, K. T. Wang and R. B. Merrifield, *Int. J. Pept. Protein Res.*, 1987, **30**, 662.

[104] J. L. Krstenansky, D. Trivedi, D. Johnson and V. J. Hruby, *J. Am. Chem. Soc.*, 1986, **108**, 1696; S. Mojsov and R. B. Merrifield, *European J. Biochem.*, 1984, **145**, 601.

[105] E. Atherton, M. Caviezel, H. Fox, D. Harkiss, H. Over and R. C. Sheppard, *J. Chem. Soc., Perkin Trans. 1*, 1983, 65; C. H. Li, D. Yamashiro, L. F. Tseng and H. H. Loh, *J. Med. Chem.*, 1977, **20**, 325.

[106] B. Scolaro, L. Gozzini, R. Rocchi and C. Dibello, *Int. J. Pept. Protein Res.*, 1989, **34**, 423; J. E. Rivier and M. R. Brown, *Biochemistry*, 1978, **17**, 1766; C. Dibello, A. Lucchiari, O. Buso, R. De Castiglione and L. Gozzini, *Gazz. Chim. Ital.*, 1986, **116**, 221.

[107] C. G. Fields, G. B. Fields, R. L. Noble and T. A. Cross, *Int. J. Pept. Protein Res.*, 1989, **33**, 298; G. B. Fields, G. C. Fields, J. Petefish, H. E. Van Wart and T. A. Cross, *Proc. Natl. Acad. Sci.*, U.S.A., 1988, **85**, 1384.

[108] R. G. Almquist, S. R. Kadambi, D. M. Yasuda, F. L. Weitl, W. E. Polgar and L. R. Toll, *Int. J. Pept. Protein Res.*, 1989, **34**, 455.

[109] D. Le-Nguyen, D. Nalis and B. Castro, *Int. J. Pept. Protein Res.*, 1989, **34**, 492.

[110] T. A. Lyle, S. F. Brady, T. M. Ciccarone, C. D. Colton, W. J. Palevda, D. F. Veber and R. F. Nutt, *J. Org. Chem.*, 1987, **52**, 3752; G. Breipohl, J. Knolle and W. König, *Klin. Wochenschr.*, 1986, **64** (suppl. 6), 16.

[111] M. T. Tosteson, J. J. Levy, L. H. Caporale, M. Rosenblatt and D. C. Tosteson, *Biochemistry*, 1987, **26**, 6627.

[112] T. Kubiak, D. B. Whitney and R. B. Merrifield, *Biochemistry*, 1987, **26**, 7849.

[113] D. B. Scanlon, M. A. Eefting, C. J. Lloyd, A. W. Burgess and R. J. Simpson, *J. Chem. Soc., Chem. Commun.*, 1987, 516.

[114] T. Fairwell, A. V. Hospattankar, B. H. Brewer Jr. and S. A. Khan, *Proc. Natl. Acad. Sci.*, U.S.A., 1987, **84**, 4796.

[115] F. Baleux, B. Calas and J. Mery, *Int. J. Pept. Protein Res.*, 1986, **28**, 22.

[116] J. J. Blake, *Int. J. Pept. Protein Res.*, 1986, **27**, 191.

[117] J. E. Gairin, H. Mazarguil, P. Alvinerie, S. Saint-Pierre, C. Meunier and J. Cros, *J. Med. Chem.*, 1986, **29**, 1913.

[118] A. S. Dutta, J. J. Gormely, A. S. Graham, I. Briggs, J. W. Growcott and A. Jamieson, *J. Med. Chem.*, 1986, **29**, 1163; *ibid.*, 1986, **29**, 1171.

[119] J. Green, O. M. Ogunjobi and R. Ramage, in *Peptides 1988* (Eds E. Bayer and G. Jung), Walter de Gruyter, Berlin, 1989, 172; O. M. Ogunjobi and R. Ramage, *Biochem. Soc. Trans.*, 1990, **18**, 1322; J. P. Briand, J. Coste, A. Van Dorsselaer, B. Raboy, J. Neimark, B. Castro and S. Muller, in *Peptides 1990* (Eds E. Giralt and D. Andreu), ESCOM Science Publishers, Leiden, 1991, p. 80.

[120] J. Schneider and S. B. H. Kent, *Cell*, 1988, **54**, 363; R. F. Nutt, T. M. Ciccarone, S. F. Brady, P. L. Darke and D. F. Veber, in *Peptides: Chemistry, Structure and Biology* (Eds J. Rivier and G. R. Marshall), ESCOM Science Publishers, Leiden, 1990, p. 825.

[121] B. Penke and L. Nyerges, in *Peptides 1990* (Eds E. Giralt and D. Andreu), ESCOM Science Publishers, Leiden, 1991, p. 158.

[122] D. Alewood, J. L. Andrews and S. B. H. Kent, ref. [121], p. 167.

[123] W. Liu, G. H. Shine and J. P. Tam, in *Peptides: Chemistry, Structure and Biology* (Eds J. Rivier and G. R. Marshall), ESCOM Science Publishers, Leiden, 1990, p. 271.

8

Solid phase oligonucleotide synthesis

M. A. White

8.1 INTRODUCTION

The chemical synthesis of DNA molecules of defined sequence was a massive challenge for chemists and a greatly desired holy grail for molecular biologists in the sixties and seventies. The introduction of solid phase methods for the synthesis of oligodeoxyribonucleotides met the challenge and now the rapid and reliable synthesis of DNA fragments up to 150 nucleotides long is routinely available to molecular biologists.

The first synthesis of a dinucleotide was reported by Michelson and Todd in 1955 [1]. The culmination of the solution chemistry approach was in 1979 with Khorana's synthesis of a tyrosine transfer RNA gene (126 nucleotides long) using a combination of chemical and enzymatic methods [2]. Such solution synthesis was a heroic effort requiring massive amounts of labour to purify the product of each elongation. Michelson and Todd used a *phosphotriester* approach, the free OH group of the phosphodiester bond being masked with a protecting benzoyl group making a phosphotriester. Khorana, on the other hand, used *phosphodiester* chemistry, which produces an ionic product, allows side reactions and correspondingly gives poor yields when long chains are synthesized.

Merrifield's introduction of peptide synthesis on a solid support [3] soon led to the first report by Letsinger and Mahadevan in 1965 of an immobilized oligodeoxyribonucleotide synthesis [4]. The manifold advantages of a solid phase approach that obviated the need for intermediate purifications attracted great interest, but the requirement for a very high yield for the coupling step meant that satisfactory methods did not develop as rapidly as might have been expected [5]. Eventually, two satisfactory approaches were refined: the phosphotriester method [6] and the phosphoramidite version of the phosphite triester method [7]. Later, a third approach was developed that used hydrogen phosphonate synthons [8-10,10a].

8.2   APPLICATIONS OF OLIGONUCLEOTIDES

The term *oligonucleotide* is often used to mean specifically oligo**deoxyribo**nucleotide, but this usage should be avoided as methods for the synthesis of oligo**ribo**nucleotides are also now available. It is preferable to use the term oligonucleotide only where appropriate to both types of oligomer. Oligonucleotide lengths are usually given as '*x*-mer' (i.e. 17-mer, 83-mer). Quantities are often expressed as 'A_{260} units' where one unit is that amount of oligonucleotide that gives an absorbance of 1 at 260 nm (in a 1 cm cell) when dissolved in 1 ml.

The familiar purine to pyrimidine base pairings of the DNA double helix (adenine to thymine [uracil in RNA]; guanine to cytosine) are at the heart of most applications of oligonucleotides. Strong duplex formation between an oligonucleotide and a part of a DNA molecule requires complete complementarity: a single mismatch substantially weakens the association. This means that 'hybridization' conditions can be established so that only completely specific association is detected. Labelled (usually with ^{32}P) oligodeoxyribonucleotides are frequently used as probes to detect the presence of particular sequences in mixtures of DNA fragments separated by electrophoresis. A typical mammalian cell has roughly one metre of DNA (some 3×10^9 nucleotides) yet single copies of sequences only 17 nucleotides long can be detected. Such methods have a plethora of uses in molecular biology and increasingly in clinical diagnosis.

Oligodeoxyribonucleotides see applications in 'protein engineering' where oligodeoxyribonucleotides are used to replace particular sections of the DNA coding for the amino acid sequence of a protein. This allows production of proteins with designed amino acid changes. More routinely, oligonucleotides are used as primers in sequencing methods and in many other procedures in molecular biology. A particularly interesting application is in the polymerase chain reaction ('PCR'). Here, specific oligonucleotides select a region of the DNA and act as primer for the enzyme DNA polymerase, which uses the parent strand as a template to produce a copy. This process is repeated and produces enormous amplifications of the DNA sequence selected. More extensive discussions are to be found in many textbooks and more specialized sources [11–13].

8.3   GENERAL STRATEGY FOR OLIGONUCLEOTIDE SYNTHESIS

Examination of segments of DNA and RNA reveals the problems of oligonucleotide synthesis (Fig. 8.1). A new phosphodiester bond must be between the 5′ hydroxyl of the growing chain and the monomer being added, and not between two monomers (or to the 2′ hydroxyl of the ribose if RNA is being constructed). This requires blocking of these hydroxyls. The protecting group used to block the 5′ hydroxyl must be removed once the synthon has been added to the chain. Oligoribonucleotide synthesis requires additionally a 2′ blocking group that is stable under the conditions required for 5′ deprotection.

The hydroxyl group of the phosphodiester bond phosphorus atom must be protected lest chain cleavage occur during the synthesis, or other side reactions (possibly even chain branching) become problems. Finally, the exocyclic amino

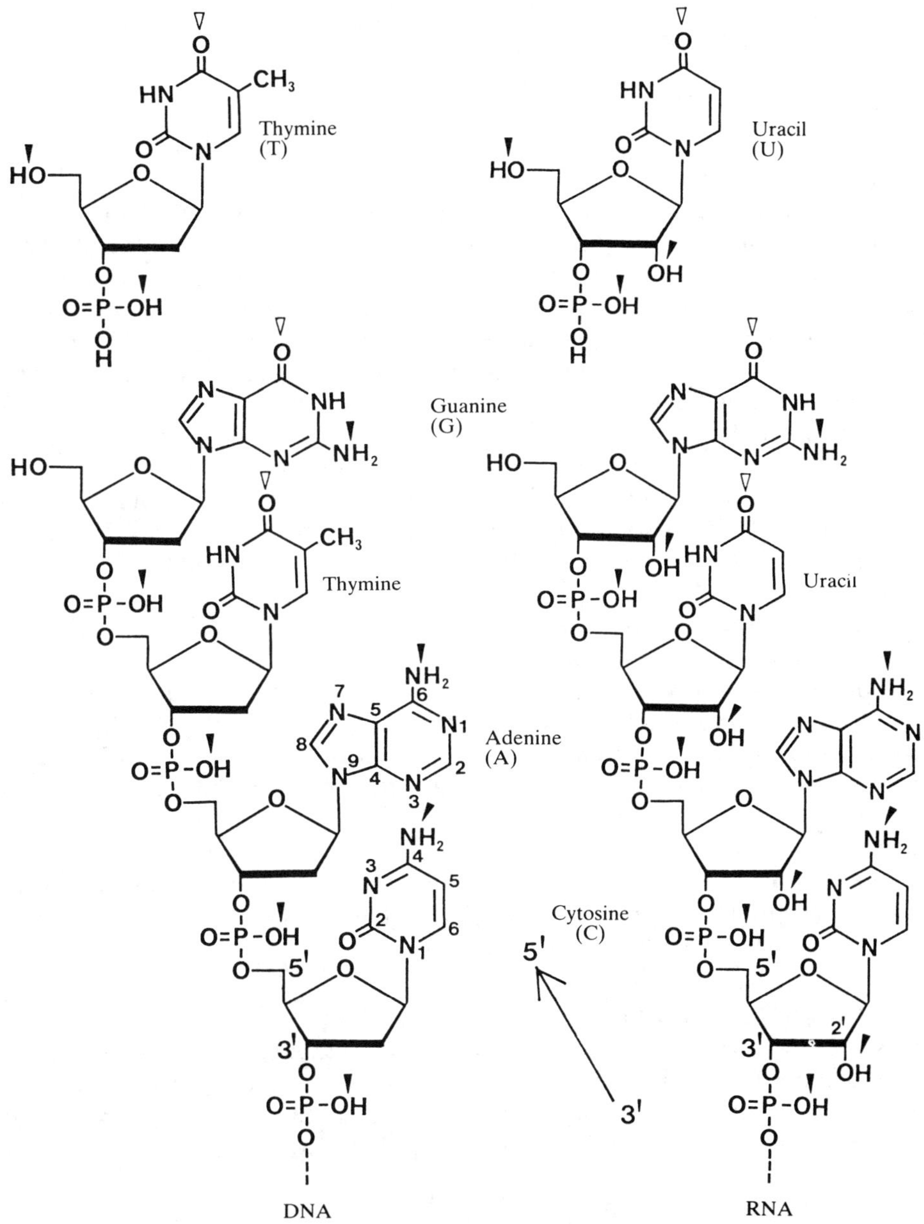

Fig. 8.1 — Segments of DNA and RNA molecules. The names of the bases and their standard abbreviations are shown. Groups requiring protection are arrowed (open arrows indicate groups *sometimes* protected). Note that in chemical synthesis the chain is elongated from 3′ to 5′ but in biological synthesis the direction is always from 5′ to 3′.

groups of the bases themselves require protection, as they would react with the activated monomer.

The various protecting groups have the additional advantage of rendering the synthons as well as the initial product soluble in the solvents required for phosphodiester bond formation (acetonitrile and/or pyridine).

8.3.1   Base protection

Virtually all of the methods for oligonucleotide syntheses use the same base protecting groups. Traditional protecting groups are the benzoyl group for adenine and cytosine and the isobutyryl group for guanine (Fig. 8.2) [5,14]. These amides are cleaved by treatment with concentrated ammonia solution at the end of the synthesis. Variant protecting groups have been introduced to allow more rapid deprotection under milder conditions [15–17]. Thymine and uracil are not usually protected [14] but O^4 may require blocking to avoid side reactions during a phosphotriester synthesis. The 4-nitrophenylsulphonylethyl group [18] or the phenyl group [19] have been employed. Additional protection of guanine is sometimes used in the phosphoramidite synthesis, as O^6 may be phosphitylated, leading to chain cleavage or the eventual introduction of 2,6-diaminopurine in place of the guanine. A β-cyanoethyl group or a p-nitrophenylethyl group can be used to prevent this side reaction [20–22].

8.3.2   Primary hydroxyl group protection

The 5′-hydroxyl must be unblocked at each cycle. The standard protecting group is the acid-labile 4,4′-dimethoxytrityl (dimethoxytriphenylmethyl) group (shown in Fig. 8.4) although the pixyl group (phenylxanthen-9-yl) (shown in Fig. 8.8) has some advantages [5,6]. Mild acid treatment (2 or 3% w/v dichloroacetic or trichloroacetic acid in dichloromethane or dichloroethane) rapidly removes these groups with the added benefit of giving a strongly coloured cation (orange for the dimethoxytrityl cation and virulent yellow-green for the pixyl cation) which may be used in quantitation. The yield of the last coupling step is reflected in the amount of protecting group released. Colour-coding of the different synthons becomes possible with the use of different triarylmethyl ethers [23].

8.3.3   Phosphodiester bond protection

The particular coupling chemistry determines the method of protection used for the phosphodiester bond. Chlorophenyl esters are usually employed in phosphotriester methods while the β-cyanoethyl group is used often in phosphoramidite and hydrogen phosphonate methods. These are discussed later.

8.3.4   Outline of solid phase synthesis

The basic pattern of a solid phase synthesis is simple. The first protected nucleoside is immobilized on a solid support such as silica gel or beads of porous glass through its 3′-hydroxyl group. The 5′-hydroxyl group is deprotected and reacts to form a phosphodiester bond with the next monomer to be added. The sequence of deprotect, couple; deprotect, couple is repeated until the chain is complete (Fig. 8.3). The product

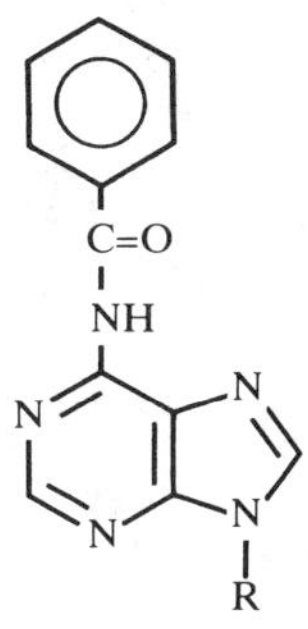

Fig. 8.2 — Protected base derivatives. N^6-Benzoyladenine (**1**), N^4-benzoylcytosine (**2**), and N^2-*iso*butyrylguanine (**3**) derivatives are standard. N^2-*iso*Butyryl-O^6-(4-nitrophenylethyl) guanine (**4**) and O^4-(4-nitrophenylsulphonylethyl)uracil (**5**) (or thymine) derivatives are sometimes employed. N^6-Di-*n*-butylaminomethyleneadenine (**6**) and N^6-phenoxyacetyladenine (**7**) derivatives are less prone to depurination under acid conditions than benzoyladenine derivatives. N^4-*iso*Butyrylcytosine (**8**), N^2-phenoxyacetylguanine (**9**), and phenoxyacetyladenine derivatives are deprotected more rapidly and under milder conditions.

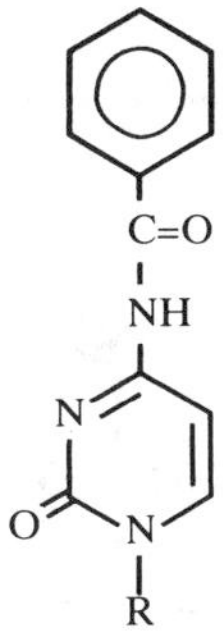

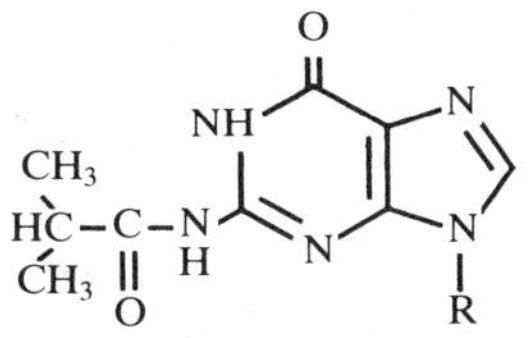

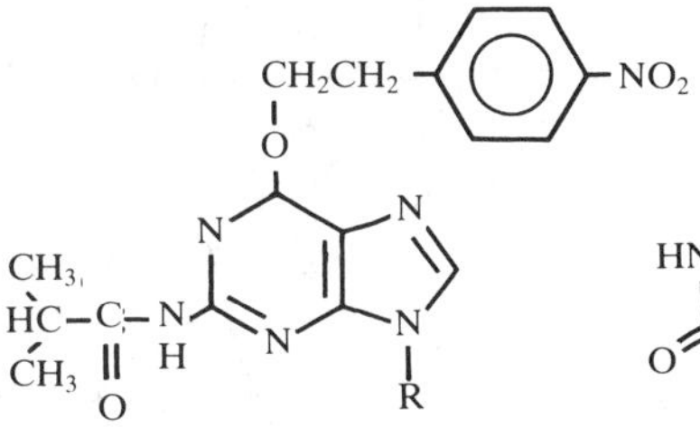

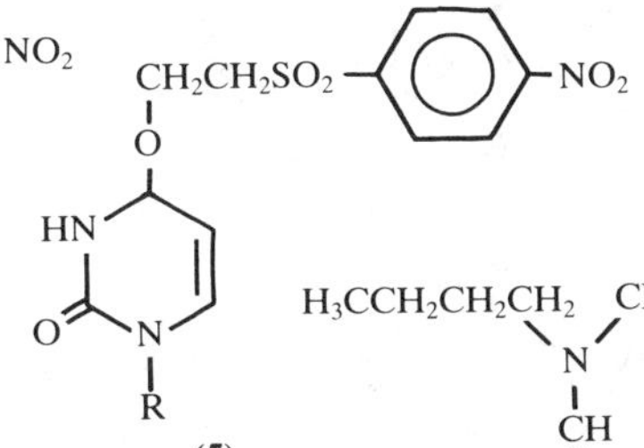

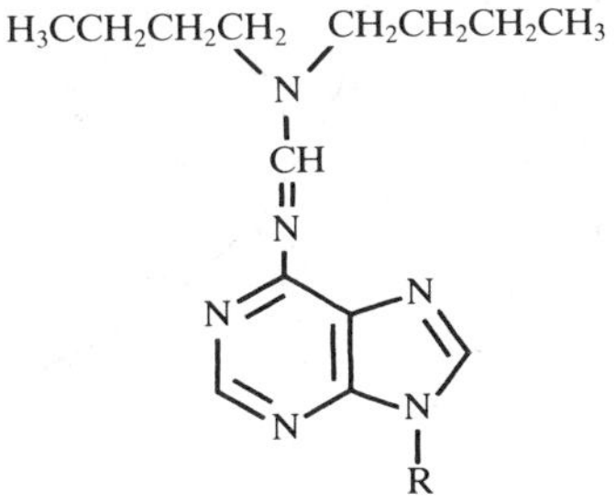

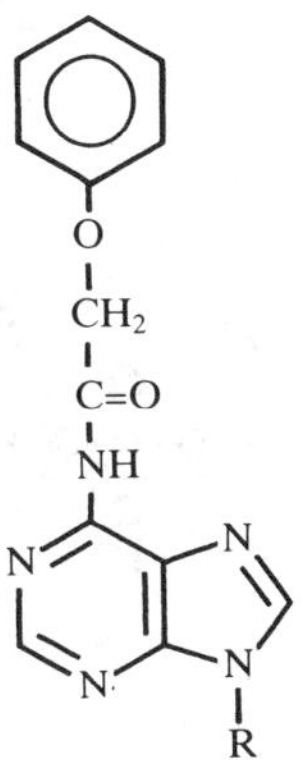

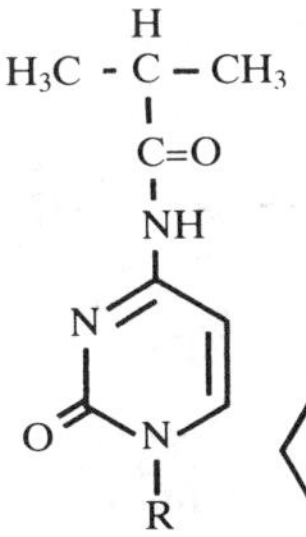

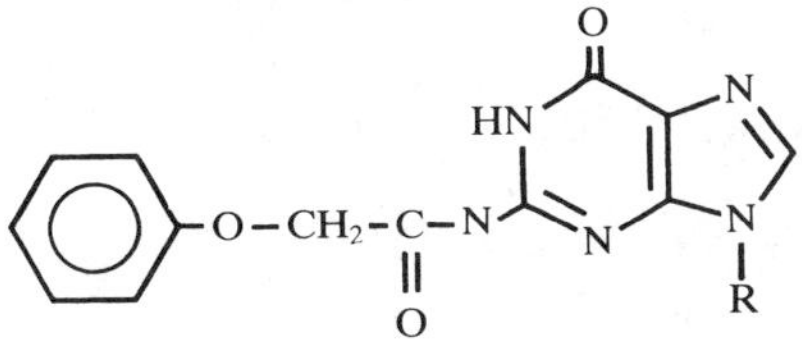

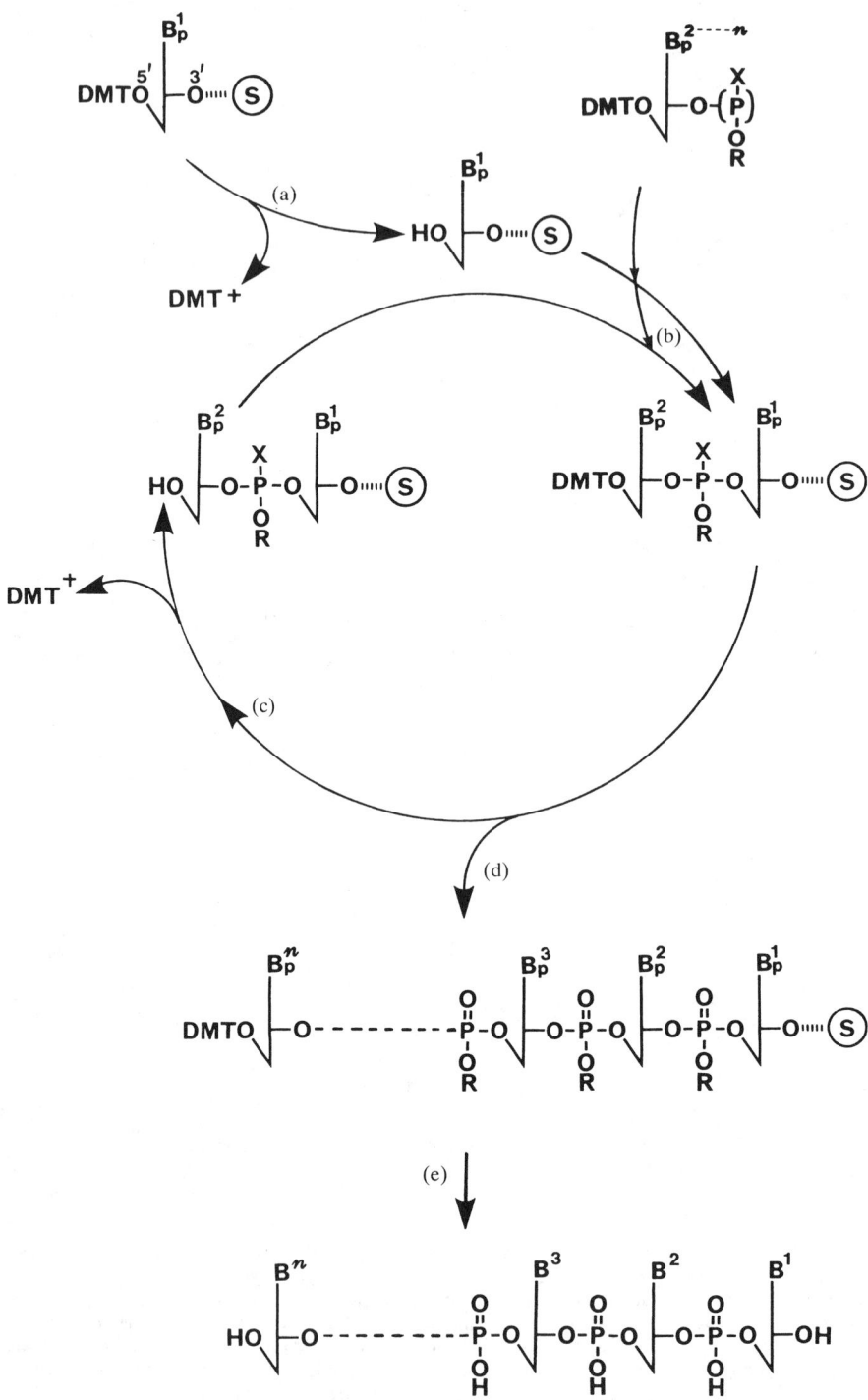

must be removed from the support, deprotected and purified. Synthesis always starts from the 3' end, as the coupling reactions used are all faster with the primary hydroxyl group than with a secondary hydroxyl group.

All of the coupling reactions in use are extremely sensitive to water, and great care must be taken to exclude water from the reagents and solvents. Although a great boon, solid phase methods do suffer from the possible accumulation of side products, which would normally be removed during a solution phase synthesis. In a solid phase synthesis, reagents are used in considerable excess (often 15- to 20-fold or greater) and the growing chain is exposed to the reagents many times as the synthesis proceeds, as well as to considerable volumes of solvents. It is therefore critical that all of the reagents be of the highest purity possible to avoid cumulative effects from impurities.

Any repetitive operation depends for its success on a high yield at each step. The length of chain that can be synthesized depends on the yield of the coupling step. The overall yield of such a process is y^n where y is the yield of the coupling step. If the coupling step gave an 80% yield, the overall yield would be 11% after 10 couplings. At a coupling step yield of 99%, the overall yield is 36% after 100 couplings. An overrall yield much below 5% may not be viable; although often only minute quantities of product are required, the product may simply vanish in the 'noise' of the workup and purification. Therefore, most of the developments in solid phase oligonucleotide synthesis have aimed at improving the yield of the coupling process. As a result, oligomers longer than 100 nucleotides are now routinely synthesized.

The apparatus for a manual solid phase oligonucleotide synthesis is basically simple. Some workers have used small funnels to retain the particles of support [7] and others use a small chromatography column for this purpose. A synthesis on the scale of 1 μmol using controlled pore glass beads as a support would have roughly 150 μl of the support. If a small column is used, it is convenient to make it semi-automatic by using a rotary valve which allows selection of the different solvents and reagents required during the cycle. The addition of each activated nucleotide can be done by injection [6]. Dry nitrogen or argon is used to exclude moisture and to drive solvent flow in flow systems. Automated systems use the same principle with the addition of automated mixing and addition of the coupling agent and synthon (see Efcavitch [24] for a discussion of automation). Additional valves may be added to permit operation of two or more columns.

Fig. 8.3 — Generalized oligonucleotide synthesis. DMT = the dimethoxytrityl group (see Fig. 8.4); B = base; B_p = protected base; Ⓢ = solid support. o⌐o is the standard abbreviation for 2'-deoxyribose.

The particular synthetic approach determines the nature of X and R on the synthon. In phosphotriester chemistry, X is = O, R is 2-chlorophenyl, and there is a third oxygen. For a phosphoramidite synthesis, X is N,N-di-iso-propylamino and R is β-cyanoethyl. A hydrogen phosphonate synthesis again has X as = O, R is H and there is an additional hydrogen.
(a) The initial 5' deprotection of the immobilized first nucleoside. (b) The COUPLING step. (c) 5'-Deprotection as part of the cycle. CAPPING may be introduced. (d) Oxidation may be part of the cycle (phosphoramidite), at the end of the synthesis (H-phosphonate), or not required (phosphotriester). (e) The final WORKUP. This involves removal of phosphate protecting groups (not required in H-phosphonate methods), cleavage from the support, deprotection of the bases and removal of the 5'-terminal protecting group.

8.4 PHOSPHOTRIESTER SYNTHESIS

Although the phosphotriester method is not the most common approach to oligonucleotide synthesis it has the advantage of simplicity, which makes it a good introduction to synthetic methods and perhaps the easiest to set up in the laboratory on a modest scale. The synthons are more readily prepared than those required for the other methods, which often makes it the method of choice when an unusual or modified base is to be introduced [25, 26]. It is possible to add a single nucleotide to a chain using phosphotriester chemistry even if phosphoramidites are used for the rest of the synthesis [27].

The most commonly used phosphotriester method is that described by Sproat and Gait [6] (Fig. 8.4). The synthons are 5'-O-dimethoxytrityl or pixyl ethers of base-protected 2-chlorophenyl phosphates in the triethylammonium salt form. Dry pyridine is used as solvent.

The coupling reagent is 1-mesitylenesulphonyl-3-nitro-1,2,4-triazole ('MSNT') (Fig. 8.5), which is thought to bring about the formation of a dinucleoside pyrophosphate intermediate (10) [28,29]. Coupling times are long (about one hour), which is a major disadvantage when long chains are being made, but the addition of a catalyst such as N-methylimidazole reduces the time to 5 to 15 min by forming a more reactive intermediate (11). Times may be further reduced using an 'intensive' phosphotriester approach at elevated temperatures [30,31]. Sproat et al. introduced intramolecular catalysis by synthesizing synthons (12) with 2-(1-methylimidazol-2-yl)phenyl groups on the phosphate in place of the usual chlorophenyl group [32]. MSNT leads to some side reactions such as sulphonation of O^6 of guanine. These are avoided in a novel method introduced by van Boom and coworkers where the active nucleotide synthon is made during the reaction from the free nucleoside (Fig. 8.6) [33].

Whichever synthon is used the method is basically the same. The first nucleoside is immobilized via a spacer arm to a support (usually beads of controlled pore glass). The nucleoside is tethered through its 3'-hydroxyl group and the synthesis proceeds from the 3' to the 5' end. The dimethoxytrityl group is removed by treatment with 3% dichloroacetic acid in dichloroethane for 30 to 60 s. The first addition is accomplished (after flushing with dichloroethane and then dry pyridine) by introduction of a mixture of the appropriate synthon in 15- to 20-fold molar excess over the amount of immobilized nucleoside (synthesis is usually on a 0.1 to 1 μmol scale), with the coupling agent in five fold excess over the synthon and if appropriate a catalytic amount of N-methylimidazole. Reagents are flushed out with pyridine, the solvent is changed to dichloroethane and an acetylating mixture of acetic anhydride and 4-dimethylaminopyridine in acetonitrile is introduced. This 'capping' step blocks the 5'-hydroxyl groups of any chains that have failed to couple. Opinions differ as to its utility: Sproat and Gait omit it [6] while others feel it is essential [5]. After the capping step, the reagents are flushed out and the newly added nucleotide is deprotected with dichloroacetic acid. This cycle is continued until the chain is complete.

The product is a fully protected oligomer still bound to the beads of support. The 5'-terminal dimethoxytrityl group is left on the chain during the workup. It may be

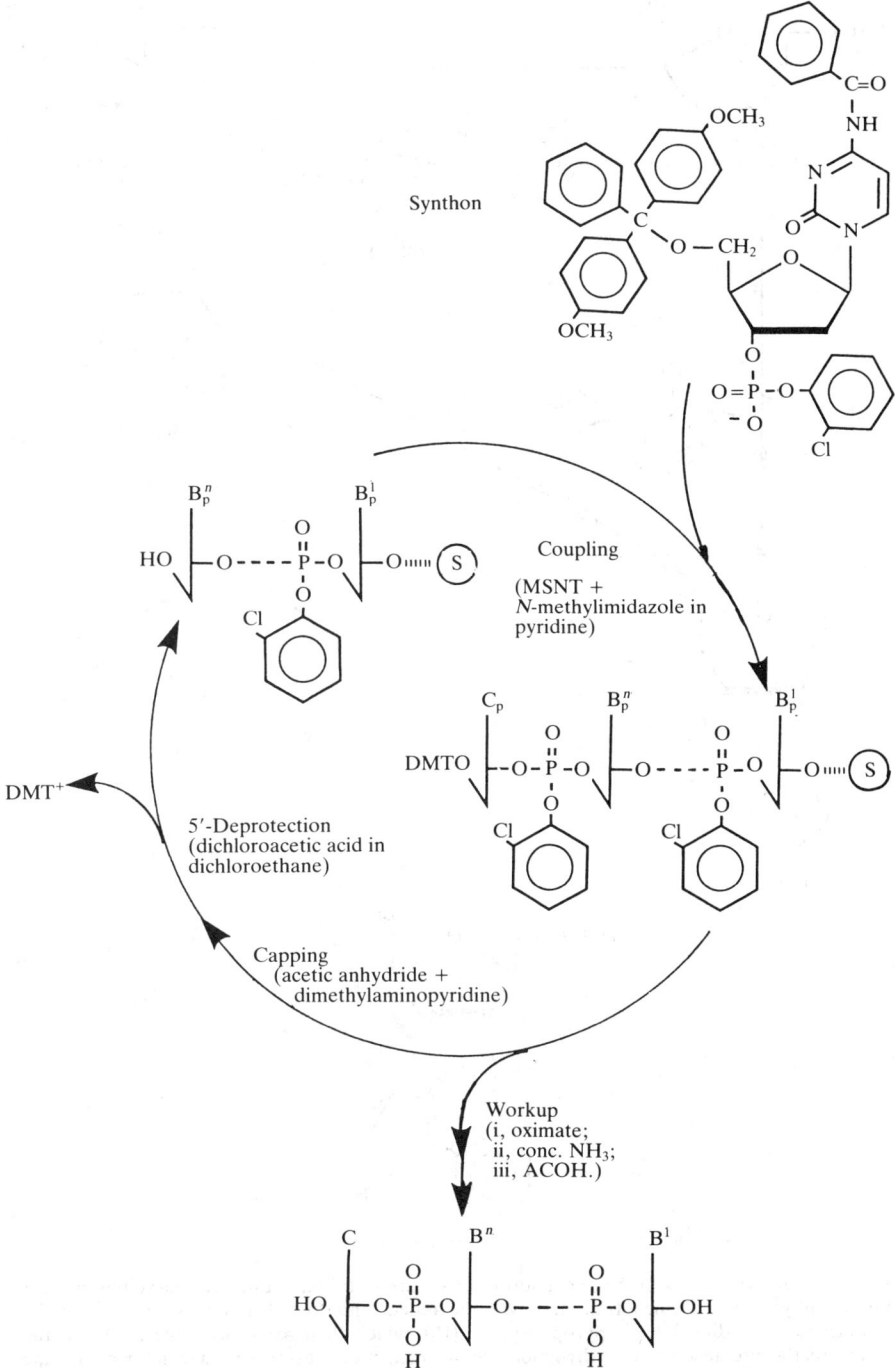

Fig. 8.4 — A phosphotriester synthesis cycle. The coupling agent is 1-mesitylenesulphonyl-3-nitro-1,2,4-triazole ('MSNT') with N-methylimidazole as a catalyst.

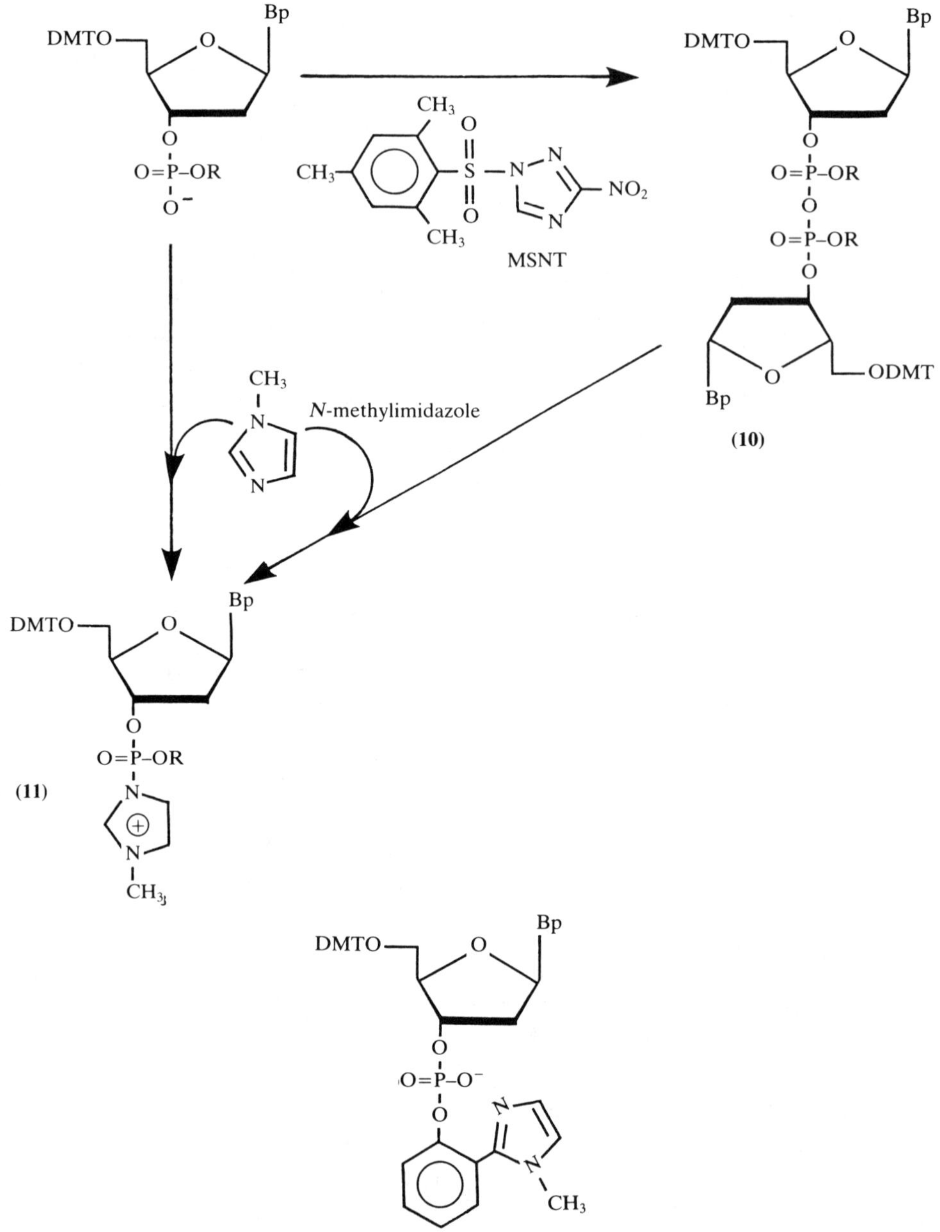

Fig. 8.5 — Intermediates in the phosphotriester synthesis. Initially a nucleotide–sulphonic acid mixed anhydride is formed from the synthon added, and this then reacts to produce the symmetrical 3',3'-dinucleoside pyrophosphate (10), which then reacts with the 5'-OH of the chain. In the presence of N-methylimidazole, a much more reactive imidazolium–nucleotide intermediate (11) is formed. The synthon (12) is an alternative which gives *intra*molecular catalysis.

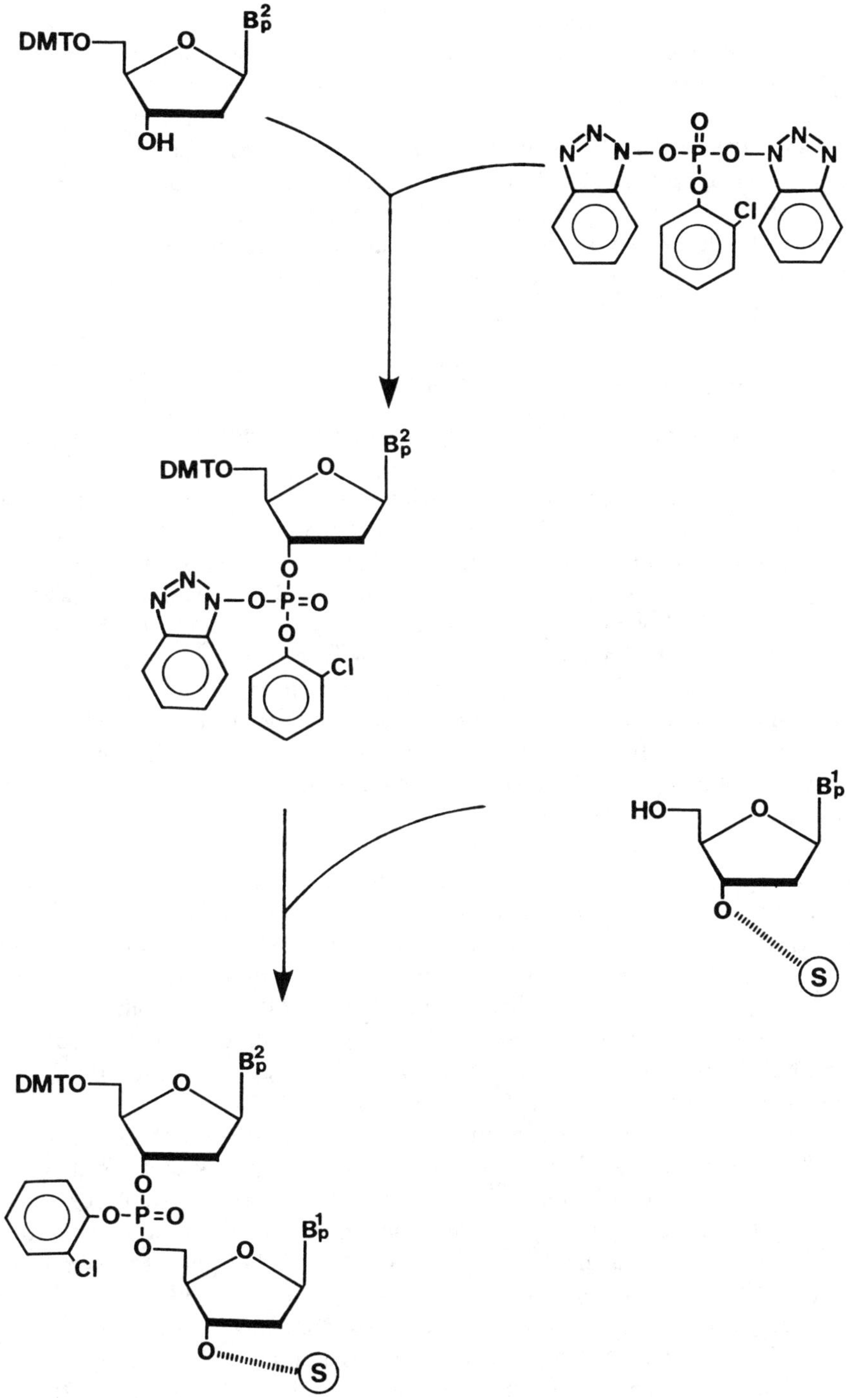

Fig. 8.6 — Hydroxybenzotriazole method for *in situ* formation of the reactive nucleotide.

removed as a final step of the workup or left on as an aid in reversed-phase hplc purification. Treatment with a basic solution of oximate anion both cleaves the succinyl ester link to the support and removes the chlorophenyl groups from the phosphates. Bases are deprotected by heating to 60° in concentrated aqueous ammonia for at least 6 h. The isobutyryl group used for guanine protection is more resistant to cleavage and some persists even after 17 h. Furthermore, the concentrated ammonia ($\simeq 35\%$ w/w) may alter unusual or modified bases incorporated into the chain. More labile protecting groups which are removed at room temperature, in shorter times and at 25 to 29% ammonia have been introduced [17,34].

8.5 PHOSPHORAMIDITE SYNTHESIS

The long coupling times required in the phosphotriester synthesis in its early stages of development led workers to investigate the use of the much more reactive phosphorus(III) compounds, and many phosphite triester methods were introduced [35]. In such methods, the newly formed phosphite triester bond is oxidized to a phosphate triester after each coupling. These systems suffered from the disadvantage of having difficult to prepare and somewhat unstable synthons. The introduction of the more stable N,N-dialkyl phosphoramidite synthons [36] solved these problems and allowed the rapid development of the phosphoramidite synthesis as the standard method which is used on all of the automated synthesizers on the market. The most commonly used phosphoramidites are β-cyanoethyl-N,N-di-isopropyl phosphoramidites (Fig. 8.7) where the β-cyanoethyl group serves to protect the hydroxyl of the phosphite and phosphate triester bonds. Phosphoramidite techniques have coupling times on the order of 3 to 5 min.

A weak acid such as tetrazole is used to produce a reactive tetrazole–phosphite intermediate [37]: the two components in acetonitrile are mixed immediately before use. Oxidation of the somewhat unstable phosphite triester bond is carried out with iodine (typically 0.2 mol 1^{-1} I_2 in 2:2:1 tetrahydrofuran: 2,6-lutidine: water) [35]. There is some phosphitylation of O^6 of guanine if this is unprotected. This can lead to phosphityl migration to the N^7 position during the oxidation, which in turn causes some loss of guanine (depurination) and consequent chain cleavage. The dimethylaminopyridine/acetic anhydride capping mixture will reverse most of the guanine phosphitylation if it is carried out *before* oxidation [20,21].

The synthesis cycle is generally similar to that described for the phosphotriester method. The first nucleoside is immobilized on a solid support such as controlled pore glass, deprotected with dichloroacetic acid in dichloromethane and treated with a mixture of tetrazole and the appropriate phosphoramidite synthon in acetonitrile. The phosphoramidite is usually in 20-fold excess over the quantity of immobilized nucleoside and the tetrazole in 80-fold excess. After excess reagents have been washed out with acetonitrile, chains that did not elongate are capped with acetic anhydride/ dimethylaminopyridine. Oxidation with aqueous iodine follows capping. The reagents are flushed out, the bed of support is washed with dichloromethane, the new 5'-hydroxyl group is deprotected with dichloroacetic acid and the cycle is then repeated.

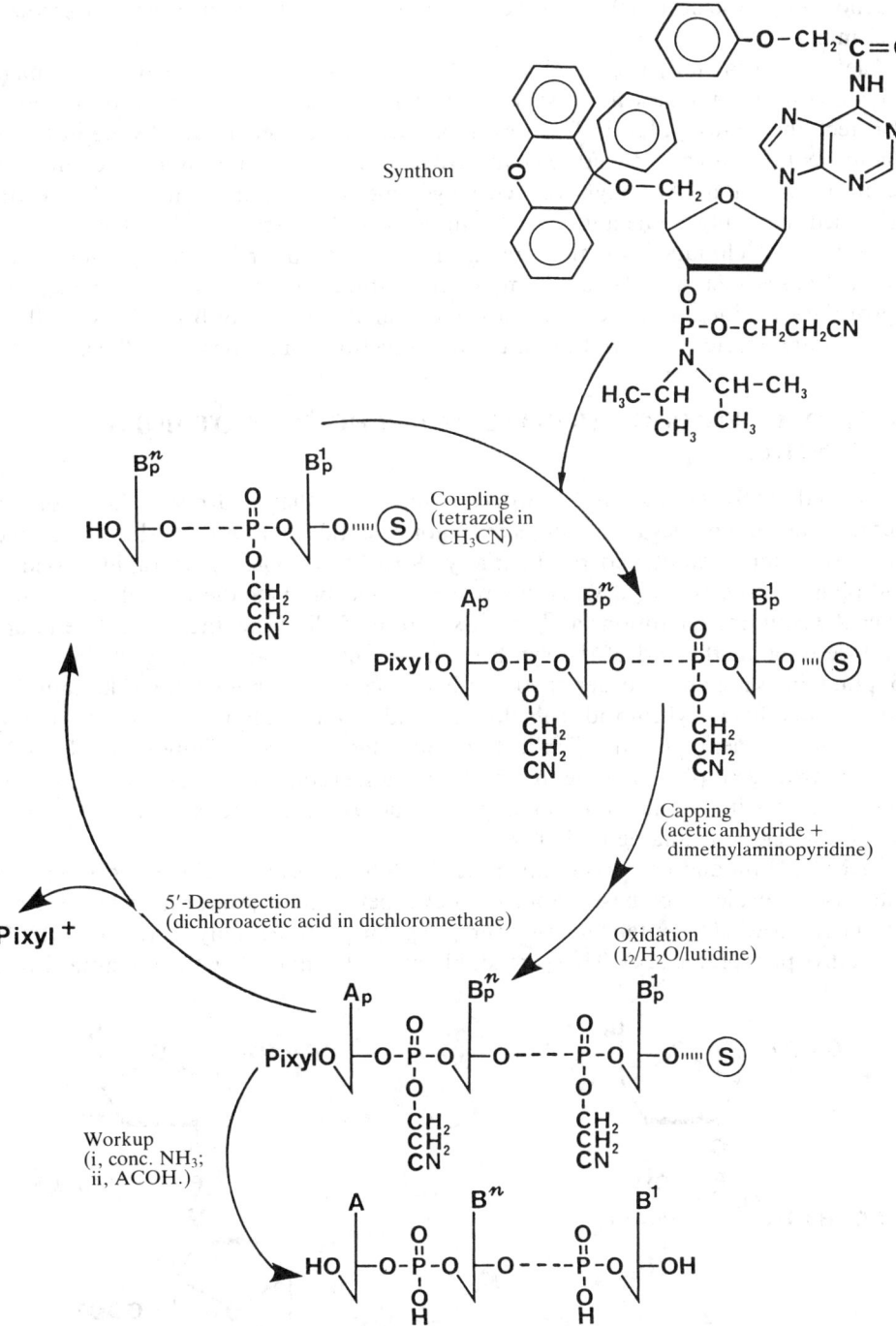

Fig. 8.7 — A phosphoramidite synthesis cycle. The coupling agent is tetrazole. The synthon is shown with 'pixyl' (9-phenylxanthen-9-yl) 5′-protection.

Depending upon scale and flow rates of reagents, the time for a full cycle varies from 7 to 17 min.

Cleavage of the fully protected product from the support is effected by treatment with aqueous ammonia, which also serves to remove the β-cyanoethyl groups and to deprotect the bases [14,38]. The result of this single step is a fully deprotected oligonucleotide with a 5'-dimethoxytrityl group, which can be removed or retained for use in purification. Although the cleavage and deprotection step is frequently performed manually, some automated synthesizers also carry out this step.

Synthesis techniques for the preparation of phosphoramidite synthons with modified bases such as O^4-alkylthymine and 6-thioguanine are now available. The appropriate position is protected with a 4-triazolo group. Substitutions at these positions are effected by variation of the deprotection conditions [38a,38b].

8.6 HYDROGEN PHOSPHONATE AND 'PHOSPHOROTHIOITE' SYNTHESES

In the mid-1980s, Garegg and coworkers suggested that hydrogen phosphonates offered some advantages as synthons in oligonucleotide synthesis, as they are stable, relatively water tolerant and react rapidly [9,10,39,40]. They were rapidly used in solid phase synthesis (Fig. 8.8), as the hydrogen phosphonate diester links are stable under the coupling conditions and thus oxidation of all of the internucleotide bonds could be done at the end of the synthesis with a considerable saving of time [41]. Coupling times are on the order of 90 s with a cycle time of about 4 min. The coupling agent is trimethylacetyl chloride (pivaloyl chloride). Some capping may be effected by the coupling reagent, but later publications introduce a normal capping step [39,40].

The hydrogen phosphonate method has undergone rapid development and improvement with applications in oligoribonucleotide synthesis and synthesis of deoxyribonucleotide analogues [41–43].

In 1987, Hata and coworkers introduced a 'phosphorothioite' method [44]. The synthons were said to be base-protected β-cyanoethyl 3'-phosphorothioites (13) but were later shown to contain no sulphur [45]. They are actually bis(nucleoside) 2-cyanoethyl phosphites (14). They are stable at $-30°$ and offer the advantage of a

(13)

(14)

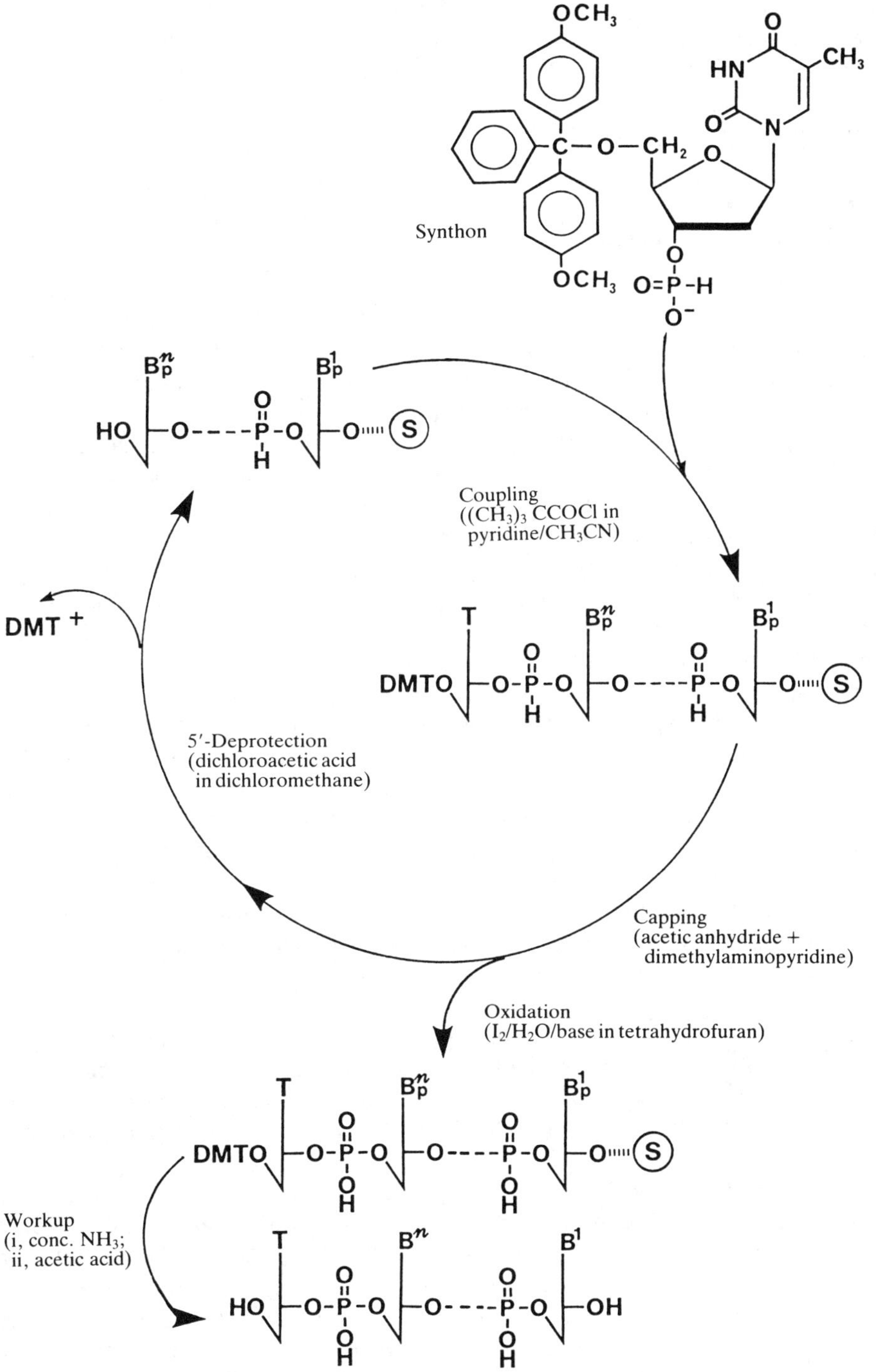

Fig. 8.8 — A hydrogen phosphonate synthesis cycle.

single step condensation and oxidation. The synthesis cycle is otherwise similar to the more established methods. Syntheses on polystyrene and controlled pore glass have been reported, the latter giving 30 s coupling times and overall yields better than 95%. Syntheses of longer oligodeoxyribonucleotides give poor yields and thus cannot compete with the other methods.

8.7   THE SOLID PHASE

8.7.1   The support base

The material selected for the support must be compatible with the reagents and solvents employed, offer a reasonable surface area for chain attachment, offer sites for convenient attachment of the first nucleoside, allow good flow rates and mass transfer for solvents and reagents, and have good mechanical properties. In addition, swelling or shrinkage when solvents are changed and as the chains grow is a disadvantage.

Polystyrene cross-linked with divinylbenzene (in the form of 'popcorn' polystyrene) was the first support used [4] and has often been used since [30,46] usually via aminoethylpolystyrene [47]. Polystyrene swells in some solvents and a rather unusual support consisting of polystyrene 'grafted' onto polytetrafluoroethylene has sometimes been employed to minimize swelling and to provide good mass exchange [48]. Perhaps because the preparation of this support requires gamma irradiation of the PTFE in the presence of styrene vapour, it has been little used.

Polysaccharides have been used in the form of the cross-linked dextran Sephadex® LH-20 (substituted with hydroxypropyl groups) [49] and cellulose [50].

Resins such as polyacrylmorpholide [51] and polydimethylacrylamide [52] have had their adherents but tend to swell in some solvents. A useful variant is polydimethylacrylamide polymerized into Kieselguhr [53]. This has mechanical strength and allows good flow rates although it does undergo size changes when solvents are changed and swells as the chains elongate. A further disadvantage is its relative basicity, which requires more concentrated acid (10% trichloroacetic) for deprotection and a subsequent washing with N,N-dimethylformamide to fully remove the acid [6,54]. It has some advantages in larger-scale syntheses of chains shorter than 25 nucleotides [55].

Silica gel is an obvious choice of support, and many groups have used it. The development of silica gel hplc media has provided a support with excellent characteristics [56]. However, there have been hints of steric effects causing problems when chains of 51 nucleotides were synthesized, which does not occur with controlled pore glass [57].

Glass is another logical support, but simple glass beads provide too little surface area to be viable [58]. Porous glass, however, has proved to be almost ideal. Gough *et al.* [59] appear to be the first to employ controlled pore glass as a support. Usually, particles of 100–200 μm diameter with a pore diameter of 50 nm are used [5,6]. Perhaps the only drawback to controlled pore glass is the relatively low loading possible (10–30 μmol g^{-1}, compared to loadings of 90–180 for polydimethylacrylamide–Kieselguhr [6] and around 200 μmol g^{-1} for silica gel [5]). Controlled pore glass is now the standard support, with the others employed only where a requirement of high loadings applies.

A rather unusual method uses cellulose acetate as a 'support' [60]. Cellulose acetate is coupled to the first nucleoside with A 4-(2-hydroxyethyl)dihydrocinnamoyl) spacer. The 'support' is soluble in pyridine but insoluble in ethanol. The authors use a homogeneous phase phosphodiester coupling and then precipitate by addition of ethanol. It remains to be seen if this periodically solid phase synthesis will be other than a novelty.

8.7.2   Attachment of the chain to the support

The first nucleoside is usually tied to the support by a succinyl ester, the other end of which forms an amide link with a complex 'long chain alkylamine' (lcaa), which is tied to the support (15). The succinate link may be cleaved by traces of basic impurities in the solvents used, but a more stable urethane variant (16) can be used, though removal of the chain from the support is more difficult [54].The more labile oxalyl link offers the advantage of rapid cleavage at the end of a synthesis under conditions that leave protected modified bases unaltered [60a]. These links suffer from the disadvantage that a different functionalized support is needed for each of the four starting nucleosides. After the synthesis, the product has a free 3'-hydroxyl group after cleavage of the succinyl or urethane link.

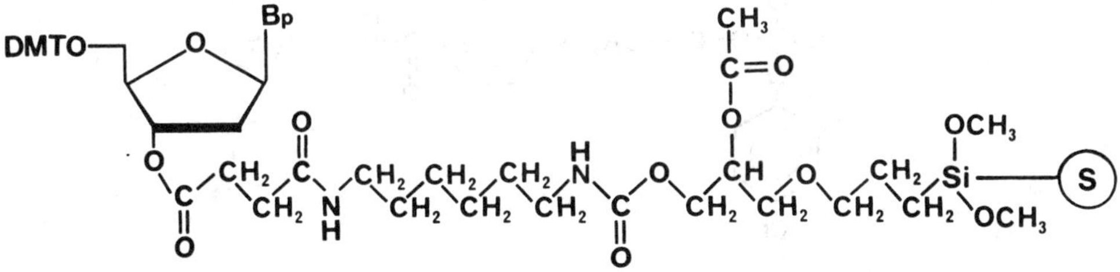

(15)

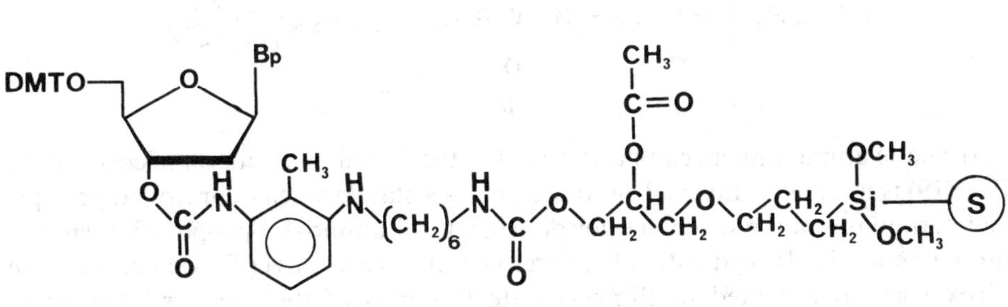

(16)

The ideas of a 'universal' support suitable for chains starting with any nucleoside was first developed by those who used the ribonucleotide uridine as a link point [50,61]. The 2'- (or 3') hydroxyl group is left free and the other hydroxyl group is protected (acetyl or benzoyl). The oligodeoxyribonucleotide is constructed on the free hydroxyl group. After the synthesis, the second hydroxyl group is unblocked under alkaline conditions, which also removes the uridine as its 2′,3′-cyclic phosphate, thereby freeing the oligodeoxynucleotide product with a free 3′-hydroxyl group.

The increasing use of oligonucleotide probes in clinical laboratories has made it desirable to develop non-radioactive label systems to circumvent the short half-life and hazards of ^{32}P. This need has led to development of methods that incorporate a 'linker' group at the 3′-end between the chain and the glass. Phosphate, thiol and amino groups have all been introduced at the 3′-position.

'CAMET' [2-(4-carboxyphenylmercapto)ethanol] linked compounds (17) are stable until the sulphide is oxidized to a sulphone or sulphoxide [62]. Once oxidized, the link is base labile and is readily cleaved to afford an oligomer with a 3′-phosphate. Terminal phosphates are also the products following use of a universal support (18) produced from substituted controlled pore glass which has been extended by reaction with methacrylic acid derivatives and β-mercaptoethanol [63]. The spacer terminates in an − OH group, which can be condensed with the synthon in a standard coupling reaction.

(17)

(18)

Aliphatic amino linkers can be attached at the 3′-end of an oligonucleotide if 'MF-CPG' (19) is used as a universal support. This is a standard lcaa-controlled pore glass, which terminates in the succinyl ester of protected 3-amino-1,2-propanediol [64]. The amino function is blocked with a 9-fluorenylmethoxycarbonyl ('Fmoc') group. The 2-hydroxyl group is linked to the succinate terminus of the lcaa, and the primary hydroxyl group is blocked with a dimethoxytrityl group. An initial acid deprotection frees the 1-hydroxyl group for coupling with the first nucleotide synthon. The final ammonia deprotection gives an oligonucleotide with a pendant aminohydroxypropyl group available for further derivatization. Asseline and Thuong [65] have also produced amine-terminated, modified oligonucleotides with a universal support (20) based on silica gel. The support (21) allows the incorporation of a similar linker terminating in a thiol group [66].

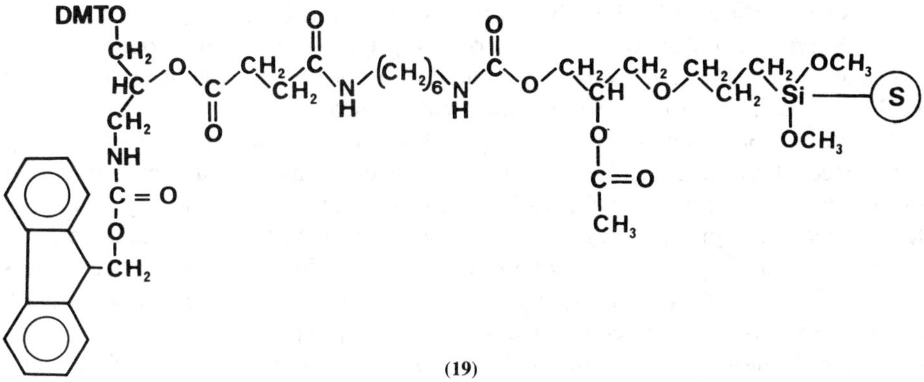

(19)

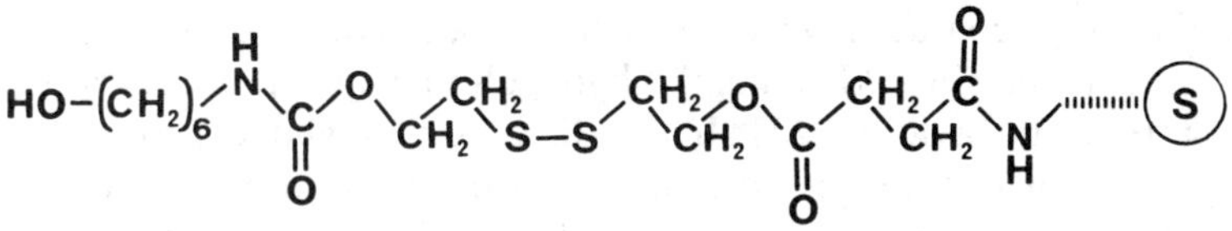

(20)

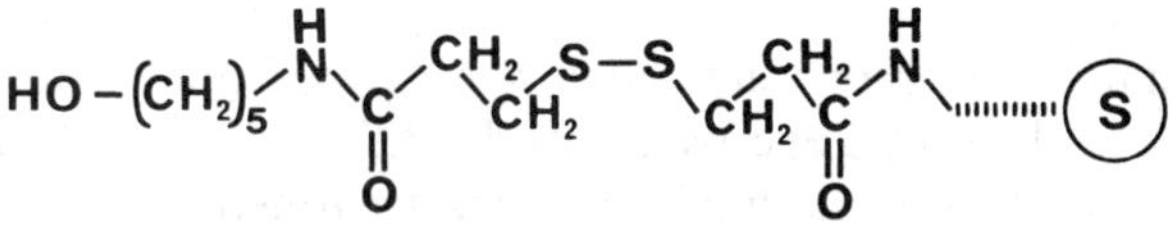

(21)

8.8 OLIGONUCLEOTIDE PURIFICATION

The protocol employed to purify the product oligonucleotide depends on the particular application. Often, molecular biologists are able to utilize minute amounts of the desired oligonucleotide without significant purification, as the complementarity allows selection and amplification. The makers of some synthesizers claim that the product may be used directly. Other molecular biological applications, chemical applications and structural studies require much higher degrees of purity.

Depending upon the base sequence and length, the synthetic method used, the exact protocol and the purity of reagents and solvents, the product oligomer will be contaminated with different amounts of chains with variously modified bases and with shorter chains, the latter resulting from failure to couple and capping as well as cleavage during the final workup. It is possible to remove uncapped failure sequences by exonuclease digestion before the product is removed from the support using snake venom phosphodiesterase, which requires a free 5′ hydroxyl [67].

The simplest methods of purification are either electrophoresis on polyacrylamide gels or thin layer chromatography [68,69]. Polyacrylamide gel electrophoresis may be used analytically if the product mixture is first labelled with ^{32}P at the 5′-end using polynucleotide kinase [67]. The bands are detected by autoradiography. In the preparative mode, the gel is placed over a UV-fluorescent surface (such as a fluorescent TLC plate). When irradiated with a UV lamp the bands are seen as shadows where the exciting light has been absorbed. The base composition may affect electrophoretic mobility of oligonucleotides [70], as does the absence of a 5′-phosphate [71]. Electrophoresis does not afford a chemically pure product, but usually suffices for most molecular biological applications.

For pure products it is better to employ hplc, using either ion-exchange, ion-pair or reversed-phase columns [72]. Ion-exchange columns separate mainly (but not completely) according to chain length: the product peak is the last major peak to be eluted when a salt gradient is run. Separations may be run in formamide-containing buffers to disrupt any base pairing between or within oligomers if there are regions of self-complementarity. Volatile buffers, which remove the requirement for desalting after the separation, have been introduced recently [73].

Reversed-phase hplc separates on the basis of base composition, which often leads to difficulties, as the product peak may be hard to identify and/or may be in the middle of a blur of poorly separated peaks. In such cases, retention of the 5′-dimethoxytrityl group affords relief since this highly hydrophobic group significantly alters the chromatographic behaviour [72]. If the dimethoxytritylated product peak from a reversed-phase run is deprotected and run again, the removal of the hydrophobic group shifts it away from any shorter chains that have cochromatographed with it. This method may fail with longer chains because of contamination with dimethoxytritylated shorter chains.

 This problem may be circumvented before removal from the support by cleavage at any sites that have lost a purine base with lysine [74] at pH 9. This removes the possible contaminants before the final ammonia treatment. Chains up to 118 bases long have been separated by this method. Reversed-phase hplc using polystyrene and dilute sodium hydroxide at 60° has been successful in dealing with chains up to 50 long which have strong secondary structure [75].

Long-chain oligonucleotides may be separated by ion-pair hplc, where both ionic and hydrophobic interactions support the separation [72,76].

Base modifications, the base composition and the incorporation of any unusual bases may be checked by digestion with snake venom phosphodiesterase and bacterial alkaline phosphatase followed by hplc analysis of the resulting nucleoside mixture [27,77].

The final touchstone of a successful synthesis is sequence analysis, although the correct sequence is very often taken on faith especially as most automated synthesizers allow thorough checking of the program. There are various sequencing methods available [68,78], but the solid phase method is very convenient [79]. The oligodeoxyribonucleotide is bound to anion-exchange paper before being subjected to base-specific degradations as in the Maxam and Gilbert method [79a].

8.9   SYNTHESIS OF OLIGORIBONUCLEOTIDES

The phosphotriester, phosphoramidite and H-phosphonate methods have all been applied in the synthesis of oligoribonucleotide [80–82]. The 2′-hydroxyl group introduces difficulties, as it must be protected during the synthesis on account of the greater lability of RNA derivatives. The 2′-protecting group must be stable under the conditions of the synthesis cycle, including the acidic 5′-deprotection step, and not be removed by the strongly alkaline conditions used for base deprotection at the end of the synthesis. Much effort has been expended on the development of a set of protecting groups that are mutually compatible in terms of their stability and removal conditions, and many variations have been published.

Two phosphoramidite approaches serve as an illustration. Thus, Iwai and Ohtsuka recommend derivatives such as **(22)**, incorporating a tetrahydrofuranyl group for 2′-protection and a levulinyl group for 5′-protection [83]. The levulinyl group is removed each cycle by a 10-min treatment with hydrazine hydrate in pyridine/acetic acid. This reagent does not affect the β-cyanoethyl group on the phosphate but does cause some slow removal of benzoyl groups used to protect adenine and cytosine. Ammonia treatment is used to deprotect the bases, and dilute HCl (pH 2) removes the tetrahydrofuranyl group. Gait and coworkers have suggested derivatives of the type **(23)**, incorporating the base-labile 9-fluorenylmethoxycarbonyl ('Fmoc') group for 5′-protection and the acid-labile 4-methoxytetrahydropyran-4-yl group for 2′-protection [84]. Removal of the 5′-Fmoc group is accomplished by a two-minute treatment with 1,8-diazabicyclo-[5,4,0]-undec-7-ene ('DBU') in acetonitrile.

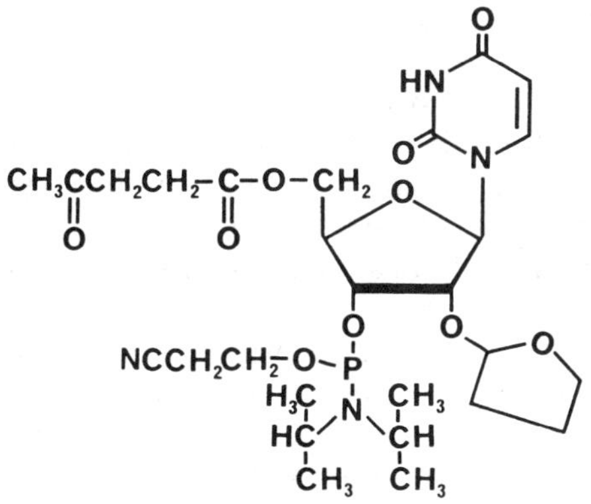

(22)

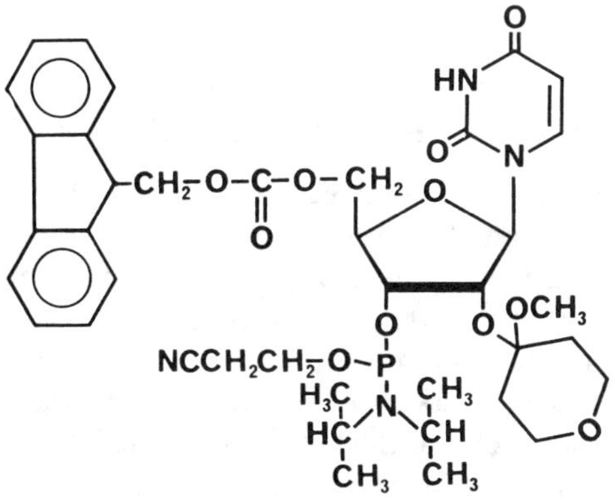

(23)

Dilute HCl (pH 2) removes the 2'-protecting groups after ammonia deprotection of the bases. The greater lability of RNA derivatives makes it desirable also to retain the 2'-OH protection during purification and storage of the oligoribonucleotide.

8.10 PROBLEMS IN OLIGONUCLEOTIDE SYNTHESIS

8.10.1 Base modification during synthesis

Synthetic DNA has been shown to have sufficient base alterations that when cloned into bacteria, various types of mutation (at levels of 0.2–0.3% per nucleotide) have been detected [85,86]. This, and the problems of synthesis of oligoribonucleotides and other labile oligonucleotides incorporating unusual or modified bases, has led to the introduction of base-protecting groups which are removed under gentler conditions [15–17,27]. Stengele and Pfleider have recently suggested the use of p-nitrophenyleth-oxycarbonyl protection of the exocyclic amino groups [87]. Treatment with DBU removes this group and the phosphate β-cyanoethyl groups while the oligomer is still bound to the support. The final cleavage is carried out with ammonia at room temperature. This version of the phosphoramidite synthesis gives improved yields when compared with the standard protocol. A coupling yield of 99.8% is reported with an overall yield of 49% for an 80-mer [87].

8.10.2 Loss of purines during synthesis

Depurination in phosphoramidite synthesis as a result of phosphitylation at guanine has been discussed [21,22]. A more general problem is that of depurination at adenine. Benzoyladenine is prone to depurination (which later results in chain cleavage) under the acid conditions used to remove the dimethoxytrityl group [6,88]. Since the growing chain is exposed to acid at each coupling cycle, depurination may be a problem if long chains rich in adenine are being made. This difficulty may be avoided by the use of a non-protic acid such as zinc bromide for removal of the dimethoxytrityl (DMT) group, but it reacts relatively slowly and the reaction rate decreases with increasing chain length [5]. Alternatively, the pixyl group is marginally more acid labile than the DMT group, and this allows diminution of the time of acid exposure [6]. Perhaps the best approach is to use alternative protecting groups such as phenoxyacetyl and isopropoxyacetyl [16,17].

8.11 PREPARATION OF OLIGONUCLEOTIDE MIXTURES

The demand for 'mixed probes' may generate problems. Frequently, an oligonucleotide probe is required to find the complete gene for a particular protein. Often the probe sequence is dictated by a short amino acid sequence obtained from a small sample of protein. The genetic code is redundant; most amino acids have more than one codon (serine has six codons, glycine four, for example). This leads to a requirement for a mixture of probes, one of which will be the 'true probe'. Each chain could be separately synthesized but the labour and reagents required by this approach may be prohibitive. However, it is possible to introduce a mixture of nucleotides in a single addition to produce a mixture of chains differing in one nucleotide. The different synthons couple at different rates, but the differences are not great and the quantities used may be altered to correct for the differences [5]. Also, use of different 5'-protecting groups allows monitoring of the actual incorporation of the different nucleotides [23].

It is possible to synthesize over 100 different oligonucleotides at once using 'segmented synthesis'. In this technique, picomolar quantities of each oligonucleotide are synthesized using a small filter paper disc for each. Four synthesis units are used — one for each nucleotide — and the discs are dried and sorted at each cycle. Both phosphotriester [89] and phosphite triester [90] methods have been used. This is a more convenient way of synthesizing mixed probes, although it does require considerable patience.

The requirement for a probe mixture may sometimes be circumvented either by using an oligonucleotide long enough to overcome one or two mismatches [91] or by incorporating a 'promiscuous' base such as hypoxanthene (the nucleoside is inosine) which will pair with all bases [92].

8.12 PREPARATION OF MODIFIED OLIGONUCLEOTIDES AND VARIANT OLIGONUCLEOTIDES

8.12.1 Introduction of a 5'-phosphate
Molecular biologists frequently need a 5'-phosphate on an oligonucleotide. This can be accomplished using the enzyme polynucleotide kinase, but it is often more convenient if it can be carried out chemically. This may be readily accomplished if the phosphate group is added at the end of the synthesis using a standard coupling reaction. Connolly employed S-triphenylmethyl-O-methoxymorpholinophosphinyl-2-mercaptoethanol [93] while Uhlmann and Engels used a p-nitrophenyl ester-blocked β-cyanoethyl-N,N-disopropyl phosphoramidite [94].

8.12.2 Introduction of label groups
Pressure to produce non-radioactive probes has led to the development of methods that either introduce a reporter group into the chain (as a modified base) or produce a 5'-amino or 5'-thiol group at the end of the chain — which is then used to couple a label group. A biotin at the 5'-end may be introduced using a biotin phosphoramidite [95]. Various groups (biotin, dinitrophenyl, dansyl, etc.) may be inserted into an oligonucleotide chain using appropriate nucleoside phosphoramidites or hydrogen phosphonates [96]. Another way of introducing amino groups (and presumably a variety of others) is to prepare oligopeptide–oligonucleotide hybrid molecules using solid phase peptide synthesis followed by solid phase oligonucleotide synthesis [97]. Multiple lysines have been incorporated into the peptide portion of such a hybrid to provide points for label attachment. This method could incorporate thiol groups if cysteine were used. Alternatively, a 5'-thiol is incorporated if a 5'-mercapto-2',5'-dideoxyribonucleoside 3'-O-phosphoramidite is coupled at the end of a synthesis [98].

8.12.3   Branched chain and variant backbone oligonucleotides

Stimulated by demonstrations that branched RNA structures (lariats) exist when RNA molecules are 'spliced' in the cell, some workers have explored solid phase methods for the synthesis of branched oligonucleotides. In one recent report a branch is introduced by linking two adjacent chains on the support by introducing a nucleoside 2',3'-bis(phosphoramidite), which links the chains 2' to 5' and 3' to 5'. A single chain is formed after the branch [99]. Cyclic deoxyribonucleotides have been made on solid supports using immobilization of the chain via the exocyclic amino group of a cytosine so that a closed circle can be synthesized [100].

Nucleases capable of cleaving DNA and/or RNA are ubiquitous, and there has been considerable effort devoted to the synthesis of oligonucleotides which are nuclease resistant. This is becoming increasingly important as the use of oligonucleotide probes becomes more common in routine diagnostic laboratories. A further stimulus to the preparation of nuclease-resistant oligomers is their potential as antiviral agents and possibly antitumour agents in the form of 'antisense' inhibitors. Oligonucleotides are constructed that are complementary to particular RNA sequences. The inhibitor forms a based-paired duplex with the target messenger RNA molecule. This duplex formation prevents synthesis of the protein coded for by the messenger RNA. If the sequence targeted is that of a specifically viral RNA, viral replication will be blocked. This is only feasible if the antisense oligonucleotide is not degraded by nucleases.

Many nuclease-resistant forms of oligonucleotides have been synthesized. Oligoribonucleotides with a methyl group at the 2'-position of the furanose ring are completely nuclease resistant [101,102]. However, the 2'-methyl group slows the phosphoramidite coupling reaction and 5-(4-nitrophenyl)-1H-tetrazole is required as activator to achieve coupling times of 6 min [101]. Oligonucleotides have also been constructed using α-anomeric synthons [103], as have oligomers with glycerol replacing the ribose [104].

Oligonucleotides connected by non-phosphate links have also been synthesized, and the formacetal link **(24)** is the simplest linkage tried. Variants on the phosphodiester link have been explored. Methyl phosphonate links **(25)**, [105], phosphoroselenoate links **(26)** [106], phosphorothioate bonds **(27)** [106–108], phosphoroamidates **(28)** [108] and phosphorodithioates **(29)** [109] have all been produced using solid

(24)   (25)   (26)

(27)   (28)   (29)

phase synthesis. Selenium or sulphur may be introduced during the oxidation step of the phosphoramidite and hydrogen phosphonate methods. The resulting phosphoro-selenoate or phosphorothioate diester bonds are chiral, and this may lead to problems in purification or applications so there are advantages to achiral links such as the phosphorodithioate diester bonds.

8.13 FUTURE DEVELOPMENTS

There is little doubt that coupling methods will see further development and improvement and that improved base-protecting groups will enter general use. It seems likely also that more interest will centre on the development of nuclease-resistant oligomers for antiviral use and use in clinical diagnostic applications.

REFERENCES

[1] A. M. Michelson and A. R. Todd, *J. Chem. Soc.*, 1955, 2632.
[2] H. G. Khorana, *Science*, 1979, **203**, 614
[3] R. B. Merrifield, *Fed. Proc.*, 1962, **21**, 412.
[4] R. L. Letsinger and V. Mahadevan, *J. Am. Chem. Soc.*, 1965, **87**, 3526.
[5] B. E. Kaplan and K. Itakura, in *Synthesis and Applications of DNA and RNA* (S. A. Narang, Ed.), Academic Press, London, 1987, p. 9.
[6] B. S. Sproat and M. J. Gait, in *Oligonucleotide Synthesis: a Practical Approach* (M. J. Gait, Ed.), IRL Press, Oxford, 1984, p. 83.
[7] T. Atkinson and M. Smith, in *Oligonucleotide Synthesis: a Practical Approach* (M. J. Gait, Ed.), IRL Press, Oxford, 1984, p. 35.
[8] P. J. Garegg, T. Regberg, J. Stawinski and R. Strömberg, *Chimica Scripta*, 1985, **25**, 280.
[9] P. J. Garegg, T. Regberg, J. Stawinski and R. Strömberg, *Chimica Scripta*, 1986, **26**, 59.
[10] B. C. Froehler, P. G. Ng and M. D. Matteucci, *Nucl. Acids Res.*, 1986, **14**, 5399.
[10a] G. M. Blackburn and M. J. Gait (Eds), *Nucleic Acids in Chemistry and Biology*, IRL Press, Oxford, 1990.
[11] S. A Narang (Ed.), *Synthesis and Applications of DNA and RNA*, Academic Press, London, 1987.
[12] J. Sambrook, E. F. Fritsch and T. Maniatis, *Molecular Cloning: a Laboratory Manual*, 2nd edn., Cold Spring Harbour Laboratory Press, Cold Spring Harbour, N.Y., 1989.
[13] J. M. Walker and E. B. Gingold (Eds), *Molecular Biology and Biotechnology*, 2nd edn., Royal Society of Chemistry, London, 1989.
[14] M. J. Gait, in *Oligonucleotide Synthesis: a Practical Approach* (M. J. Gait, Ed.), IRL Press, Oxford, 1984, p. 1.
[15] C. Chaix, D. Molko and R. Téoule, *Tetrahedron Lett.*, 1989, **30**, 71.
[16] B. Uznanski, A. Grajkowski and A. Wilk, *Nucl. Acids Res.*, 1989, **17**, 4863.
[17] J. C. Schulhof, D. Molko and R. Téoule, *Nucl. Acids Res.*, 1987, **15**, 397.

[18] C. A. A. Claesen, A. M. A. Pistorius and G. I. Tesser, *Tetrahedron Lett.*, 1985, **26**, 3859.

[19] M. V. Rao and C. B. Reese, *Nucl. Acids Res.*, 1989, **17**, 8221.

[20] R. T. Pon, N. Usman, M. J. Damha and K. K. Ogilvie, *Nucl. Acids Res.*, 1986, **14**, 6453.

[21] J. S. Eadie and D. S. Davidson, *Nucl. Acids Res.*, 1987, **15**, 8333.

[22] I. K. Furrance, J. S. Eadie and R. Ivarie, *Nucl. Acids Res.*, 1989, **17**, 1231.

[23] E. F. Fisher and M. H. Caruthers, *Nucl. Acids Res.*, 1983, **11**, 1589.

[24] J. W. Efcavitch, in *Macromolecular Sequencing and Synthesis* (D. H. Schlesinger, Ed.), Alan R. Liss, New York, 1988, p. 221.

[25] A. Ono and T. Ueda, *Nucl. Acids Res.*, 1987, **15**, 3059.

[26] T. Hayakawa, A. Ono and T. Ueda, *Nucl. Acids Res.*, 1988, **16**, 4761.

[27] C. A. Smith, Y. Z. Xu and P. F. Swann, *Carcinogenesis*, 1990, **11**, 811.

[28] V. A. Efimov, S. V. Reverdatto and O. G. Chakhmakhcheva, *Nucl. Acids Res.*, 1982, **10**, 6675.

[29] V. A. Efimov, O. G. Chakhmakhcheva and Y. A. Ovchinnikov, *Nucl. Acids Res.*, 1985, **13**, 3651.

[30] R. Charczuk and C. Tamm, *Helv. Chim. Acta*, 1987, **70**, 717.

[31] A Szemzõ, J. Szécsi, J. Sági and L. Ötvös, *Tetrahedron Lett.*, 1990, **31**, 1463.

[32] B. S. Sproat, P. Rider and B. Beijer, *Nucl. Acids Res.*, 1986, **14**, 1811.

[33] E. de Vroom, H. C. P. F. Roelen, C. P. Saris, T. N. W. Budding, G. A. van der Marel and J. H. van Boom, *Nucl. Acids Res.*, 1988, **16**, 2987.

[34] J. C. Schulhof, D. Molko and R. Téoule, *Tetrahedron Lett.*, 1987, **28**, 51.

[35] M. H. Caruthers, in *Synthesis and Applications of DNA and RNA* (S. A. Narang, Ed.), Academic Press, London, 1987, p. 47.

[36] S. L. Beaucage and M. H. Caruthers, *Tetrahedron Lett.*, 1981, **22**, 1859.

[37] B. H. Dahl, J. Nielson and O. Dahl, *Nucl. Acids Res.*, 1987, **15**, 1729.

[38] N. D. Sinha, J. Biernat and H. Köster, *Nucl. Acids res.*, 1983, **24**, 5843.

[38a] Y. Z. Xu and P. F. Swann, *Nucl. Acids Res.*, 1990, **18**, 4061.

[38b] Y. Z. Xu, Q. Zheng and P. F. Swann, *Tetrahedron Lett.*, 1991, **32**, 2817.

[39] A. Andrus, J. W. Efcavitch, L. J. McBride and B. Giusti, *Tetrahedron Lett.*, 1988, **29**, 861.

[40] B. L. Gaffney and R. A. Jones, *Tetrahedron Lett.*, 1988, **29**, 2619.

[41] J. Stawinski, R. Strömberg, M. Thelin and E. Westman, *Nucl. Acids Res.*, 1988, **16**, 9285.

[42] O. Sakatsume, H. Yamane, H. Takaku and N. Yamamoto, *Tetrahedron Lett.*, 1989, **30**, 6375.

[43] M. Matteucci, *Tetrahedron Lett.*, 1990, **31**, 2385.

[44] M. Fujii, H. Nagai, M. Sekine and T. Hata, *Tetrahedron Lett.*, 1987, **28**, 1435.

[45] H. Nagai, T. Fujiwara, M. Fujii, M. Sekine and T. Hata, *Nucl. Acids Res.*, 1989, **21**, 8581.

[46] V. A. Efimov, A. A. Buryakova, S. V. Reverdatto, O. G. Chakhmakhcheva, Y. A. Ovchinnikov, *Nucl. Acids Res.*, 1983, **11**, 8369.

[47] A. R. Mitchell, S. B. H. Kent, B. W. Erickson and R. B. Merrifield, *Tetrahedron Lett.*, 1976, 3795.

[48] V. K. Potapov, V. P. Veiko, O. N. Koroleva and Z. A. Ahabarova, *Nucl. Acids Res.*, 1979, **6**, 2041.

[49] H. Köster and K. Heyns, *Tetrahedron Lett.*, 1972, 1531.

[50] R. Crea and T. Horn, *Nucl. Acids Res.*, 1980, **8**, 2331.

[51] K. Miyoshi, R. Arentzen, T. Huang and K. Itakura, *Nucl. Acids Res.*, 1980, **8**, 5473.

[52] M. L. Duckworth, M. J. Gait, P. Goelet, G. F. Hong, M. Singh and R. C. Titmas, *Nucl. Acids Res.*, 1981, **9**, 1691.

[53] M. J. Gait, H. W. D. Matthes, M. Singh, B. S. Sproat and R. C. Titmas, *Nucl. Acids Res.*, 1982, **10**, 6243.

[54] B. S. Sproat and D. M. Brown, *Nucl. Acids Res.*, 1985, **13**, 2979.

[55] C. Mingati, K. N. Ganesh, B. S. Sproat and M. J. Gait, *Anal. Biochem.*, 1985, **147**, 63.

[56] M. D. Matteucci and M. H. Caruthers, *J. Am. Chem. Soc.*, 1981, **103**, 3185.

[57] S. P. Adams, K. S. Kavka, E. J. Wykes, S. B. Holder and G. R. Galluppi, *J. Am. Chem. Soc.*, 1983, **105**, 661.

[58] H. Köster, *Tetrahedron Lett.*, 1972, 1527.

[59] G. R. Gough, M. J. Brunden and P. T. Gilham, *Tetrahedron Lett.*, 1981, **22**, 4177.

[60] K. Kamaike, Y. Hasegawa, I. Masuda, Y. Ishido, K. Watanabe, I. Hirao and K. I. Miura, *Tetrahedron*, 1990, **46**, 163.

[60a] R. H. Alul, C. N. Singman, G. Zang and R. L. Letsinger, *Nucl. Acids Res.*, 1991, **19**, 1527.

[61] G. R. Gough, M. J. Brunden and P. T. Gilham, *Tetrahedron Lett.*, 1983, **24**, 5321.

[62] E. Felder, R. Schwyzer, R. Charubala, W. Pfleiderer and B. Schulz, *Tetrahedron Lett.*, 1984, **25**, 3967.

[63] W. T. Markiewicz and T. K. Wyrzykiewicz, *Nucl. Acids Res.*, 1989, **17**, 7149.

[64] P. S. Nelson, R. A. Frye and E. Liu, *Nucl. Acids Res.*, 1989, **17**, 7187.

[65] U. Asseline and N. T. Thuong *Tetrahedron Lett.*, 1990, **31**, 81.

[66] K. C. Gupta, P. Sharma, S. Sathyanarayana and P. Kumar, *Tetrahedron Lett.*, 1990, **31**, 2471.

[67] M. S. Urdea and T. Horn, *Tetrahedron Lett.*, 1986, **27**, 2933.

[68] R. Wu, N. H. Wu, Z. Hanna, F. Georges and S. Narang, in *Oligonucleotide Synthesis: a Practical Approach* (M. J. Gait, Ed.), IRL Press, Oxford, 1984, p. 135.

[69] R. McGookin, in *Methods in Molecular Biology*, vol. 4 (J. M. Walker, Ed.), Humana Press, Clifton, New Jersey, 1988, p. 215.

[70] H. Frank and H. Köster, *Nucl. Acids Res.*, 1979, **6**, 2069.

[71] S. P. Becerra, S. D. Detera and S. H. Wilson, *Anal. Biochem.*, 1983, **129**, 200.

[72] L. W. McLaughlin and N. Piel, in *Oligonucleotide Synthesis: a Practical Approach* (M. J. Gait, Ed.), IRL Press, Oxford, 1984, p. 117.

[73] J. M. Munholland, K. A. Bright and R. N. Nazar, *Anal. Biochem.*, 1989, **178**, 320.

[74] T. Horn and M. S. Urdea, *Nucl. Acids Res.*, 1988, **16**, 11559.

[75] M. W. Germann, R. T. Pon and J. H. van de Sande, *Anal. Biochem.*, 1987, **165**, 399.

[76] T. R. Floyd, S. E. Cicero, S. D. Fazio, T. V. Raglione, S. H. Hsu, S. A. Winkle and R. A. Hartwick, *Anal. Biochem.*, 1986, **154**, 570.

[77] J. S. Eadie, L. J. McBride, J. W. Efcavitch, L. B. Hoff and R. Cathcart, *Anal. Biochem.*, 1987, **165**, 442.

[78] W. Ansorge, A. Rosenthal, B. Sproat, C. Schwager, J. Stegemann and H. Voss, *Nucl. Acids Res.*, 1988, **16**, 2203.

[79] A. Rosenthal, S. Schwertner, V. Hahn and H. Hunger, *Nucl. Acids Res.*, 1985, **13**, 1173.

[79a] R. Wu and R. Yang, in *Synthesis and Applications of DNA and RNA* (S. A. Narang, Ed.), Academic Press, London, 1987, p. 137.

[80] E. Ohtsuka and S. Iwai, in *Synthesis and Applications of DNA and RNA* (S. A. Narang, Ed.), Academic Press, London, 1987, p. 115.

[81] O. Sakatsume, M. Ohsuki, H. Takaku and C. B. Reese, *Nucl. Acids Res.*, 1989, **17**, 3689.

[82] S. Yamakage, M. Fujii, H. Takaku and M. Uemura, *Tetrahedron*, 1989, **46**, 5459.

[83] S. Iwai and E. Ohtsuka, *Nucl. Acids Res.*, 1988, **16**, 9433.

[84] C. Lehmann, Y. Z. Xu, C. Christodoulou, Z. K. Tan and M. J. Gait, *Nucl. Acids Res.*, 1989, **17**, 2379.

[85] W. H. McClain, K. Foss, K. L. Mittelstadt and J. Schneider, *Nucl. Acids Res.*, 1986, **14**, 6770.

[86] M. A. Wosnick, R. W. Barnett, A. M. Vincentini, H. Erfle, R. Elliott, M. Sumner-Smith, N. Mantei and R. W. Davies, *Gene*, 1987, **60**, 115.

[87] K. P. Stengele and W. Pfleiderer, *Tetrahedron Lett.*, 1990, **31**, 2549.

[88] V. Amarnath and A. D. Brown, *Chem. Rev.*, 1977, **77**, 183.

[89] H. W. D. Matthes, W. M. Zenke, T. Grundström, A. Staub, M. Wintzerith and P. Chambon, *EMBO J.*, 1984, **3**, 801.

[90] J. Ott, and F. Eckstein, *Nucl. Acids Res.*, 1984, **12**, 9137.

[91] R. Lathe, *J. Mol. Biol.*, 1985, **183**, 1.

[92] F. Seela and K. Kaiser, *Nucl. Acids Res.*, 1986, **14**, 1825.

[93] B. A. Connolly, *Tetrahedron Lett.*, 1987, **28**, 463.

[94] E. Uhlmann and J. Engels, *Tetrahedron Lett.*, 1986, **27**, 1023.

[95] A. J. Cocuzza, *Tetrahedron Lett.*, 1989, **30**, 6287.

[96] A. Roget, H. Bazin and R. Téoule, *Nucl. Acids Res.*, 1989, **17**, 7643.

[97] J. Haralambdis, K. A. Pownall, L. Duncan, M. Chai and G. W. Tregear, *Nucl. Acids Res.*, 1990, **18**, 501.

[98] B. S. Sproat, B. Beijer, P. Rider and P. Neuner, *Nucl. Acids Res.*, 1987, **15**, 4837.

[99] M. J. Damha and S. Zabarylo, *Tetrahedron Lett.*, 1989, **30**, 6295.

[100] S. Barbato, L. De Napoli, L. Mayol, G. Piccialli and C. Santcroce, *Tetrahedron Lett.*, 1987, **28**, 5727.

[101] B. S. Sproat, A. I. Lamond, B. Beijer, P. Neuner and U. Ryder, *Nucl. Acids Res.*, 1989, **17**, 3373.

[102] B. S. Sproat, B. Beijer and A. Iribarren, *Nucl. Acids. Res.*, 1990, **18**, 41.

[103] F. Morvan, B. Rayner, J. L. Imbach, S. Thenet, J. R. Bertrand, J. Paoletti, C. Malvy and C. Paoletti, *Nucl. Acids Res.*, 1987, **15**, 3421.

[104] N. Usman, C. D. Juby and K. K. Ogilvie, *Tetrahedron Lett.*, 1988, **29**, 4831.

[105] A. V. Lebedev, G. R. Wenzinger and E. Wickstrom, *Tetrahedron Lett.*, 1990, **31**, 851.

[106] K. Mori, C. Boiziau, M. Matsukura, C. Subasinghe, J. S. Cohen, S. Broder, J. J. Toulmé and C. A. Stein, *Nucl. Acids Res.*, 1989, **17**, 8207.

[107] P. C. J. Kamer, H. C. P. F. Roelen, H. van den Elst, G. A. van der Marcel and J. H. van Boom, *Tetrahedron Lett.*, 1989, **30**, 6757.

[108] R. Eritja, V. Smirnov and M. H. Caruthers, *Tetrahedron*, 1990, **46**, 721.

[109] A. Grandas, W. S. Marshall, J. Neilsen and M. H. Caruthers, *Tetrahedron Lett.*, 1989, **30**, 543.

9

Immobilized biocatalysts

J. M. Woodley

9.1 INTRODUCTION

One of the most exciting developments in biology within recent years has been the application of biological techniques to solve industrial, diagnostic and analytical problems that were previously approached by chemical means. The unique activity of many proteins has led to much promise for these techniques, and this is particularly true in the case of enzymes, proteins with a structure such that they can bind reactants specifically and catalyse useful organic reactions. It is likely that most, if not all, organic reactions can be catalysed by enzymes, of which some two thousand are known and several hundred available commercially [1]. Since the 1960s, considerable effort has been spent in trying to use these catalysts to carry out reactions to produce compounds of use not only to the biotechnological growth areas of pharmaceuticals, food products and agricultural chemicals but now also to the more traditional organic chemicals market. The use of a biocatalyst may therefore be viewed as an alternative means of catalysing an organic reaction.

Biological catalysis is focused around the use of enzyme systems. The catalyst may be an individual isolated enzyme or, alternatively, a group of enzymes acting collectively, within a whole animal, plant or more likely microbial cell. Isolation of an enzyme from its original host cell may prove an expensive process, particularly for an intracellular enzyme. Other enzymes may be membrane bound or need to be associated with other cellular functions (e.g. cofactor regeneration) such that their isolation is impractical or economically disadvantageous for the particular reaction they catalyse. Whole cell catalysis may also capitalize upon the metabolic pathway of a cell using several enzymes in series to convert in one stage what an organic synthesis may take several reaction steps to achieve.

Therefore, choosing between an enzymic and a whole cell catalyst is, in general, predetermined by either the nature of the enzyme or the reaction. There have been several reviews highlighting the considerations required to make such a choice [2,3]. The organic chemist may also find employing a commercially available enzyme to catalyse a reaction simpler than using whole cells (with the necessary fermentation required to produce the catalyst).

However, whether a whole cell or an isolated enzyme is employed as the reaction catalyst there are several key features that distinguish biological from chemical catalysis of an organic reaction [3a].

9.2 BIOLOGICAL CATALYSIS

Although many compounds of interest to organic chemists can be obtained via conventional chemical synthesis, these techniques are often not selective. Perhaps the most important of the features that distinguish bioligical from chemical catalysts is their high degree of selectivity (reaction, regio- and stereoselectivity). For example, steroids (of key importance to the pharmaceutical industry) may be regioselectively oxidized [4] or reduced [5] by biocatalysts. In contrast, chemical catalysis of such reactions is complicated by the many reaction sites on the reactant molecules.

Resolution of stereoisomers to produce optically pure products is also a key feature of biocatalyst selectivity. Particularly useful are those catalysts that can act on a wide range of reactants but still catalyse in a stereospecific manner. Thus, the nature of the particular selective catalytic properties of a given biocatalyst is of crucial importance to its application. Unwanted isomers may have undesired properties (a particular problem for pharmaceutical products). Moreover, the production of inactive isomers also represents a loss of reactant material and increases the process inventory per unit of active product. Since separation of structurally similar compounds is difficult and expensive, it is particularly important to use a selective catalyst where possible. The biocatalytic resolution of stereoisomers has already been used successfully on an industrial scale [6]. This finds particular application in the production of agrochemicals and pharmaceuticals, many of which are currently racemic mixtures with only one active enantiomer. Currently, half of all drugs are chiral, with a third of these optically pure. It is predicted that by the year 2000, two-thirds of all drugs will be chiral, with three-quarters of these optically pure. Already six of the top fifteen selling drugs are marketed as single enantiomers. This trend will increasingly demand the use of biological catalysts [7,8,8a].

Using biocatalysts in a specific manner may also produce fewer byproducts, resulting in higher yields, more control of the reaction and considerably reduced purification procedures downstream of the synthesis. On an industrial scale this leads to substantial cost savings. In addition, since a particular enzyme generally catalyses one specific moiety, protecting the other functional groups present is often unnecessary and therefore may further simplify an otherwise complex multistep organic synthesis.

The second key characteristic of biocatalysts is the defined environment within which they work (i.e. a narrow band of pH, temperature and concentration). This is also useful for reaction control, but of far greater benefit is their ability to operate at near ambient temperature, atmospheric pressure and neutral pH, with reaction rates of similar magnitude to those achieved by chemical catalysts at more extreme conditions. This can make for process energy savings and reduced capital costs. Although it is commonly believed that all biocatalysts operate under these mild conditions, this is not true. Microbial strains have been isolated suitable for catalysis under conditions of alkali, acid, extremes of temperature and extremes of pressure [9]. However, in all cases the operational band of the catalyst (whether whole cell or enzyme) is restricted to the extreme conditions. Although biocatalysts have seen little application, relative to conventional chemical catalysts (for example, precious metals or metal oxides) many of these distinguishing features are advantageous. Furthermore, some chemical catalysts carry penalties (for example, special equipment for acid or alkali catalysts) which are not seen with biocatalysts.

There are also some problematic features of biocatalysts that do not lend themselves to everyday use by the organic chemist. The most important of these is that biological catalysts operate in an aqueous reaction medium. This is problematic because most compounds of interest to the organic chemist, either in the laboratory or in industry, have low water solubilities, leading to inefficient reactions. Since the mid-1970s, techniques have been devised to overcome this limitation via the use of water–organic solvent mixtures [10,11]. In the extreme it has proved possible to use enzymes in anhydrous organic solvent, hydrated merely by the few water molecules necessary to sustain catalytic activity, thereby shifting equilibria such that synthetic reactions become thermodynamically favoured over hydrolytic routes [12,12a]. In less extreme cases, approximately equal volumes of organic and aqueous phases may be used in order to operate microbially catalysed reactions, presenting reactant(s) and product(s) to the catalyst in whichever phase they are predominantly soluble [13,13a].

Enzymes are proteins with molecular weights ranging from about 20,000 to several million, and as a result they depend on their natural structure to provide the special interactions that form the catalytic site(s). Substantial disturbance of the protein structure therefore leads to a loss of catalytic stability. Hence adverse pH conditions or elevated temperatures may cause denaturation of the protein and a resultant loss in the observed activity. Enzymes may also be poisoned by chemicals that bind irreversibly at the catalytic site. Similarly, microbial catalysts may be damaged by adverse conditions (e.g. shear effects on mycellial organisms). This limited catalyst life is the second major drawback in the use of biological catalysts. Consequently, until recently it has been difficult to use biocatalysts as 'off the shelf' laboratory reagents, which has precluded their widespread use. Thus, the characteristic of high sensitivity to their environment, which makes them so useful in the control of organic syntheses, is also highly problematic for their application. In Table 9.1 the features of biological catalysts are listed, as outlined in the foregoing discussion.

Several methods have been proposed to improve the stability of these catalysts so that chemists may take advantage of their special features. Protein engineering and recombinant DNA technology are now offering new routes to stable biocatalysts. However, there is still considerable research required to achieve this goal. A more pragmatic approach and one which has already achieved some success as a method of biocatalyst stabilization is incorporation of an enzyme, or more recently whole cells, within or upon a solid support. Here, therefore, is the ultimate example of the use of a solid support as an enabling technology for the application of a catalyst in organic synthesis.

This chapter outlines the development of immobilized biocatalyst technology and its use in organic synthesis.

Table 9.1 — Features of biological catalysts

Stereo-, regio- and reaction-specific catalysis
Defined operating environment (pH, temperature, pressure)
Defined operation, frequently under mild conditions
Non-robust nature of catalyst
Catalyst requirement for water
Catalyst produced in unlimited quantities
Possibility to catalyse a series of reactions in one step
Readily degradable catalysts (i.e. non-polluting)
Non-toxic catalysts

9.3   BIOCATALYST IMMOBILIZATION

The immobilization of a biocatalyst may be defined as a technique that entraps a single, or several, catalytically active enzyme(s) within a reaction system such that they are prevented from entry into the mobile phase carrying reactant(s) and product(s). Immobilization relies upon a phase boundary between the catalyst-rich phase and the mobile phase. This may be achieved via presentation of the reactant(s) and/or product(s) in the gas phase or in an appropriate organic liquid phase. Discussion here is confined to those methods of immobilization that involve the entrapment of the biocatalyst either within or on a solid support, which is precisely analogous with other solid-supported catalytic techniques for use in organic synthesis as exemplified in this volume. Microbial catalysts trapped in slime layers and catalytically active enzymes bound to soil particles are examples of such solid-supported biocatalysts to be found in nature. Use is made of these phenomena by the waste treatment and agricultural-based industries, respectively. However, using biological catalysts in organic chemistry (whether for laboratory or industrial-scale syntheses) requires more precisely defined support materials.

Historically, the first attempt to create a solid-supported biocatalyst was the adsorption of invertase onto charcoal in 1908 [14]. This had only limited success, however, and it was not until the 1960s that immobilization became an active area of research [15]. In 1969, Chibata and his coworkers at the Tanabe Seiyaku company in Japan pioneered an industrial immobilized biocatalytic process — the production of L-amino acids using immobilized aminocyclase [16]. Since then, several industrial processes have come to rely on solid-supported enzymes, and more recently whole microbial cell preparations, to carry out reactions.

Immobilization of a biocatalyst inevitably changes the native characteristics of the catalyst. The nature of the support and the method of attachment of the biocatalyst to it are both crucial issues and are examined in section 9.5. Some typical support materials are given in Table 9.2.

Although the original rationale for biocatalyst immobilization was to improve catalyst stability and operational life, there are many other consequences for reaction, reactor and process. These are best illustrated with particular applications and are

Table 9.2 — Some common supports for biocatalysts

Acrylamide	Maleic anhydride copolymers
Affinity absorbents	Metal oxides
Agar	Nylon
Agarose	Pectate
Albumin	Phenolic resins
Alginate	Polyacrylates
Alumina	Polystyrene
Carageenan	Polyurethane
Ceramics	Sephadex
Chitosan	Silica
Collagen	Zirconia
Dextran	
Glass	
Ion-exchange resins	
Magnesia	

emphasized in section 9.6, but a summary is given in Table 9.3. One crucial result is that choices between whole cell and isolated enzyme catalysts may be affected. The use of an individual enzyme in non-viable, though intact, cells is usually only justified if, for example, increased stability of the catalyst can counteract the lower catalytic activity that results from immobilization of large quantities of inactive biomass. However, such effects can be of considerable advantage. Prokaryotes and yeast cells are more easily prepared, immobilized and controlled than cells of the higher, eukaryotic species such as plant and animal cells. Consequently, the remainder of this chapter focuses on immobilized enzyme and microbial catalysts, of greatest use to the organic chemist.

In the following section the most common methods of immobilization are discussed. An enormous volume of literature has been written on this subject, and only the principles are described. Something in the region of one hundred immobilization techniques have been proposed [16–22] and in the last five years nearly one thousand patents on these methods have been applied for (about half from Japan). The techniques may be broadly categorized into four classes, dependent primarily on the nature of the catalyst attachment to the solid support.

9.4   TECHNIQUES FOR IMMOBILIZATION

9.4.1   Adsorption onto solid supports

The simplest of all biocatalyst immobilization procedures is adsorption of an enzyme onto a solid support (Fig. 9.1). Practically, an enzyme solution is stirred in the presence of the support for a few minutes and subsequently the remaining enzyme in solution is removed by washing. Ion-exchange resins are the most common support

Table 9.3 — Consequences of solid-supported biocatalysis

Potential advantages
  Catalyst operating life extended
  Aqueous phase rheology improved
  Easier catalyst separation from product-rich stream
  Retention of catalyst in reactor
  High catalyst concentration in reactor
  Control of catalyst microenvironment
  Complete and rapid removal of catalyst from reaction
  Alternative reactor operation possible

Potential disadvantages
  Cost of catalyst production increased
  Mass transfer limitations
  Design variables increased
  Loss of catalytic activity during immobilization via:
    — irreversible heat damage
    — irreversible pH change damage
    — irreversible free radical poisoning
  Difficulties for reactions with pH changes
  Difficulties for reactions producing gas
  Difficulties for reactions with product inhibition

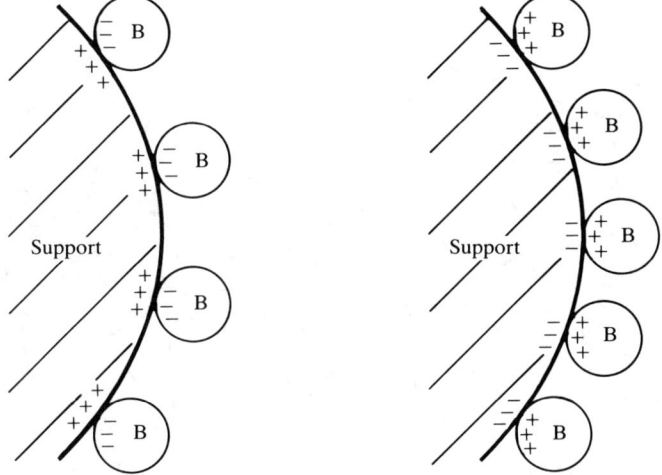

Fig. 9.1 — Schematic of biocatalyst adsorption. Ⓑ represents a biocatalyst.

materials for this technique. Since most proteins are readily adsorbed onto ion-exchange resins, these have become the most common support materials. Industrially, both anion-exchangers (e.g. diethylaminoethylcellulose: DEAE-cellulose) and cation-exchangers (e.g. carboxymethylcellulose: CM-cellulose) are used.

Electrostatic interactions between a carrier and whole cell catalysts also result in adsorption of biocatalyst onto a solid support. Theoretically, adsorption is a reversible process, but some irreversible binding may occur under non-optimal conditions.

Advantages of immobilization by adsorption include its simplicity and the mild treatment of the enzyme during immobilization as well as the large range of variously charged and shaped solid supports available. However, finding the optimal conditions for adsorption is largely a trial-and-error method and potential leakage is also a problem with this technique. Some applications, where high product purity is required, cannot tolerate contamination of this kind.

9.4.2   Entrapment

An alternative method of immolbilization is to entrap the biocatalyst within a solid support (Fig. 9.2), which must obviously have the right properties to allow reactant to diffuse into the active site and product to diffuse back out to the bulk solution. Here, biocatalyst (cell or enzyme) is added to a monomer in solution. Subsequently, polymerization is initiated, either thermally or chemically, and the catalyst becomes entrapped within a polymeric gel. Both covalent gels (e.g. polyacrylamide cross-linked with N,N'-methylenebisacrylamide) and non-covalent gels (e.g. calcium alginate) may be used.

This is a widely used technique of biocatalyst immobilization. However, acryla-mide gels are polymerized via free radical initiation, which may damage whole cell catalysts. Another caution is that alginate entrapped catalyst may be disrupted via the presence of phosphate ions (for example, present in the form of buffer). Entrapment

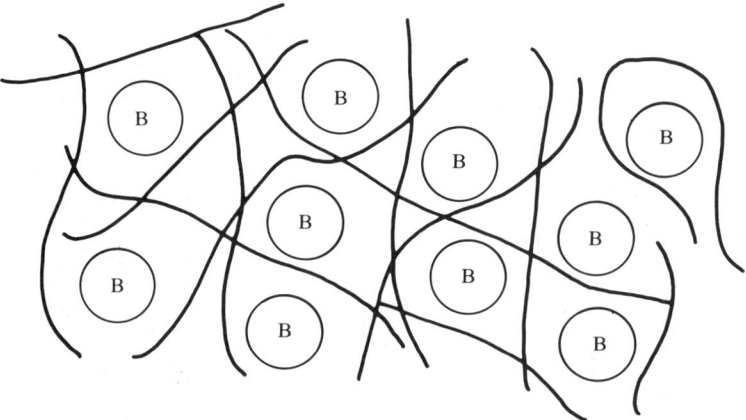

Fig. 9.2 — Schematic of biocatalyst entrapment.

does lead to good operational stability, and operational half-lives in acrylamide gel up to 180 days have been observed [23]. This method is also very versatile. For example, photo-cross-linkable resin prepolymers have been used to entrap whole cells of *A. simplex* [24] and *N. rhodocrous* [25] for catalysis of specific steroid dehydrogenations.

9.4.3 Covalent attachment to solid supports

A more complex immobilization method involves the covalent attachment of enzymes to solid supports (Fig. 9.3) their amino or carboxyl groups. This is more normally a multistep procedure involving support activation followed by subsequent enzyme attachment. Examples of support materials include steel, charcoal, cellulose, polymers, porous glass, silica and ceramics. Immobilization by this means may well place constraints on the enzyme conformation. It is widely believed that the ability of an enzyme to act as a catalyst is dependent both upon the conformation of the enzyme and on its ability to change shape during catalysis. Hence, the specificity and selectivity of an enzyme may be altered by the attachment to the support. Consequently, binding of the functionalized support to residues at the enzyme active site or to groups required for allosteric control or sub-unit interaction should be avoided.

Covalent attachment of whole cells to solid supports is also possible using the amino or carboxyl groups on the cell surface, although it is not widely employed, as there may be significant cell damage.

This method of immobilization is rather more complex and difficult than other methods and therefore a greater loss of activity is often observed upon immobilization. However, it does result in a strong, stable attachment, yielding catalysts with excellent operational stabilities.

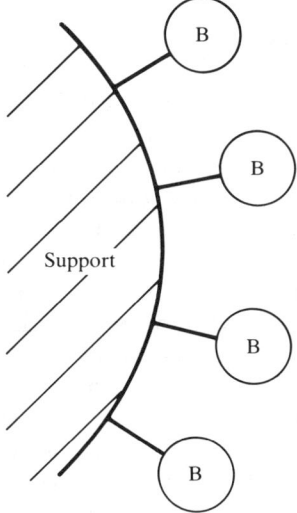

Fig. 9.3 — Schematic of biocatalyst immobilization by covalent binding.

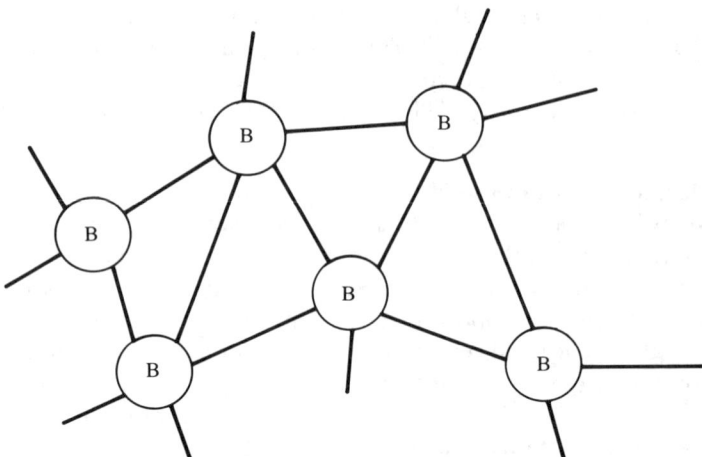

Fig. 9.4 — Schematic of biocatalyst immobilization by cross-linking.

9.4.4   Cross-linking with bifunctional reagents

Enzyme amino (or less frequently carboxyl) groups can be cross-linked to either molecules of enzyme or other proteins (Fig. 9.4). Alternatively they may cross-link with other reagents. The most common cross-linkers are glutaraldehyde, dimethyl adipimidate, dimethyl suberimidate and aliphatic diamines. As with the other three methods of immobilization, this method is applicable to the support of whole cells as well as isolated enzymes.

9.5   IMPLICATION OF INTRODUCING A SOLID SUPPORT

9.5.1   Immobilization procedure

There are three distinct ways in which the presence of a solid support can affect the subsequent catalytic performance of a biocatalyst. The first is via the method of immobilization of the biocatalyst onto or into the solid support. Methods employing harsh conditions (in particular of pH or temperature) may negate many of the advantages of immobilization by damaging the catalyst in the process (e.g. during covalent attachment to a support). Mild conditions of immobilization are therefore crucial. In a similar manner, immobilization must not block the active site by the presence of solid such that reactant access is reduced or altered (steric hindrance).

Selecting a particular immobilization method is not easy since there are many considerations to examine, including the nature of the reaction, catalyst and also type of reactor to be used. Irreversible damage to a biocatalyst support via mechanical damage (most likely from abrasion) can occur if a weak material is chosen for use in a high-shear environment. Compressive strengths of a variety of polymer beads used for entrapment have been examined [26] and vary from polyacrylamide and carageenan, which are easily ruptured, to epoxy resins, requiring one hundred times the force to rupture a single bead.

Similarly, highly elastic and compressible gels are inappropriate for use in packed-bed reactors where hydraulic performance is critical. Physico-chemical breakage of the catalyst–support link, if reversible in the reactor, may lead to leakage of cells and enzyme in addition to loss of activity. A detailed comparison of different immobilization methods — adsorption (onto duolite and celite), covalent binding (glutaraldehyde, carbodiimide and periodic acid) and entrapment (within hydrophobic photo-cross-linkable prepolymers of different chain lengths) — for triacylglyceride hydrolysis by lipase has been made [27]. Entrapment methods yielded the highest activity (30% of the free counterpart), but covalent binding yielded a catalyst particle with excellent operational and stability properties.

Other attempts at systematic comparison of support methods have failed to produce general conclusions [28,29]. Some factors that need to be considered when choosing a support are listed in Table 9.4. In general, choosing an appropriate method of catalyst support is dependent upon the particular application. It is also determined by the scale of operation. Ready-activated, commercially available, supports provide the most convenient laboratory support for immobilizing enzymes. For the larger scale, relatively simple methods are favoured that use particles through which fluids pass easily (e.g. ion-exchangers, phenolic resins and porous inorganic supports).

9.5.2 Partitioning

The second way in which the presence of the solid support will affect catalytic activity is via partitioning of reaction components between the catalyst support and the bulk liquid phase surrounding it. This may be based upon the relative hydrophobic or hydrophilic nature of the support material relative to the molecules that will transfer to and from the catalytically active site in the support. Consequently, concentrations of reactant or product within the support may not reflect those within the bulk solution. This can lead to particularly interesting results when solid-supported biocatalysts are used for the conversion of hydrophobic, poorly water-soluble reactants [24,25]. A clear discussion of these effects has been given previously [19] but it is important to emphasize that changes in operational strategy do not affect this

Table 9.4 — Factors affecting choice of support and immobilization method

Hydrophobicity of reaction components
Ionic nature of reaction components
Catalyst (whole cell or isolated enzyme) to be used
Changes in pH on reaction
Inhibitory nature of reaction components
Catalytic activity per unit mass
Nature of catalyst isolation at end of reaction
Presence and nature of organic solvents
Reactions involving changes of state
Rheological properties of reaction fluids

kind of behaviour. The effects are a direct reflection of the reactants, products and chosen support material.

9.5.3 Mass transfer

In contrast to the effects of partitioning on catalyst performance, mass transfer of reactant to, and of product from, the active site of an immobilized biocatalyst is highly dependent upon operational parameters. This is the third way in which the presence of solid support affects catalytic activity.

Whole cells or enzymes may be bound within a support (e.g. whole cells entrapped in a gel) or on the surface (e.g. enzymes covalently bound to a non-porous solid). In both cases, mass transfer of components between the bulk solution and the outside of the support are important. It may largely be determined by the particle hydrodynamics (i.e. the shape of the support and velocity of the fluid moving past it). However, even more critical is that diffusional effects limit the transfer of components within the support itself. Diffusional limitation of this type is commonly observed. It is dependent upon both the particle size and the porosity of the support as well as on catalyst concentration. Methods to alleviate diffusional limitation include increasing the reactant concentration in the bulk solution, decreasing the support particle size, using supports with larger pore sizes, decreasing the particle catalyst concentration as well as using surface bound biocatalysts on very small supports. The considerations required to evaluate these mass transfer (and partitioning) effects on the reaction kinetics have been well documented previously for both chemical and biological catalysts [19,30].

Although small support beads are advantageous to remove mass transfer limitations, they may ultimately yield problems for catalyst recovery from solution. In practical reactors, support particles vary from 0.18 to 4 mm, dependent upon catalyst and application.

In some cases, mass transfer limitations may be used to advantage, e.g. by reducing inhibitory concentrations of reactants. In other cases they may confer disadvantageous activities, e.g. reduced reaction rate via high concentrations of inhibitory product local to the catalytic site in a support particle. It is particularly important to evaluate these effects alongside partitioning of components between the bulk solution and the support.

On an industrial scale, the rate of reaction is a key performance parameter, but in the laboratory, final reactant conversion is equally important for simplification of the subsequent product purification downstream of the synthesis. It is noteworthy that high concentrations of product within catalyst supports (via partitioning or mass transfer limitations) may also limit conversion due to reaction equilibrium being attained within the particle before that in the bulk solution.

Diffusional limitations can also have an important impact on the pH in the locale of the catalytic site. Reactions resulting in a change in pH (e.g. ester hydrolysis) may develop pH gradients within the support. The ionic nature of the support may also influence concentrations of H^+ ions. Some kinetic models have been developed to take into account pH gradients within particles containing entrapped biocatalysts [31, 32]. The pH sensitivity of most biological catalysts makes this an important implication of solid-supported biocatalysis. In summary, unless the effects are clearly understood, using a solid-supported biocatalyst for a reaction producing or consuming H^+ ions may be highly disadvantageous.

9.5.4 Reactor and process design

Many organic reactions both on an industrial and on a laboratory scale are performed in stirred tank reactors, catalyst concentration being determined by the required reactor productivity. However, when a solid-supported biocatalyst is introduced to such a reactor, too high a catalyst concentration will result in a loss of productivity due to support damage via particle collision and abrasion. In general, therefore, an upper limit of 10% immobilized biocatalyst is used in a stirred tank reactor.

In contrast, a packed bed [33,34] of biocatalyst beads can be very concentrated. However, new demands are now made upon the support since a compressible gel blocks the flow of fluid transporting reactant(s) to, and product(s) from, the catalytically active site in the reactor. Fluidized beds [35] (suspending large biocatalyst particles in motion) and membrane bioreactors [36] (supporting enzyme in membrane sheets or fibres) can also be used with insoluble biocatalysts. Table 9.5 indicates some of these possibilities.

Immobilization of a catalytic cell or enzyme may confer several advantages for overall process design. In particular, primary catalyst separation can be based upon physical exclusion at the point of product removal, since the solid-supported catalysts are usually of large enough size. Rigid, well structured particles are especially preferred, as they prevent blinding of filters. Removing catalyst from the reacting liquid in this way also means that a reaction can be stopped virtually instantaneously, which may be highly beneficial for reaction control (for example, in the synthesis of an intermediate compound where subsequent reaction of the desired product may be a problem). It is also then possible to reuse the recovered catalyst by contacting with fresh reactant. A further advantage of immobilization is that the product-rich fluid is

Table 9.5 — Reactor types for reactions using biocatalysts

Reactor	Biocatalyst form		
	Soluble	Insoluble	
		Fixed	Suspended
Stirred tank	X		X
Bubble column	X		X
Fluidized bed			X
Packed bed		X	
Trickle bed		X	
Membrane	X	X	

cleaner from protein and cell debris, which can cause problems for purification with non-supported biocatalysis.

9.6   USES OF IMMOBILIZED BIOCATALYSTS

9.6.1   Industrial applications

The enormous growth of interest in immobilized biocatalysts has so far resulted in relatively few industrial processes using these catalysts. This is mainly due to the expense of developing the catalyst, and hence application is very market-dependent. Cost–benefit analyses are required to permit rational choices to be made between chemical and biological catalysis, and whether the catalyst should be supported on a solid or not [37].

The most important worldwide industrial solid-supported biocatalytic process is the production of high fructose corn syrup (HFCS) [38]. HFCS is produced via the conversion of glucose into fructose using glucose isomerase. The relative sweetness of fructose is more than twice that of glucose. HFCS has now replaced sucrose as a sweetener in many drinks, and world production is now more than three million tonnes per annum [39]. HFCS can be made at an even lower price when starch-containing waste is used as the reactant.

The other industrial solid-supported biocatalytic processes of note are the production of L-amino acids using L-amino acylase or amino acid-dehydrogenase and the production of 6-aminopenicillanic acid using penicillin acylase [40]. Table 9.6 indicates the type of catalysts used for these conversions. In these cases the economics are such that application of a biological catalyst is feasible. Each of these reactions was operating on an industrial scale well over a decade ago.

It is only recently that several other processes have been commercialized. One of these is the 1,3-regiospecific interesterification of fats by lipase used to manufacture cocoa butter substitutes. The immobilized lipase packed bed reactor is fed with hexane, in which reactants and small amounts of water are dissolved [41,42]. This

Table 9.6 — Example of industrial immobilized biocatalysts

Catalyst	Support	Reaction product
L-Amino acylase	Heteropolar bound on sephadex	L-Amino acids
Glucose isomerase	Cross-link whole cells Adsorption on silica Cross-link/entrap in gelatin	HFCS
L-Aspartase	Entrap whole cells in carageenan Entrap whole cells in polyacrylamide	L-Aspartic acid

exemplifies many of the recent process developments that are increasingly using solid-supported biocatalysts in the presence of organic solvent.

Nitto Chemical Industries in Japan [43] have recently commercialized the production of acrylamide from acrylonitrile using immobilized *corynebacterium*. This is the first process to produce a commodity chemical using an immobilized biocatalyst, but it would be quite wrong to suppose that the application of immobilized biocatalysts will move towards the commodity chemical sector. In general, 60–80% of the process cost for large tonnage chemicals is attributable to the reactant and raw materials. Consequently, immobilized biocatalysts should see a slow, but steady, growth in the fine chemicals' sector, often using organic-solvent-based media. This growth should be proportional to the market demand for optically pure products and environmentally compatible processes, which should occur increasingly in the next decade.

9.6.2   Laboratory organic syntheses

Application of solid-supported biocatalysts in the organic synthetic laboratory has not been widespread. It has been maintained that biological catalysts work only in bulk aqueous solutions, with difficult control and poor operational life. Furthermore, many of the enzymes are expensive and microbial preparations require specialist microbiological knowledge.

Although biocatalytic techniques should be examined with a certain caution, their application could be far more widespread in organic syntheses. They have the potential for assisting a laboratory organic synthesis in the following ways:

(a)   reduction in the number of synthetic steps — complex multistage syntheses can be carried out using the metabolic pathway of a whole cell in one step;

(b)   synthesis involving reactants or products that are pH or temperature labile;

(c)   synthesis of optically pure chiral molecules;

(d)   synthesis where high purity is difficult to achieve;

(e)   regioselective synthesis;

(f)   synthesis where the desired reaction may be competing with a parallel reaction.

Contrary to popular belief, immobilized biocatalysts do catalyse reactions involving hydrophobic compounds. Several laboratory-scale syntheses have already used immobilized biocatlysts in the presence of organic solvents [44]. Examples include enantioselective acylation of racemic alcohols by immobilized carboxylesterase in the presence of methyl propionate [45], asymmetric reduction of ketones by immobilized alcohol dehydrogenase in the presence of isopropyl alcohol [46], oxidative conversion of phenols to δ-quinones by immobilized polyphenol oxidase in the presence of chloroform [47] and the synthesis of optically active (R)-cyanohydrins by immobilized mandelonitrilelyase in the presence of ethyl acetate [48]. Lipases are particularly

Table 9.7 — Common laboratory organic reactions performed by biocatalysts

---

Hydrolysis and synthesis of:
— esters
— amides (especially peptides)
Reduction of:
— carbonyl compounds to alcohols
Oxidation of:
— alcohols to carbonyl compounds
Formation of:
— epoxides from alkenes
— lactones from ketones
Hydroxylation of:
— carbon skeletons at non-activated positions

---

active in organic solvent media, but the optimal choices of organic solvent and operating conditions to effect enantioselective resolution still require more research.

Table 9.7 lists the most common reactions for which immobilized biocatalysts have been used on a laboratory scale. Application of these catalysts has capitalized upon their ability to catalyse under mild conditions, to produce optically active molecules [49,50,50a], and to functionalize organic molecules in conventionally non-activated positions.

Techniques are now moving towards the use of solid-supported biocatalysis for the hydrolysis of esters of chiral and prochiral acids and alcohols, stereoselective and enantioselective reduction of ketones, hydroxylation of carbon atoms remote from functional groups, and more recently for polymerization.

In a few syntheses, supporting a biocatalyst on a solid is not beneficial. For example, reactions producing gaseous products, those undergoing a net consumption or production of H^+ ions, and those producing inhibitory products are not likely to benefit greatly. However, for the majority of reactions, providing the biocatalyst in a solid-supported form assists with storage, handling and transport of the catalyst and should consequently be beneficial to their wider application. It is likely that the next decade will see a greater growth in biocatalytic techniques for use in laboratory organic synthetic chemistry than in industrial applications.

9.7 CONCLUDING REMARKS

The increased interest in the use of biocatalysts for organic synthesis is reflected not only in the scientific literature [51–54] but also in several symposia in recent years [55–59].

There are three reasons for this increased interest by organic chemists.

• There is a need to produce organic processes that are environmentally compatible. Biological catalysts (whether supported on a solid or not) may be produced in unlimited quantities and are readily degraded subsequent to use, in contrast to conventional chemical catalysts. They also avoid the use of acidic or basic solutions and the requirement to dispose of the resultant salts. Moreover, biological catalysis is inherently a low-energy method of carrying out a reaction.

- There is increasing need for production of optically pure products whenever possible. Unwanted isomers are wasteful of resources and may be undesirable to pharmaceutical or agrochemical activity. Biological catalysts (whether supported on a solid or not) are capable of reaction selectivity, regioselectivity and stereoselectivity to varying degrees.

- It has recently become known that biological catalysts can function effectively in the presence of organic solvents and accept hydrophobic reactants as well as many other reactants that might not be encountered in nature.

It is likely that industrial applications will continue to expand at a slow rate. In the 1970s, much was promised for immobilized biocatalyst technology and its application, that was never realized. In part this was due to inherent problems of biocatalyst stability, but in part it was also due to rather exaggerated and over-stated claims about the technology.

The current revolution of biocatalysis in organic media is already seeing industrial application but will have far more widespread implications for the chemist conducting laboratory organic syntheses, who now has another catalytic tool to employ. Solid-supported biocatalysis too is reaching a level of maturity such that biocatalysts may be used much as other reagents. For example, pig liver esterase bound covalently onto oxirane acrylic beads is sufficiently stable that it can be used for many months with no loss of activity, stored at $7°C$ [50].

REFERENCES

[1] K. Kieslich, (Ed.), *Biotechnology. Vol 6A: Biotransformations*, Verlag Chemie, New York, 1984.

[2] S. Birnbaum, P-O. Larsson and K. Mosbach, *Chem. Anal.*, 1983, **66**, 679.

[3] M. D. Lilly, in *Biotechnological Applications of Proteins and Enzymes* (Eds Z. Bohak and N. Sharon), Academic Press, London, 1977, Chapter 8.

[3a] A. L. Margolin, *Chemtech.*, 1991, **21**, 160.

[4] R. Z. Kazandjian, J. S. Dordick and A. M. Klibanov, *Biotechnol. Bioeng.*, 1986, **28**, 417.

[5] C. Laane, R. Hilhorst, R. Spruijt, K. Dekker and C. Veeger, in *Flavins and Flavoproteins*, Walter de Gryten, West Berlin, 1984, pp. 879–892.

[6] V. Svedes and I.U. Galaev, *Russ. Chem. Rev. (Engl. Transl.)*, 1983, **52**, 1184.

[7] R. A. Sheldon, H. E. Shoemaker, J. Kamphuis, W. H. J. Boesten and E. M. Meijer, in *Stereoselectivity of Pesticides; Biological and Chemical Problems* (Eds. E. J. Ariëns, J. J. S. van Rensen and W. Welling), Elsevier, Amsterdam, 1988, Chapter 14.

[8]   P. E. Sonnet, *Chemtech*, 1988, **18**, 94.

[8a] D. H. Deutsch, *Chemtech*, 1991, **21**, 157.

[9]   J. A. Baross and J. W. Deming, *Nature*, 1983, **303**, 423.

[10] M. D. Lilly and J. M. Woodley, in *Biocatalysts in Organic Syntheses* (Eds J. Tramper, H. C. van der Plas and P. Linko), Elsevier, Amsterdam, 1985, pp. 179–192.

[11] M. D. Lilly, A. J. Brazier, M. D. Hocknull, A. C. Williams and J. M. Woodley, in *Biocatalysis in Organic Media* (Eds C. Laane, J. Tramper and M. D. Lilly), Elsevier, Amsterdam, 1987, pp 3–17.

[12] A. M. Klibanov, *Chemtech*, 1986, **16**, 354.

[12a] J. S. Dordick, *Enzyme Micro. Technol.* 1989, **11**, 194.

[13] J. M. Woodley, in *Biocatalysis* (Ed. D. A. Abramowicz), Van Nostrand Reinhold, New York, 1990, p. 337.

[13a] M. D. Lilly, G. A. Dervakos and J. M. Woodley, in *Opportunities in Biotransformations* (Eds L. G. Copping, R. E. Martin, J. A. Pickett, C. Bucke and A. W. Bunch), Elsevier, London, 1990, p. 5.

[14] L. Michaelis and M. Ehrenreich, *Biochem. Z.*, 1908, **10**, 283.

[15] C. Bucke and A. Wiseman, *Chem. Ind.*, 1981, 234.

[16] I. Chibata (Ed.), *Immobilized Enzymes*, Kodonsha, Tokyo, 1978.

[17] O. R. Zaborsky, *Immobilized Enzymes*, CRC Press, Cleveland, 1973.

[18] K. Mosbach (Ed.), *Methods in Enzymology 44*, Academic Press, New York, 1974.

[19] M. D. Trevan, *Immobilized Enzymes*, Wiley, New York, 1980.

[20] C. Bucke, *Phil. Trans. R. Soc. Lond.*, 1983, **B300**, 369.

[21] J. F. Kennedy and J. M. S. Cabral, in *Solid-Phase Biochemistry* (Ed. W. H. Scouten), Wiley, New York, 1983, pp 245–393.

[22] A. Rosevear, *J. Chem., Tech. Biotechnol.*, 1984, **34B**, 127.

[23] F. B. Kolat, *Process Biochem.*, 1981, 2.

[24] K. Sonomoto, A. Tanaka., T. Omata, T. Yamane and S. Fukui, *Eur. J. Appl. Microbiol. Biotechnol.*, 1979, **6**, 325.

[25] A. Tanaka and K. Sonomoto, *Chemtech*, 1990, **20**, 112.

[26] J. Klein and F. Wagner, in *Proc. First Eur. Congr. Biotechnol.*, pt 3, Verlag Chemie, New York, 1978, pp. 142–164.

[27] Y. Kimura, A. Tanaka, K. Sonomoto, T. Nihara and S. Fukui, *Eur. J. Appl. Microbiol. Biotechnol.*, 1983, **17**, 107.

[28] J. Konecny, *Enzyme Engineering*, 1978, **3**, 11.

[29] I. Chibata and T. Tosa, *Appl. Biochem. Biotechnol.*, 1976, **1**, 329.

[30] C. Webb, G. M. Black and B. Atkinson (Ed), *Process Engineering Aspects of Immobilized Cell Systems*, Pergamon, Oxford, 1986.

[31] J. E. Bailey and M. T. C. Chow, *Biotechnol. Bioeng.*, 1974, **16**, 1345.

[32] J. Engasser and C. Horvath, *Biochem. Biophys. Acta*, 1974, **358**, 178.

[33] A. R. Macrae, *J. Amer. Oil Chem. Soc.*, 1983, **60**, 291.

[34] G. Bell, J. R. Todd, J. A. Blain, J. D. E. Patterson and C. E. Shaw, *Biotechnol. Bioeng.*, 1981, **23**, 1703.

[35] R. B. Lieberman and D. F. Ollis, *Biotechnol. Bioeng.*, 1975, **17**, 1401.

[36] M. M. Hoq, T. Yamane, S. Shimizu, T. Funada and S. Ishida, *J. Amer. Oil Chem. Soc.*, 1984, **61**, 776.

[37] P. B. Poulsen, *Biotechnol. and Genetic Eng. Rev.*, 1984, **1**, 121.

[38] S. Furusaki, *J. Chem. Eng. Japan*, 1988, **21**, 219.

[39] H. Dellweg, in *Grundlagen und Verfahren*, VCH, Weinheim, 1987, pp. 223–265.

[40] P. Dunnill and M. D. Lilly, *Process Biochem.*, 1971, **6**, 29.

[41] M. H. Coleman and A. R. Macrae, U.K. Patent 1,577,933, 1980; *Chem. Abstr.*, **87**:166366.

[42] A. R. Macrae, in *Biocatalysts in Organic Syntheses* (Eds J. Tramper, H. C. van der Plas and P. Linko), Elsevier, Amsterdam, 1985, pp. 195–208.

[43] I. Watanabe, U.S. Patent 4,343,900, 1982; *Chem. Abstr.*, **95**, 59933.

[44] C-S. Chen and C. J. Sih, *Angew. Chem., Int. Edn. Engl.*, 1989, **28**, 695.

[45] G. M. Ramos Tombo, H-P. Schaer, X. Fernandez, J. Busquets and O. Ghisalba, *Tetrahedron Lett.*, 1986, **27**, 5707.

[46] J. Grundwald, B. Wirtz, M. P. Scollar and A. M. Klibanov, *J. Am. Chem. Soc.*, 1986, **108**, 6732.

[47] R. Z. Kajandjian and A. M. Klibanov, *J. Am. Chem. Soc.*, 1985, **107** 5448.

[48] F. Effenberger, T. Ziegler and S. Förster, *Angew. Chem., Int. Edn. Engl.*, 1987, **26**, 458.

[49] J. B. Jones and J. F. Beck, in *Application of Biochemical Systems in Organic Chemistry, Pt. 1* (Eds J. B. Jones, C. J. Sih and D. Perlman), Wiley-Interscience, New York, 1976, pp. 112–401.

[50] A. Akiyama, M. Bednarski, M-J. Kim, E. S. Simon, H. Waldmann and G. M. Whitesides, *Chem. in Britain*, 1987, **23**, 645.

[50a] H. G. Davies, R. H. Green, D. R. Kelly and S. M. Roberts, *Crit. Rev. Biotechnol.*, 1990, **10**, 129.

[51] J. B. Jones, *Tetrahedron*, 1986, **42**, 3351.

[52] P. Gramatica, *Chimicaoggi*, 1989, **7**, 43.

[53] C-H. Wong, *Science*, 1989, **244**, 1145.

[54] A. J. Pratt, *Chem. in Britain*, 1989, **25**, 282.

[55] R. Porter and S. Clark (Eds), *Enzymes in Organic Synthesis*, Pitman, London, 1985.

[56] J. Tramper, H. C. van der Plas and P. Linko (Eds), *Biocatalysts in Organic Syntheses*, Elsevier, Amsterdam, 1985.

[57] C. Laane, J. Tramper and M. D. Lilly (Eds), *Biocatalysis in Organic Media*, Elsevier, Amsterdam, 1987.

[58] D. A. Abramowicz (Ed.) *Biocatalysis*, Van Nostrand Reinhold, New York, 1990.

[59] L. G. Copping, R. E. Martin, J. A. Pickett, C. Bucke and A. W. Bunch (Eds), *Opportunities in Biotransformations*, Elsevier, London, 1990.

[60] K. Laumen, E. H. Reimerdes and M. Schneider, *Tetrahedron Lett.*, 1985, **26**, 407.

Part IV

Individual applications

10

Hydrogenation and hydrogenation catalysts

M. E. Fakley and **F. King**

10.1 BACKGROUND

The hydrogenation of unsaturated functional groups is important throughout industrial organic chemistry, and both homogeneous (soluble) and heterogeneous (insoluble) catalysts may be employed. Homogeneous catalysts are often highly selective in the laboratory, but the complexities of catalyst separation and recycle often mitigate against their use at larger scale. Therefore, in practice, most industrial reductions eventually require heterogeneous catalysts, which although sometimes less selective, generally prove simpler to handle and scale up.

Nevertheless, the selection of a heterogeneous catalyst for a particular reaction can appear daunting. The objective of this chapter is to provide examples and guidance for the organic chemist wishing to perform selective reductions, with illustrations from the recent chemical literature. A number of standard texts are already available [1–5], and references to key studies are also provided. It is not intended, however, to describe the preparation of heterogeneous catalysts, which is beyond the scope of this brief chapter and is amply covered elsewhere [6]. The proceedings of the 1988 and 1990 International Symposia on Heterogeneous Catalysis and Fine Chemicals have appeared recently [7].

Heterogeneous hydrogenation catalysts may contain Group VIII transition metals, i.e. Ru, Co, Rh, Ir, Ni, Pd, or Pt, or other base metals such as Cu and Fe may be used. Sometimes the metals are used in combination, optionally incorporating promoters, e.g. Cr, Ba, Mn. Precious metal catalysts are often preferred because they can operate under milder conditions and hence allow better control of selectivity.

In general, heterogeneous catalysts for organic transformations may be grouped into four distinct categories.

(a) Raney type: in which the active metal is alloyed with aluminium and activated with sodium hydroxide solution prior to use, e.g. Raney nickel [8].
(b) Impregnated: in which the active metal(s) plus any promoter(s) are impregnated onto a suitable support material, e.g. Pd on carbon, Ni on alumina. This method is employed for precious metals and low metal loadings in general.

(c) Precipitated: in which the active metal(s) plus any promoter(s) are co-precipitated with a support and calcined to give a catalyst with a high metal loading but in the oxide form.

(d) Unsupported metals or metal oxides, e.g. Pd black, PtO_2.

Catalysts of types (b) and (c) are often supplied in a pre-reduced form, which lowers the temperature of reduction and permits activation under mild conditions. In a recent study, a pre-treatment of a Ni/Al_2O_3 catalyst using $KBH_4/NH_4OH/MeOH$ permitted hydrogenations of an ethynyl carboxylate to the corresponding ethyl carboxylate to be performed at temperatures as low as 50°C [9].

Catalysts may be supplied in several physical forms, e.g. powders, granules, pellets or extrudates, depending upon the type of catalyst, the reaction and the reactor type to be employed. Care should be exercised when specifying physical form, as this can be a major determinant of both activity and selectivity.

Although, from a theoretical viewpoint, a novel reduction may require a uniquely formulated catalyst, practical considerations often result in the selection of a commercially available catalyst (e.g. Raney nickel, Ni/Al_2O_3, Pd/C, Cu chromite, etc.). In some cases, it has proved necessary to modify a standard catalyst by a subsequent treatment (e.g. additional impregnation). Catalyst performance is optimized by careful control of the reaction parameters, e.g. temperature, pressure, catalyst concentration; by choice of reaction solvent; or in some cases by selective poisoning.

The availability of standard laboratory pressure reactors from a number of suppliers (Appendix I), and growing awareness of the value of heterogeneous catalysts, has brought about a re-examination of many industrial reductions. The traditional 'nascent' hydrogenation process (Fe filings plus hydrochloric acid) has almost entirely been displaced by Raney nickel catalysis. However, with the ever increasing concerns about the handling of finely divided metals and the effluents generated during activation, there is a general move towards fixed-bed supported catalysts (e.g. Ni/Al_2O_3).

10.2   PRACTICAL CONSIDERATIONS

Successful study of the reduction of functional groups using heterogeneous catalysts requires attention to a number of practical matters.

(a) The hydrogen supply should ideally be of at least the same purity as is available at scale, should be available over a range of pressures, should be safely stored — preferably outside the building — and should have the flow limited by a suitable restrictor.

(b) A range of laboratory reactors may be used. Stirred or rocking autoclaves are used for batch-wise experiments or to model stirred reactor vessels. If fixed-bed catalysts for gas phase or trickle beds are to be specified, several types of laboratory reactor are available. All such reactors should be located within suitably ventilated areas and the appropriate hazard and operability studies performed prior to startup.

(c) Reactor heating needs to be accurately controlled. Ideally, reactors should be jacketed to ensure even heating, and good temperature control gained by the use of internal thermowells if possible.

(d) Agitation is important. Stirred autoclaves should be fitted with high shear stirrers to aid hydrogen solubility.

(e) Provision for sampling the reaction mixture as the reaction proceeds should be made. For sampling from a stirred autoclave, a dip pipe fitted with suitable pressure valves is often adequate. Care should be exercised when sampling from any pressurized system.

(f) A range of common commercial catalysts (see Appendices II and III) will be required. They should be stored away from inflammable solvents and any potential catalyst poisons (e.g. organo-sulphur, -chlorine, -nitrogen, -phosphorus, or -arsenic compounds). Failure to do so may rersult in inconsistent or irreproducible results.

Spent catalysts should be separated from reaction mixtures and re-tested in the reaction if catalyst recycling is necessary. Highly reduced catalysts may be pyrophoric and should be disposed of with care. Catalysts containing precious metals should be stored for metal recovery, preferably keeping separate the different metals.

It is essential that the raw materials (reaction substrates and solvents) used in the laboratory are of the same origin and purity as those to be used in any subsequent scale-up. Fast, accurate analytical procedures for monitoring the composition of the reaction mixture are desirable.

10.3 NEW HETEROGENEOUS HYDROGENATION CATALYSTS

Heterogeneous catalysts are available from commercial suppliers as standard grades or may be 'tailored' to suit specific reactions or process conditions. Major developments in hydrogenation catalysis generally arise from an intensive study of a particular reaction system in which both the catalyst formulation and the process conditions are optimized synergistically. Since much of this work is performed in industrial laboratories, only a fraction is published in the open literature, and so it may appear, to the causal observer, that the subject has advanced little over the past two decades. This is far from the truth, as either an inspection of *Chemical Abstracts* or a brief review of recent papers illustrates.

Various chromium-doped Raney nickel catalysts show increased activity towards the hydrogenation of acetophenone (Fig. 10.1) and glucose [10], but the doping does not influence the hydrogenation of cycohexene [11].

A description has appeared of a new type of fixed-bed Raney catalyst, which is prepared by extruding a Raney alloy powder with a polymer and plasticizer which are subsequently removed by calcination prior to caustic leaching of the aluminiium [12,13]. Higher activity than Ni/SiO_2, $Ni/SiO_2-Al_2O_3$ and granular Raney nickel is observed for both vapour-phase and trickle flow toluene hydrogenation [12].

Sulphided catalysts, traditionally used for large-scale hydrotreating of sulphur-containing petroleum feeds, have been found highly selective for transformation of

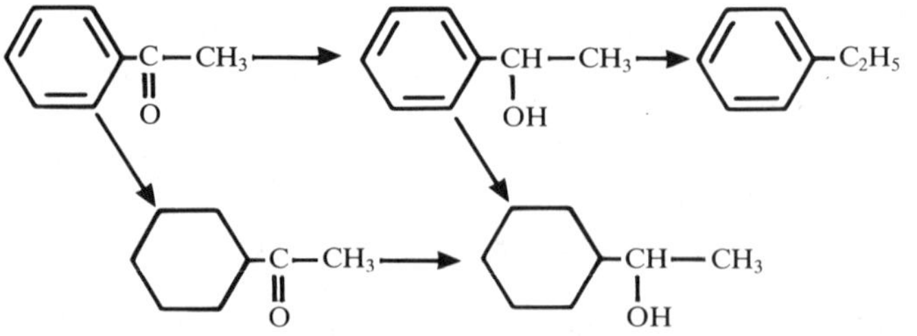

Fig. 10.1 — Stages in the hydrogenation of acetophenone.

ketones, alcohols and amines into hydrocarbons (i.e. for selective hydrogenolysis of $C-O$ and $C-N$ bonds) and for the selective deprotection of ketals and thioketals (Fig. 10.2) [14].

10.4 FUNCTIONAL GROUP REDUCTIONS

It is possible to hydrogenate most organic functional groups over suitable supported catalysts. The catalyst of choice is usually that which combines a good rate of hydrogenation (often measured as the rate of hydrogen uptake) with a high selectivity to the desired product. Side reactions are minimized by the proper selection of catalyst and reaction conditions.

10.4.1 Carbon–carbon bonds

Selective hydrogenation of acetylene in ethylene or propylene streams is performed industrially using a range of Pd/Al_2O_3 catalysts, and there are examples of the application of the reaction in organic synthesis. The saturation of olefins and the selective conversion of di-olefins to mono-olefins are important for petrochemical feedstocks, and supported nickel catalysts are often used. Comparisons of the activity and selectivity of catalysts often include the hydrogenation of simple alkenes. The complete saturation of pure unsaturated hydrocarbons is less important industrially.

10.4.1.1 Carbon–carbon triple bonds

The hydrogenation of the triple bond occurs in many synthetic reaction sequences. Usually the intention is to achieve high yields of the *cis*-olefin. Supported palladium catalysts such as $Pd/Pb/CaCO_3$ (Lindlar) or $Pd/Cd/C$ are the most effective. Few

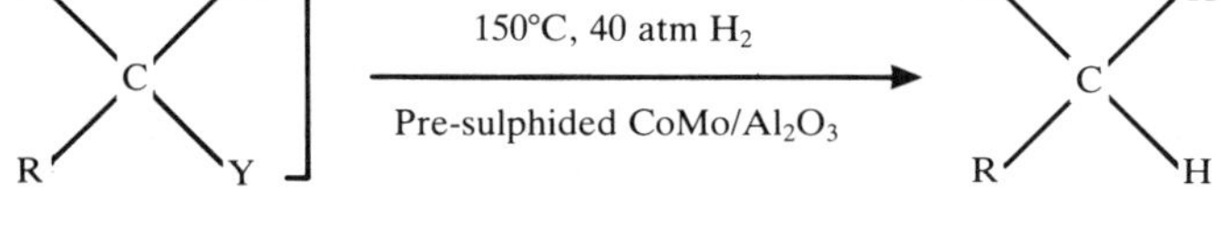

(R = various; X, Y = O, S)

Fig. 10.2 — Hydrogenolysis of ketals and thioketals.

examples of the hydrogenation of alkynes have been published in recent years [4]. However, the selective reduction of an ester of an alkynylcarboxylic acid to the corresponding saturated ester using a carefully de-passivated Ni/Al_2O_3 catalyst has been described (equation (10.1)) [9].

$$R-C\equiv C-C-C-OCH_3 \xrightarrow[\text{De-passivated Ni/Al}_2O_3]{50°C, 3.3 \text{ atm H}_2} R-CH_2CH_2C-OCH_3 \quad (10.1)$$

10.4.1.2   Carbon–carbon double bonds

Unhindered olefins are very readily hydrogenated, but the migration of the double bond during the hydrogenation process is a common problem. The rate of hydrogenation of olefins occurs in decreasing order of activity over platinum metals Pd > Rh > Pt > Ru. Palladium is some 50 to 200 times more active than nickel. Although the tendency for double bond migration decreases in the order Pd > Ni ≫ Rh ≫ Ru, Os > Ir, Pt, the reaction variables also play a major part. Raney and supported nickel catalysts are often selected as an economic compromise for low-cost products. The detailed mechanisms of olefin hydrogenation are still unknown. However, recently it has been suggested that a *cis*-concerted mechanism operates in the liquid phase hydrogenation of several styrene derivatives, $PhCH = CHR$ ($R = H$, CH_2OH, CHO, COMe, CO_2H and CO_2Me) at 25–60°C under a starting pressure of 4.1 atm on Ni/Al_2O_3, Ni/SiO_2 or $Ni/AlPO_4$ catalysts [15].

10.4.2   Cyclic structures

The saturation of aromatic rings can readily be achieved with supported metal catalysts, and in practice the preferred metals are Pt, Rh or Ni. The major difficulties encountered in aromatic hydrogenation are the retention of other functional groups, regioselectivity and the control of stereo-chemistry. Careful selection of metal catalyst and reaction conditions can permit reduction of substituted aromatics without unwanted hydrogenolysis.

10.4.2.1   Aromatic rings

Industrial catalysts for hydrogenation of benzene to cyclohexane (an important step in the route to adipic acid and hexamethylenediamine for Nylon) include Raney nickel and supported nickel or platinum. Saturation of substituted aromatics, performed for the production of solvents, uses $Ni/W/Al_2O_3$, sometimes in combination with a catalyst containing a high loading of Ni — particularly if the feed is contaminated with sulphur. Ring saturations of several functionalized benzene derivatives are also industrially important (e.g. phenol to cyclohexanone (equation (10.2)) [16] and aniline to cyclohexylamine).

$$\text{(phenol)} + 3H_2 \xrightarrow[\text{Promoted Pd/Al}_2O_3]{180°C, H_2 \text{ pressure}} \text{(cyclohexanone)} \quad (10.2)$$

The hydrogenation of mono-substituted benzenes RC_6H_5 (R = Et, Ph, cyclo-C_6H_{11}, $PhCH_2$, cyclo-$C_6H_{11}CH_2$) and of *ortho-* and *para*-substituted phenols has been studied by a batch method at 340°C and 70 atm H_2 over a sulphided $NiO/MoO_3/\gamma$-Al_2O_3 catalyst [17]. The rates of hydrogenation for phenols are always higher than for the corresponding alkylbenzenes. Pd and Ru/Al_2O_3 have shown good selectivity and activity for the hydrogenation of *N*-alkylanilines to the corresponding *N*-alkycyclohexylamines [18]. The yields of *N*-ethylcyclohexylamine and *N,N*-di-methylcyclohexylamine are 94% and 90% respectively.

10.4.2.2   *Polycyclic rings*
The selective saturation of a single aromatic ring in a polyaromatic system is frequently required in synthesis. The degree of hydrogenation of a particular molecule depends primarily upon structure but may be influenced by choice of catalyst and reaction conditions. Pd is usually much more selective than Pt, Rh or Ru, and the choice of support can be critical [4].

10.4.2.3   *Heterocyclic rings*
Heterocyclic compounds can either become saturated or undergo ring fission under hydrogenation conditions. Generally, pyrroles, pyridines, pyrimidines, quinolines and furans undergo saturation. The oxygen-containing rings are more easily reduced and suffer hydrogenolysis, as do any heterocyclic structures which are strained, activated by unsaturation or contain a particularly weak bond. Whether ring hydrogenation or ring fission of a furan occurs depends upon the substrate, the catalyst and the reaction conditions. Pd, Rh and Ru catalysts are used, and higher temperatures favour hydrogenolysis. The hydrogenation of pyridines is reported over many catalysts — Ni, Pd and Ru are employed industrially. Certain substituents can halt the hydrogenation at the tetrahydro stage. A study of the hydrogenation of isoquinoline to decahydroisoquinolines over Raney nickel catalysts at 190–220°C under hydrogen at 1–10 atm reveals the major reaction pathway to be via sequential hydrogen addition [19]. Some simultaneous reactions are found to contribute at pressures above 50 atm [19].

10.4.3   **Carbonyl groups**
Aldehydes and ketones are readily hydrogenated to the corresponding alcohol. Aliphatic carbonyls generally give higher selectivities than aromatic carbonyls, which often suffer dehydroxylation of the product. In the laboratory, aldehydes and ketones can be hydrogenated to alcohols over Pt, Rh, Ni or Ru catalysts under mild conditions [2]. The choice of reaction solvent significantly affects the rate of carbonyl reduction, and small quantities of acid or base may also influence both rate and selectivity [4]. At large scale, these hydrogenations are usually performed using either Ni or Cu catalysts, but the published data are scarce. For example, butanals can be hydrogenated to the corresponding butanols either with Ni catalysts at 2–3 atm and 115°C or with Cu catalysts at about 30 atm and 160°C. The relative reactivities of 19 saturated acyclic and cyclic ketones over copper chromite catalysts in the vapour phase at 185–240°C and 20–100 atm total pressure have been reported [20]. A linear

correlation between the logarithm of reactivities and the Taft E_s coefficients for eight methyl alkyl ketones indicates the predominance of steric control [20].

Hydrogenation of unsaturated α,β-unsaturated aldehydes is a more difficult problem. Recently, cinnamaldehyde and crotonaldehyde have been hydrogenated to unsaturated alcohols with selectivities of 96% and 84% respectively at 50-90% conversion over a Co/SiO_2 catalyst [21]. Studies using various platinum group mono- and bimetallic catalysts for the former case indicate that the factors controlling selectivity to cinnamyl alcohol are the choice of metal and the nature of the support. In particular, Pt and Ru have tunable selectivity while Pd and Ir are non-selective and selective respectively [22].

In the hydrogenation of glucose with Raney-type Ni–M bimetallic catalysts (M = Cr,Fe,Co,Cu,Mo), those containing Cr and Mo are twice as active as unpromoted Ni whilst neither added Co nor Cu changes the catalytic properties [23]. Enantioselective hydrogenations of α-keto esters to α-hydroxyesters, important chiral intermediates for the synthesis of biologically active compounds, have been reviewed [24]. A recent investigation using Pt catalysts [25] demonstrates that thermal pretreatment of the catalyst in hydrogen at 400°C gives 15-20% higher optical yields, whilst the effects of chinchona alkaloids are complex and variable.

10.4.4   Nitro, nitrile and imine groups

The conversion of a nitro group into an amine is one of the most facile and important hydrogenation reactions. The reaction can take place over a range of metals including Pd, Pt, Rh, Ru and Ni. The hydrogenation of nitrobenzene to aniline is frequently used by precious metal catalyst manufacturers as a standard quality control test. The reduction of an aromatic nitro group occurs in preference to all other functional groups with the exception of double or triple bonds.

The mechanism of the reaction of aliphatic or aromatic nitro compounds is not simple and can proceed through the formation of nitroso-, azo-, diazo- and hydroxy-amino-intermediates [2]. Selective catalytic hydrogenation of nitrobenzene to phenyl-hydroxylamine (PHA) is of considerable industrial importance because PHA can undergo rearrangement to a variety of useful products and Pt/C catalysts have been investigated under a wide range of conditions [26]. The solvent has a significant effect on selectivity, which can be correlated with dielectric constant. The selectivity of substituted nitroaromatics also correlates with the electron-releasing power of the substituent functional groups.

The reduction of halonitroaromatics to the corresponding amines is becoming increasingly important, because the traditional route using iron/hydrochloric acid is less environmentally acceptable. Probably the best catalysts to date are modified supported Pt group metals. For example, a modified Pt/C catalyst has been reported that does not promote dehalogenation, requires no process additives, and is highly selective [27]. However, the catalyst selectivity is still dependent on the halogen substituent and decreases in the order F > Cl > Br. The succcessful conversion of a series of nitroalkanes into the corresponding amines, with retention of configuration, has been reported (Fig. 10.3) [28]. Although usually performed at elevated temperatures and pressures over Raney Ni or Pt, transfer hydrogenation using ammonium

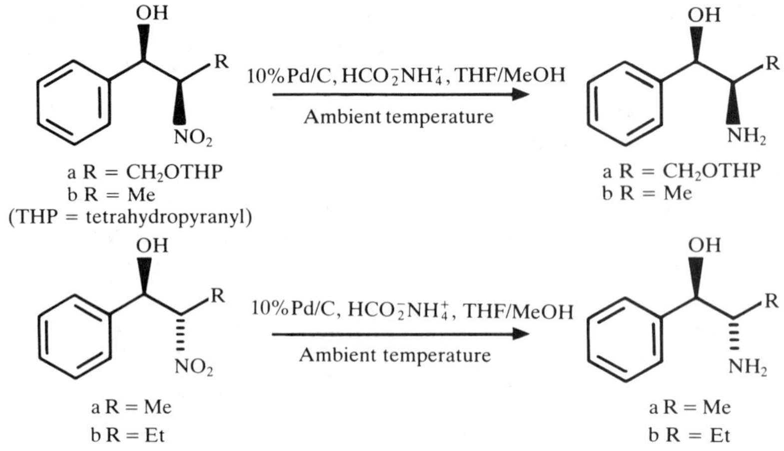

Fig. 10.3 — Transfer hydrogenation of chiral nitro compounds.

formate achieves the desired transformation under ambient conditions over Pd/C in THF/methanol. *p*-Phenylenediamine analogues can be prepared from 4-nitrodiphenylamine by transfer hydrogenation using formic acid and Raney Ni [29]. The reaction is applicable to the reduction of the condensation products of a nitro compound with an alcohol or ketone.

Few publications concerning nitrile reduction have appeared recently despite the high industrial importance of, for example, adiponitrile conversion to hexamethylene-diamine (HMD), for which a comparison of two activated Ni/Al_2O_3 catalysts with Raney Ni has appeared [9]. In this case the yield of HMD does not benefit from catalyst pre-treatment. The effect of dopant on Raney nickel catalysts has been studied using 2- and 3-butenenitriles as model reactants [30]. Cr-doped Ni does not influence the reduction of 2-butenenitrile but has a negative effect on the 3-butenenitrile isomer. Highly dispersed Rh/MgO is a very selective and efficient catalyst for the partial hydrogenation of aliphatic α,ω-dinitriles to the corresponding α-aminonitriles [31].

The reductive amination of alcohols to the corresponding amines is very important on the large scale, and the reaction has been found to proceed via imine intermediates. The kinetics of reductive amination of 2-propanol with aniline on a Cu/Cr catalyst show that the overall rate is determined by both the condensation of acetone with aniline and the subsequent hydrogenation of the ketimine intermediate [32] and is accelerated by the addition of aluminosilicate. In contrast, a kinetic study into the effect of hydrogen in catalytic amination of alcohols has shown that: (a) the abstraction of an α-hydrogen from the reactant alcohol determines the reaction rate to all amination products; (b) hydrogen has no marked influence on the overall amination rate; (c) hydrogen improves the selectivity by suppressing simultaneous disproportionation of reactant and product amines; and (d) that hydrogen is necessary to prevent catalyst deactivation [33].

10.4.5 Hydrodehalogenation

Selective hydrodehalogenation is a valuable pathway to important intermediates in organic synthesis because the carbon–halogen bond is easily cleaved by hydrogen, particularly in the presence of Pd [4]. Supported Pt and Rh are less active but allow ring saturation of molecules containing a benzyl halogen without dehalogenation. Very good yields of 3,5-dichloroaniline or 3,5-dichlorophenol are obtained by the hydrodechlorination of the tri-, tetra-, and pentahalogenated precursors in the liquid phase with Pd/C catalysts [34]. Selective hydrodechlorination of polychlorophenols requires a Lewis acid and HCl whilst that of polychloroanilines does not need a Lewis acid.

10.4.6 Hydrogenolysis

Hydrogenolysis is the hydro-cleavage of any bond within a molecule to produce two discreet carbon-containing molecules. Pd is the preferred catalyst for the hydrogenolysis of benzyl alcohols, ethers, esters, acetals and phosphates. The most important industrial example of hydrogenolysis is the reduction of esters to alcohols, usually performed over Cu-containing catalysts under forcing conditions. Recent work demonstrates that, in some cases (e.g. butyl butanoate to butanol), the reaction can proceed under surprisingly mild conditions, and the mechanism of ester hydrogenolysis over Cu catalysts has been investigated [35]. The cleavage of benzilic acid dimeric ester to yield over 90% diphenylacetic acid proceeds under ambient conditions with a Pd/C catalyst (equation (10.3)) [36].

$$\text{[structure]} \xrightarrow[\text{10\% Pd/C, EtOAc}]{\text{Ambient temp., 1 atm H}_2} 2 \text{ Ph}_2\text{CHCOOH} \qquad (10.3)$$

The liquid phase hydrogenolysis of β-lactams over Raney Ni is claimed to be a new, efficient route to 1,3-aminoalcohols [37]. A variety of natural products can also be successfully hydrogenolysed by catalytic transfer hydrogenation using Pd/C and cyclohexene [38]. Catalytic hydrogenolysis of the C6-halogen bond of halopenicillanates can be achieved with Pd/CaCO$_3$ or Rh/Al$_2$O$_3$ (equation (10.4)) [39].

$$\text{[structure]} \xrightarrow[\substack{25°C, \text{ Pd/CaCO}_3 \\ \text{or Rh/Al}_2\text{O}_3}]{\text{H}_2(\text{D}_2), 1 \text{ atm}} \text{[structure]} \qquad (10.4)$$

X,Y = I,Br,Cl

10.5   FUTURE TRENDS

The true value of heterogeneous hydrogenation catalysts continues to extend into all branches of organic chemistry. For example, a dimethyl phosphate protecting group for the tryptophan moiety in peptide synthesis has been chosen because of its stability to Pd-catalysed hydrogenation [40]. The long cherished goal of heterogenizing highly selective homogeneous catalysts by their attachment to functionalized polymers [41] is apparently yet to bear fruit in an operating commercial process. The high cost of such catalysts may be a stumbling block to their use at volumes higher than laboratory scale.

In addition to the obvious goal of increasing catalyst performance (in terms of activity and selectivity, i.e. productivity), two major trends in the application of heterogeneous catalysts are emerging. Firstly, further research is being directed at the development of more efficient processes for the introduction of stereoselectivity [24]. Essentially, the aim is to modify the catalyst surface with a sacrificial chiral molecule or a physical pre-treatment. Although initially successful, under the reaction conditions catalyst surfaces seem to re-structure, eliminating the initial effect. The application of such techniques to catalytic systems that can operate under very mild conditions would therefore appear to offer the best likelihood of achieving the desired induction of enantiomeric selectivity. The effect of crystallite size and pore diffusion have been shown to affect hydrogenation of methyl acetoacetate to optically active methyl 3-hydroxybutanoate over modified Ni/SiO_2 catalysts [42]. Optical yields in the hydrogenation of ethyl acetoacetate are in the range 1.6–7.1% with a series of R,R-(+)-tartaric acid-treated, bimetallic Cu/Pd and Ni/Pd catalysts [43], which compares with 9.9–21.0% over similarly modified Raney nickel [44].

Secondly, the reduction of the environmental impact of synthetic processes using metallic catalysts is being addressed by moving from batch reactors using powders towards fixed-bed or closed-loop, reactors. The actual quantity of catalyst handled may sometimes be reduced by switching to a precious metal catalyst. Although there may be a trade-off against costs, particularly as such catalysts often need recovery, the main benefit may be to operate under significantly milder conditions. The decontamination of spent catalysts prior to disposal or recovery is also increasingly important.

10.6   APPENDICES

10.6.1   Appendix I. Hydrogenation equipment suppliers

10.6.1.1   *Laboratory reactors*
Autoclave Engineers/George Meller Ltd
Berghof Autoclaves/Techmate Ltd
Buchi Autoclaves/Ken Kimble & Company Ltd
Parr Stirred & Non-stirred Autoclaves, Pressure Vessels
Scientific & Medical Products Ltd
Pressure Products Industries, Inc./F. Kenyon (Process Equipment) Co.

10.6.1.2   Gas-handling equipment, gas mixtures, etc.
Gas and Equipment Limited
Electrochem Ltd

10.6.2   Appendix II. Some suppliers of hydrogenation catalysts
BASF Catalysts
Calsicat
Crosfield Catalysts
Degussa
Engelhard (Incorporating Harshaw)
Girdler Sud-Chemie
Heraeus
ICI Catalysts
Johnson-Matthey
Procatalyse

10.6.3   Apppendix III. Hydrogenation catalysts available from suppliers of laboratory chemicals
Raney Copper, active catalyst [*Chem. Abstr.* Registry Number 7440-50-8]
Copper chromite [122053-18-8]
Copper chromite, barium promoted [12053-18-8]
Raney Nickel, active catalyst [7440-02-0]
Nickel–aluminium alloy
Nickel on Kieselguhr
Nickel on silica/alumina
Palladium black [7440-05-3]
Palladium on active carbon (1–10% loadings)
Palladium on alumina, various
Palladium on barium carbonate
Palladium on barium sulphate
Palladium on calcium carbonate
Palladium on calcium carbonate, poisoned with lead (Lindlar)
Palladium(II) oxide
Platinum powder [7440-06-4]
Platinum sponge [7440-06-4]
Platinum on activated carbon (1–10% loadings)
Platinum on alumina, various
Platinum(IV) oxide [1314-15-4] (Adam's catalyst)
Platinum(IV) oxide monohydrate [52785-06-5] (Adam's catalyst)
Rhodium black [7440-16-6]
Rhodium on alumina, various
Rhodium on carbon, various
Ruthenium black [7440-18-8]
Ruthenium on alumina, various
Ruthenium on carbon, various
Ruthenium(IV) oxide [12036-10-1]

REFERENCES

[1] R. L. Augustine, *Catalytic Hydrogenation, Techniques and Applications in Organic Synthesis*, Marcel Dekker, New York, 1965.

[2] R. L. Augustine, *Catal. Rev. Sci. Eng.*, 1976, **13**, 285.

[3] M. Freifelder, *Catalytic Hydrogenation in Organic Synthesis: Procedures and Commentary*, Wiley, New York, 1978.

[4] P. N. Rylander: (a) *Catalytic Hydrogenation in Organic Syntheses*. Academic Press, New York, 1979; (b) *Catalytic Processes in Organic Conversion*, p. 1 in *Catalysis Science and Technology* (Eds J. R. Anderson and M. Boudart), Springer-Verlag, New York, 1983, Vol. 4; (c) *Hydrogenation Methods*, Academic Press, New York, 1985.

[5] H. M. Colquhoun, J. Holton, D. J. Thompson and M. V. Twigg, *New Pathways in Organic Synthesis*, New York, 1984.

[6] L. L. Hegedus (Ed.), *Catalyst Design, Progress and Perspectives*, Wiley, London, 1987.

[7] (a) M. Guisnet, J. Barrault, C. Bouchoule, D. Duprez, C. Montassier and G. Perot (Eds), *Heterogeneous Catalysis and Fine Chemicals, Proceedings of an International Symposium, Stud. Surf. Sci. Catal.*, Elsevier, New York, 1989, Vol. 41; (b) M. Guisnet, J. Barrault, C. Bouchoule, D. Duprez, G. Perot, R. Maurel and C. Montassier (Eds), *Heterogeneous Catalysis and Fine Chemicals II, Stud. Surf. Sci. Catal.*, 1991, Vol. 65.

[8] T. Yoshino, T. Abe, S. Abe and I. Nakabayashi, *J. Catal.*, 1989, **118**, 436.

[9] M. G. Scaros, H. L. Dryden Jr., J. P. Westrich, O. J. Goodmonson and J. R. Pilney, *Catal. Org. React. Chem. Ind.*, 1988, **33**, 419.

[10] J. M. Bonnier, J. P. Dalmon and J. Masson, *Appl. Catal.*, 1988, **42**, 285.

[11] T. Koscieski, J. M. Bonnier, J. P. Dalmon and J. Masson, *Appl. Catal.*, 1989, **49**, 91.

[12] W-C. Cheng, L. J. Czarnecki and C. J. Pereira, *Ind. Eng. Chem. Res.*, 1989, **28**, 1764.

[13] W-C. Cheng, C. B. Ludsager and R. M. Spotnitz, U.S. Patent 4 826 799 (1989), *Chem. Abs.*: **111**:141608b.

[14] C. Moreau, R. Durand, P. Graffin and P. Geneste, p. 139 in Ref. [7(a)].

[15] F. M. Bautista, J. M. Campelo, A. Garcia, R. Guardeno, D. Luna and J. M. Marinas, *J. Chem. Soc., Perkin Trans. II*, 1989, 493.

[16] I. Dodgson, K. Griffin, G. Barberis, F. Pignataro and G. Tauszik, *Chem. Ind. (London)*, 1989, 830.

[17] C. Aubert, R. Durand, P. Geneste and C. Moreau, *J. Catal.*, 1988, **112**, 12.

[18] I. Palkovics, I. Gemes and P. Szonyi, *Hung. J. Ind. Chem.*, 1989, **17**, 113.

[19] H. Okazaki, M. Sveda, Y. Ikefuji and R. Tamaru, *Appl. Catal.*, 1988, **43**, 71.

[20] J. Jenck and J. E. Germain, *J. Catal.*, 1980, **65**, 133.

[21] Y. Nitta, Y. Hiramatsu and T. Imanaka, *Chem. Express*, 1989, **4**, 281.

[22] (a) A. Giroir-Fendler, D. Richard and P. Gallezot, p. 171 in Ref. [7(a)]; (b) P. Fouilloux, p. 123, in Ref. [7(a)].

[23] J. Court, J. P. Damon, J. Masson and P. Wierzchowski, p. 189 in Ref. [7(a)].

[24] M. Bartok (Ed.), *Stereochemistry in Heterogeneous Metal Catalysis*, Wiley, New York, 1985, pp. 511–524.

[25] (a) H. U. Blaser, H. P. Jalett, D. M. Monti, J. F. Reber and J. T. Wehrli, p. 153 in Ref. [7(a)]; (b) H. U. Blaser, H. P. Jalett, D. M. Monti and J. T. Wehrli, *Appl. Catal.*, 1989, **52**, 19.

[26] S. L. Karwa and R. A. Rajadhyaksha, *Ind. Eng. Chem. Res. Dev.*, 1987, **26**, 1746.

[27] (a) G. G. Ferrier and F. King, *Plat. Met. Rev.*, 1983, **27**, 2; (b) K. Griffin and J. Anderson, *Perf. Chem.*, 1988, 19.

[28] A. G. M. Barrett and C. D. Spilling, *Tetrahedron Lett.*, 1988, **29**, 5733.

[29] A. A. Banerjee and D. Mukesh, *J. Chem. Soc., Chem. Commun.*, 1988, 1275.

[30] J. L. Dallons, G. James and B. Delmon; (a) p. 115 in Ref. [7(a)]; (b) *Catal. Today*, 1989, **5**, 257 .

[31] F. Mares, J. E. Galle, S. E. Diamond and F. J. Regina, *J. Catal.*, 1988, **112**, 145.

[32] J. Dlouhy and J. Pasek, *Coll. Czech. Chem. Commun.*, 1989, **54**, 326.

[33] A. Baiker, p. 283 in Ref. [7(a)].

[34] G. Cordier, p. 19 in Ref. [7(a)].

[35] M. A. Kohler, N. W. Cant, M. S. Wainwright and D. L. Trimm, *Proc. Int. Congr. Catal., 9th* (Eds M. J. Phillips and M. Ternan), Chem. Inst. Can., Ottawa, 1988, **3**, 1043.

[36] P. Strazzolini, A. G. Giumanini and G. Verardo, *Synth. Commun.*, 1987, **17**, 1919.

[37] M. Bartok, L. Gera, G. Gondos and A. Molner, *J. Mol. Catal.*, 1988, **49**, 103.

[38] A. Bianco, P. Passacantilli and G. Righi,, *Tetrahedron Lett.*, 1989, **30**, 1405.

[39] E. L. Setti, D. U. Belinzoni and O. A. Mascaretti, *J. Org. Chem.*, 1989, **54**, 5731.

[40] H. A. Guillaume, J. W. Perich, J. D. Wade, G. W. Tregear and R. B. Johns, *J. Org. Chem.*, 1989, **54**, 5731.

[41] C. U. Pittman Jr., 'Catalysis by Polymer-supported Transition Metal Complexes', in *Polymer-supported Reactions in Organic Synthesis* (Eds P. Hodge and D. C. Sherrington), Wiley, New York, 1980.

[42] Y. Nitta and T. Imanaka, *Bull. Chem. Soc. Jpn*, 1988, **61**, 295.

[43] T. I. Kuznetsova, I. P. Murina, A. A., Vedenyapin, V. M. Akimov and E. I. Klabunovski, *React. Kinet. Catal. Lett.*, 1988, **37**, 363.

[44] A. Boerner, H.-W. Krause and K. Kortus, *React. Kinet. Catal. Lett.*, 1988, **36**, 103.

11

Designing micro-reactors — a personal account

P. Laszlo

11.1 INTRODUCTION

Preparation of speciality chemicals receives ever increased priority in the industrialized countries. The technologically more advanced countries are witnessing a gradual shift from production of basic chemicals (such as ethylene oxide or vinyl chloride) to the synthesis of fine chemicals while developing countries are becoming responsible for increasing proportions of the production of simple molecules as a result of the availability of cheap labour and fossil hydrocarbons (oil or natural gas) as sources both of energy and of raw materials. Other reasons for the rise of fine chemistry in developed countries are the production overcapacity of units set up during the 70s, which pulled down the prices of heavy chemicals, and indirect pressure from ecologists, through national governments and through the EEC, leading to stricter and stricter regulations for environmental protection.

This shift in emphasis increases the need to devise reactions and processes that are quantitative, proceed under mild conditions (i.e. ambient temperature and pressure), and are economical of reagents, catalysts, and solvents. It is very important that they be selective also, since very often separation procedures are a bottleneck. A new process, if it is to be implemented nowadays, must take place with close to 100% yield and with better than 85% selectivity.

The obvious model is the admirable efficiency of enzymes. These biological catalysts, generally proteic (RNA is an exception, perhaps of profound significance for the abiotic formation of primeval biomonomers and polymers), give quantitative yields. Reactions are usually specific: a single product is formed, to the extent of 100%. Furthermore, the reactions take place in aqueous solution, at atmospheric pressure and at body temperature. These performances contrast with typical conditions for present-day industrial chemistry, where many reactions proceed at elevated temperatures and pressures, often in the gas phase.

The challenge to chemistry is thus to emulate the high performances of enzymes and to avoid environmental pollution, ideally by devising processes that occur in water and whose output, besides the desired product in quantitative yield, consists ideally of only water, dioxygen, and dinitrogen!

One approach is to imitate enzymes, to take a leaf from the album of Nature so to say, and to try to duplicate the enzymatic works. Enzymes are micro-reactors. At the scale of about a nanometre, they fulfill various functions comparable to those of an industrial reactor, whose dimensions are of the order of metres. Enzymatic micro-reactors start by binding the substrate, and they bind it — to quote a wonderfully fertile concept of Linus Pauling [1] — in a stressed state. Enzymes force a substrate into a conformation packed with energy, hyperreactive, that prepares it for the transformation that it will undergo.

This is the backdrop for our contribution to the new methodologies of organic synthesis. Very early on we opted for using *clays* as reaction supports and as catalysts. Clays were chosen because of their favourable features as micro-reactors, to be described in section 11.2. This account is a personal case history of our work in the area and documents the variety of sources that have influenced its course.

The work drew its origin from a successful study in prebiotic chemistry, i.e. in the spontaneous formation of biomolecules in the absence of life, which convinced me that clays can serve as extremely efficient catalysts. It was tempting to explore how general this property was and we opted, in my laboratory at the University of Liège, for oxidation reactions. We found that the surface of a clay is capable of stabilizing a reagent which is otherwise fierce and dangerous. Thus, we could hope to set up a whole new class of oxidizing reagents. Such an ability at stabilization of hyperreactive species is an important characteristic of clays, a real asset. Another promising aspect made clay minerals attractive as micro-reactors: their ability at concentration of organic matter within their interlamellar spaces. This concentration factor had played a role in the prebiotic reactions that we had observed.

We came quickly to realize that clays had yet another favourable feature: they force chemical reactions to occur on a planar surface rather than in a three-dimensional reaction volume. This leads to large accelerations, by up to five or six orders of magnitude in reaction rates. Clay micro-reactors also increase the activity of ions congregated near their surface, which accelerates ionic reactions.

Activation of a clay surface is easily affected. Dehydration in an oven, at a temperature low enough that it does not totally destroy the internal structure of the clay, increases the catalytic activity. The explanation is that removal of water from surface silanol groups O_3Si-OH creates radical centres, of the O_3Si-O type. These are efficient oxidizing centres.

Thus, the main themes of this chapter have been stated: ability of clays to concentrate molecules about to be transformed; stabilization offered to highly reactive species; fast diffusional kinetics on clay surfaces; acceleration of ionic reactions by clay polyelectrolytes; ease with which dehydrative activation can be performed. These various assets correspond to the successive stages in the following narrative.

11.2 FROM PREBIOTIC CHEMISTRY TO PRESENT-DAY FRONTIERS OF CHEMISTRY

Who else but Desmond Bernal, the British crystallographer, could have come up, in the mid-1940s, with the notion of clays as efficient micro-reactors for organic chemistry? This Renaissance man, a scholar well-informed in many fields of knowledge, had the intuition of a clay niche for the spontaneous formation of the first organic molecules under abiotic circumstances. He based it on three areas of his expertise or beliefs: his competence about silicates, many structures of which he had determined by X-ray diffraction; his knowledge of biopolymers — he had been one of the very few pioneers in the crystallographic study of proteins using X-rays; and his Marxist beliefs, which made him put aside any creationist scheme.

In a lecture that Bernal gave [2], he drew attention to a whole series of features that made clays good candidates as supports for prebiotic chemistry. He stressed their swelling aptitude, which enables clays to store the organic matter formed within and so shield it from the destructive influence of ultraviolet radiation from the Sun. He pointed to their huge specific surfaces, up to $750 \, \text{m}^2 \, \text{g}^{-1}$ (of the order of magnitude of the area of a small town per kilogram of dry clay powder). He stated that the lamellar structure of clays allows not only concentration but also compartmentalization of various chemical species to keep them apart from one another. He also considered that the ordered structure of clays was an asset, together with their ability to serve as polymerization templates.

Microscopic examination shows that clay particles are highly organized. Most often, the clay platelets are stacked vertically on top of one another, in the manner of a deck of playing cards. Thus, kaolinite (china clay) has the microscopic aspect of flakes with roughly hexagonal shapes. Certain types of clay (smectites), once dehydrated, can regain water layers and swell. The water molecules intercalate between the parallel plates. The clay sheets thus gain translational mobility, which explains their plasticity, a major source of attraction of such materials.

X-ray diffraction is the choice means for measuring the distrance between clay sheets. This so-called *basal spacing* (denoted as d_{001}) is about 1 nm in a totally dehydrated smectite, a totally collapsed clay in which the sheets come into van der Waals contact. Rehydration increases it, usually by discrete amounts. After a monomolecular layer of water has been reinserted, there comes in a second, a third layer, etc., increasing the basal spacing to ca. 4–5 nm, the exact value depending on the number of interlamellar water layers.

Clay structure is thus lamellar. It consists of parallel layers formed by silicate SiO_4 tetrahedral modules, and by aluminate AlO_6 octahedral modules. Kaolin clays consist of one such tetrahedral layer linked, through oxygen bridges, to one such octahedral layer (Fig. 11.1; for further details see also Chapter 1).

Another important clay family is that of *montmorillonites*. They are 2:1 clays, in which one octahedral layer is sandwiched between two tetrahedral layers (Fig. 11.2; see also Chapter 1).

In practice, substitutions occur. Here and there, a trivalent aluminium atom in the octahedral layer is replaced by a divalent metallic ion, such as Fe(II), Mn(II), etc. Likewise, in the tetrahedral layer, tetravalent silicon is replaced by a trivalent metal, such as Al(III) or Fe(III). Accordingly, the valences of the attached oxygens are no longer saturated. Local excesses of negative charge are set up and interstitial metallic ions come into the interlamellar space to maintain electroneutrality. In natural clays, these are typically uni- and divalent cations such as Na^+, K^+, NH_4^+, Ca^{++}, Mg^{++},

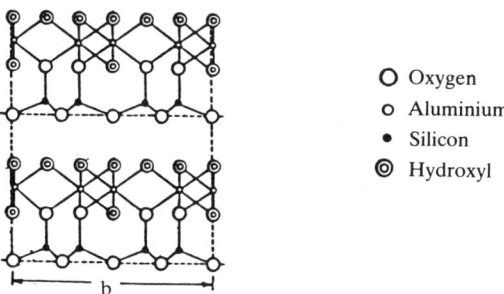

Fig. 11.1 — Schematic structure of a kaolinite (1:1) clay.

and have a key role for the cohesion of the clay particles. They are the glue holding individual platelets together — their overall negative charge would otherwise lead to mutual repulsion. Swelling can be viewed as increased hydration of these interlamellar ions.

When a substitutional defect arises in the octahedral layer, the attendant increase in negative charge is *localized* on the coordinated oxygens, next to the site of isomorphous substitution. Conversely, when the site of substitution is in one of the two tetrahedral layers for a montmorillonite, the local pile-up of negative charge can be neutralized easily through migration of a proton through the hydrogen-bond network of silanol groups: the electronic excess is therefore *delocalized* over the whole surface. Accordingly, for other reasons as well, the counterions congregated near the clay surface are merely *condensed*, rather than being *bound* to specific sites on the surface.

The platelets of course do not have indefinite extension. A typical size is 1 nm. At the edges, there are dangling, broken bonds between oxygen and the silicon and aluminium atoms. This exposes coordinatively unsaturated silicons in the tetrahedral layer, and coordinatively unsaturated aluminiums in the octrahedral layer. Both types, the latter especially, are potential catalytic sites.

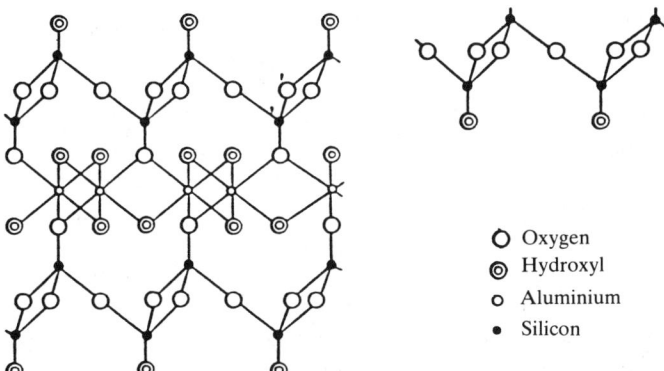

Fig. 11.2 — Schematic structure of a montmorillonite (2:1) clay.

Our initial involvement with these clay minerals arose in the 1970s from a fascinating scientific problem concerned with the self-assembly in saline aqueous solutions of a nucleotide, 5'-guanosine monophosphate (5'-GMP). The genetic code, as we know it, has GGG (i.e. a trimer of 5'-GMP!) coding for the synthesis of glycine, one of the most archaic amino acids on our planet [3]. We were witnessing the spontaneous formation of oligomers (octamers) of 5'-GMP. We speculated that there might have been a causal relationship. In an attempt to find some experimental support for such an idea, we studied the possibility of forming a nucleotide such as 5'-GMP from a nucleoside and a source of phosphate, and of forming in turn the nucleoside from condensation of a nucleic base with ribose sugar. To simulate the prebiotic environment, we naively opted for a clay template, and chose a montmorillonite clay for the purpose. The results were positive. In the presence of the clay, we obtained small amounts of the condensation products, the nucleotide as well as the nucleoside.

This work will not be published since there can be no rigorous proof of the relevance of the results to prebiotic molecular evolution, but since a clay had the proven ability to catalyse a possible prebiotic-type reaction, it appeared likely that clays might have the ability more generally to catalyse reactions of interest to present-day chemists. At the time (early 1980s) the community of organic chemists had one overwhelming priority: *to try to emulate enzymatic efficiency.* This paradigm of contemporary advances in the methodology of organic synthesis can be viewed as *a research programme* in the sense of Imre Lakatos [4]: that advancement of science results from Darwinian selection between competing research programmes.

Buoyed by our enthusiasm for the potential of clay minerals, we decided to embark on a programme of design of chemical reactions which would proceed selectively and in quantitative yield under mild conditions using only low-cost reactants, thereby emulating enzyme-catalysed reactions.

11.3   THE RELEVANCE OF ROCKETRY AND DRUNKENNESS TO THE DESIGN OF A NEW OXIDIZING REAGENT

During the sixties, the group of Professor Addison [5] at the University of Nottingham, England, produced a most impressive amount of work on *covalent* metallic nitrates. They showed that once dehydrated, certain metallic nitrates, with a general formula $M(NO_3)_n$, do not admit any longer of an ionic formulation as $M^{n+}(NO_3^-)_n$. They display true covalent bonds between the nitrato ligands and the metal. Addison gave these requirements for metallic nitrates to be strong oxidants: bidentate coordination of the nitrato groups; covalent bonding; availability of stable lower oxidation states for the metal. Addison and his group also showed that vibrational spectroscopy could differentiate readily between the various modes of attachment of the nitrato group to the metal (Fig. 11.3). The Pentagon became interested in these compounds for their potential as rocket propellants.

We became interested in ferric nitrate on account of its low toxicity, low cost and potential oxidizing properties. The commercial product, however, exists as the nonahydrate and is ionic — which seemed to defeat our purpose. Fortunately, an old

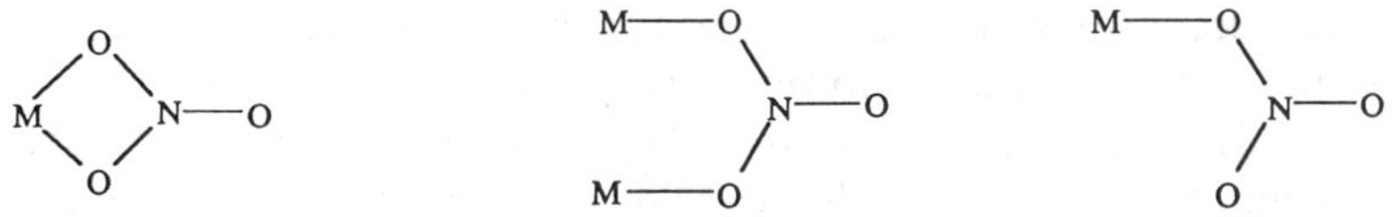

Fig. 11.3 — Various types of covalent bonding between metals and nitrate groups.

publication by Naumann [6] together with more recent work by Naldini [7] suggested that anhydrous material could be obtained by evaporating to dryness an acetone suspension of the commercial, hydrated, ferric nitrate. We obtained in this manner an oil whose composition corresponded to an anhydrous ferric nitrate solvated by two molecules of acetone. Its infrared spectrum suggested bidentate coordination and it even oxidized the nujol paraffin that had served to make the mull on which we had recorded our infrared spectrum!

We learned afterwards that *this acetone disolvate of covalent ferric nitrate has one-third the explosive power of TNT*! In order to use such a dangerous species in the laboratory, it had first to be tamed. We followed the model of Alfred Nobel, who made dynamite by stabilizing nitroglycerine on a finely divided solid support, kieselguhr (diatomaceous earth). We impregnated the covalent ferric nitrate, as it was being formed in the acetone medium, on a swelling clay, the K10 acidic montmorillonite.

The choice of a clay was governed by probability theory. For reaction to occur, the two reacting species must encounter each other. The probability of encounter is related to the mathematical problem of the drunkard wandering haphazardly in an attempt to get home [8]. In normal three-dimensional space, the probability of his success is low [9], but in two-dimensional space, the probability of his eventually arriving at the desired point is unity. According to Delbrück the ability of enzymes to constrain reactants to diffuse on a surface rather than in a volume is one of the primary reasons for their efficiency [10].

Although some solids (e.g. silica gel) have such rough surfaces that the effective dimensionality of their surfaces is close to 3, lamellar solids such as graphite and clays have effective dimensions only slightly greater than 2. In order to have a cheap solid with reproducible properties to stabilize anhydrous ferric nitrate we chose the bulk commercial clay codenamed K10. This is a natural clay, which has been subjected to calcination and treatment with mineral acid.

We nicknamed the ferric nitrate impregnated clay 'clayfen' [11]. It proved to possess a wide reactivity spectrum: oxidation of alcohols into aldehydes and ketones, of thiols into thionitrites, oxidative cleavage of bis (thioacetals), nitration of phenols, etc. [11]. This body of work was performed during the 1980–1985 period.

11.4 CONCENTRATION EFFECTS BY CLAYS

11.4.1 Concentration of organics by tactoids
As early as 1949, Desmond Bernal drew attention to the ability of clays at concentration of organic matter [2]. Clay platelets are smooth with an effective

dimension very close to 2, which does not seem to be conducive to the accumulation of organic matter at their surface. However, they are able to pile up to provide the microscopic equivalent of a house of cards. These aggregates are named tactoids, and they embody empty spaces. Fripiat has shown that these tactoids increase in number but not in size when the amount of a clay suspended in water is increased [12]. The aggregation number is small; in general, only a few clay platelets (2–6) make up these tactoids [12].

The presence of easily accessed cavities internal to the tactoids is another big asset of clays. In this manner, real clay particles behave just like microporous solids such as silica gel. Organic molecules can accumulate and spend long residence times in the pockets that trap them. Accordingly, their collisional encounters are made more highly probable and more frequent.

A great many structures in the physical world are self-similar on various scales. They resemble themselves when observed with a different magnification, whether smaller or greater. Their representative mathematical function can be characterized with a single parameter, named the fractional (or fractal) dimension. Clays often have fractal shapes, and Thorez, a mineralogist whose expertise is with clay minerals, has devised a unifying model, which he calls the *clay integron* [13]. This provides, just like the Mandelbrot fractals, a unitary description. It provides an analogy between the presence of oil in a sedimentary rock formation and the existence within tactoids of compartments capable of housing organic molecules.

11.4.2   Concentration of ions at clay interfaces

The planar surface of a montmorillonite 2:1 clay is not electrically neutral on account of the deviation from a strict aluminosilicate composition. Here and there a divalent cation, Fe(II), Mg(II), Mn(II), ..., replaces aluminium(III) in the octahedral layer. Likewise, an occasional Al(III) is found instead of Si(IV) in the tetrahedral layer. The latter substitutions are responsible for negative charge delocalized on the oxygens attached to the clay surface. In order to maintain electroneutrality, counterions are present. The distribution of the counterions in a direction perpendicular to the plane of the platelets is not uniform. The positively charged ions — typically Na^+, K^+, NH_4^+, Ca^{++}, Mg^{++}, in a natural clay — congregate near the charged surface. The aluminosilicate sheet, covered with electrical charges, all negative, is a *polyelectrolyte*. The condensation of the counterions at the interface increases their concentration, relative to the mean concentration, by factors up to 1000 or more. The explanations for this finding are on the one hand, coulombic attraction, and, on the other hand, the need to shield from one another the negative charges on the surface by way of positive ions interposed. Such ionic screening is the way by which plants use their root system to extract from the soil (of which clays are a prime constituent [14]) their nutritional electrolytes.

One of the consequences of the accumulation of electrical charge at the clay interface is the awesome magnitude of the corresponding electrostatic fields, up to 10^{10} V m^{-1} according to our most recent calculations. This is a favourable factor for chemical reactions performed on clay platelets or within clay tactoids serving as micro-reactors. Such intense electrical fields help electrons to flow from one reaction partner (the donor) to the other reaction partner (the electron acceptor). The electrical fields can be calculated by computer simulations of the clay interface (at the Poisson–Boltzmann level of approximation or with more refined, Monte Carlo, approaches). Such numerical models have enabled us to interpret multinuclear magnetic resonance measurements and to provide an explanation for the very high surface acidity of clays — an important property for the catalytic activity of K10 montmorillonite.

We initially used very crude models such as that by Manning [15], but we have steadily improved our approach.

Presently, we perform approximate solutions of the Schrödinger equation for small atomic aggregates (or clusters) that stand for an aluminosilicate structural module interacting with external ions and neutral molecules. Such quantum calculations demand computers among the most powerful, with vectorial processing. However, used together with various nuclear magnetic resonance methods they help us to gain a better understanding of the structure and microdynamics of the adsorbates — the water molecule is of prime importance — at the interface of a clay. This has helped us to understand the very high surface acidity (Bronsted) of clay minerals like montmorillonites, of which K10 is an example. The surface acidity of K10 is high enough to place it between concentrated nitric acid and oleums of sulphuric acid. It is due to polarization of water molecules in the first clay hydration layer. These are pinched between the negatively charged aluminosilicate sheet and the positively charged counterion. In this 'push–pull' manner, electronic density migrates from the water oxygen, thus freeing protons [16].

11.5   CATALYSIS BY CLAYS

Numerous chemical reactions that are subject to general acid catalysis can and do benefit from the high surface acidity of clays (see Chapter 4). One example is that of a cyclohexadienone-to-phenol conversion (a Cope rearrangement) (equation (11.1)).

$$(11.1)$$

A careful kinetic study had been performed more than 30 years ago, with sulphuric acid as the catalyst [17]. The half-life of the starting material was five *hours*, at a reaction temperature of 80°C. When sulphuric acid is replaced with the K10 acidic montmorillonite — without any special care — the half-life becomes five *hours* at room temperature (20°C). If the clay is dried before use (6 or 7 h in an oven at 120–130°C), the reaction becomes yet more rapid: the half-life becomes five *minutes* at 20°C. Finally, if the clay is doped with a Lewis acid such as Fe^{3+} ions and dried, the half-life becomes five *seconds* at 20°C [18]. Overall, from 5 h at 80°C to 5 s at 20°C, the gain is five or six powers of ten acceleration.

Owing to the presence of counterions in their interlamellar galleries, clays are ion exchangers. This feature is well known in the cosmetics industry. Clays are the active ingredients in a number of shampoos, of skin lotions; they buffer acidity variations. Replacement of Na^+ ions by Fe^{3+} ions in the K10 montmorillonite, alluded to above, is another example.

Their ability as *ion exchangers* is yet another asset for the clay micro-reactors. Since very many different divalent and trivalent cations, including those deriving from transition metals, can be exchanged into a clay, such an exchangeable clay can be doped with a wide diversity of Lewis acids. In this manner, the clay micro-reactors can be made to combine catalysis by protic acids such as HCl with catalysis by Lewis acids such as $AlCl_3$ or $FeCl_3$.

An industrially important case is that of the Friedel–Crafts reaction [19]. It allows one to substitute an aromatic ring. Thus, benzene can be turned into toluene. And toluene can be alkylated in turn into the xylenes. Unfortunately, under conditions of homogeneous catalysis, the catalyst — aluminium chloride, say — is destroyed by the secondary reaction products. Water is such a secondary product resulting from use of an alcohol (methanol for a methylation) as an alkylating agent. As soon as water molecules are formed, they attach themselves to the *3p* vacant orbital on the metal and they poison the catalyst (equation (11.2)).

$$AlCl_3 + H_2O \longrightarrow Al(OH)Cl_3^{(-)} + H^{(+)} \tag{11.2}$$

As a consequence, *stoichiometric* (rather than *catalytic*) amounts of the Lewis-acidic catalyst are required. This is a serious drawback. Especially in this environmentally conscious time, large amounts of aluminium-rich industrial wastes are a big concern. By exploiting ionic condensation to bring up the *local concentration* in Lewis acids near the walls of the micro-reactors to elevated values, we have been able to remove this handicap: we have been able to decrease the amount of catalyst necessary by up to a factor 2000 in weight! Thus, a *true catalyst* of this century-old major reaction of organic chemistry has been achieved [20].

11.6   THE CATALYTIC SITES

11.6.1   The nature of the sites

So far, in this description of clay micro-reactors, only *diffuse* catalytic activity has been mentioned: whether dispersed on the whole surface of aluminosilicate sheets (and giving rise to their Bronsted acidity), or spread into the whole interlamellar volume for clays exchanged with Lewis acidic cations. There are also all-important localized sites. One type has been described already; it involves the oxygen basic sites in the octahedral layer. These sites are adjacent to the positions in the lattice where a divalent ion has replaced aluminium(III). A second type of important active site is that resulting from *isomorphic substitution* by an oxidizing metal in a high oxidation state such as iron(III) or manganese(VI). A third type, that has received a fine recent description [21], consists of those superficial sites that bear oxygen atoms with unpaired electrons. Whether they are isolated, or linked into a peroxo $O-O$ bond, or converted into superoxide centres O_2^-, they function as powerful oxidizing centres. A fourth type involves edge sites, where the sheets break off. These sites consist of coordinatively unsaturated and thus hyperreactive aluminium atoms. Pinnavaia has drawn attention to their prime importance for the catalyatic activity of clays [22].

The simplest means to activate a clay as a catalyst is through thermal dehydration. We have been doing this for a decade now. As a result, we believe that, while most silanol groups O_3SiOH are converted into $O_3SiOSiO_3$ siloxane groups, a small minority may be converted into the highly reactive silyloxy radicals $O_3SiO\cdot$. A similar process in quartz crystals, giving rise to $Al^{3+}\ldots\cdot O-SiO_3$ units, accounts for the grey colour of smoky quartz. Likewise, the presence of $Fe^{3+}\ldots\cdot O-SiO_3$ centres is responsible for the violet colour of amethysts.

11.6.2   Designed catalytic reactions

For the reasons outlined above, clays can and do oxidize organic matter. Aromatic rings, in phenols for instance, are oxidized through electron transfer from the aromatic donor to a metallic centre as an acceptor. We have made use of this property to devise an efficient nitration of phenols [23] by 'clayfen'. The radical cation derived from the phenol on contact with the modified clay couples with an NO_2 radical generated by the slow decomposition of the anhydrous ferric nitrate impregnated on the clay.

Unexpectedly, this study of phenol nitration gave us a rewarding entry into Diels–Alder chemistry! At the beginning of the eighties, the group led by Bauld had devised a catalysis of the cycloaddition reaction between unactivated dienes and dienophiles in which the active ingredient was an Ar_3N^+ radical cation. This enabled the reaction to be run at ambient temperature (instead of 200°C) in a few minutes or hours (instead of a few days).

A drawback was the cost of the catalyst, about US $80 per 5 g at the time. Since radical cations, we thought, were the intermediates in nitration of phenols, we decided to try to apply them to catalysis of the Diels–Alder reaction, by running the two reactions (the cycloaddition and the aromatic nitration) in one pot. The result surpassed our wildest dreams: the efficiency was superior to that of the Bauld catalyst, and our catalyst was (literally) dirt-cheap. We went on to other applications of clays to improved Diels–Alder methodologies [23], accomplishing very high stereo-selectivities (better than a 20:1 *endo : exo* ratio) in like manner to the aqueous Diels–Alder's that had been pioneered by Breslow, but in an organic solvent, dichloromethane. In this way, we succeeded in making the very unreactive furan dienophiles react at ambient pressure (instead of the 15 kbar that are otherwise necessary).

The possibilities of modified clays as catalysts for oxidation reactions are virtually untapped yet. A superb application has been proposed already, to detoxification of dioxin [24]. Dibenzo-*p*-dioxin forms a radical cation and polymerizes, under very mild conditions, on smectites doped with copper(II). This insolubilizes and immobilizes the foul chemical, which is thus trapped inside the clay particles.

11.7 CONCLUSIONS. THE SIGNIFICANCE OF CLAYS IN A WIDER CONTEXT

The catalytic sites resulting from isomorphous substitution in clays are also responsible for the oxidation of organic matter by soils and sediments. In the first case, it is preceded by microbiological oxidations that convert into humus ca. 1 kg carbon per square metre in times that vary from a few weeks under the tropics to a few months in more temperate climates. Mineralization proper, with full oxidation into carbon dioxide and water as the end result, is a much slower process (thousands of years). The reaction partners are atmospheric oxygen as the oxidant, iron and manganese isomorphically substituted as electron acceptors, and organic matter as the electron donor [14].

In this manner, the carbon cycle is completed. Organic matter was very likely formed at the contact of silicates and aluminosilicates, in primeval times when planet Earth was accreted from the primitive nebula. Organic matter finishes its decomposition, also at the contact of the same mineral types, of which clays are the most important component. Science joins up with myth (such as that of Osiris; Fig. 11.4): organic molecules have first appeared on clays, billions of years ago; they decompose and mineralize on the same supports and catalysts.

ACKNOWLEDGEMENTS

The editor has had the kindness to give me the leeway of writing this personal account. I shall abuse it by closing with an enumeration and an encapsulation of the lively personalities that were involved. I was fortunate indeed to have such co-workers. Let me cite first and foremost Dr André Cornélis, whose chemical intuition is amazing; he is one of the few persons who, when in France, will seek an Italian restaurant, and, when in Japan, will look for French cuisine, while in Belgium he will guide one to an Oriental restaurant. An early contributor who brought us our first samples of clay was Keith Pannell, from the University of Texas at El Paso. A peripatetic scientist — one of the few persons who travels more than I do — he is a craftsman of organometallic chemistry. We had the good fortune that he elected our laboratory to be his mailbox while on sabbatical leave. World politics helped us. We had a number of brilliant, hard-working, superbly trained scientists from Eastern Europe. This Warsaw Pact contingent included the Hungarian Dr Maria Balogh, and the Poles, Drs Janusz Baran and Waclaw Kolodziejski. They were attracted by the opportunity to do chemistry under more favourable material circumstances, both personally (the hard currency they could take home afterwards) and professionally (availability of equipment, of glassware, and of chemicals). I shall always remember Waclaw's reaction on a day when there was a general strike in Belgium (which did not prevent him from working his normal 18 hours in the lab): 'these people are crazy. They don't know how lucky they are. What do they strike about?'.

Another recruit from Eastern Europe was Eugenio Polla. His passport was that of a Yugoslav, but nearly everything else was Italian. Eugenio kept to himself. (He had a tendency to keep *for* himself much of the glassware also, which frayed some tempers

Fig. 11.4 — Clays in the moulding of men. The Egyptian god, Khnemou, moulds the first man on a potter's wheel. Behind him stands god Thoth, who marks the duration of human existence. Khnemou is a primaeval divinity. Originally a water-god associated with the Nile and its overflowing, he became also the maker of the egg from which arose the Sun, the creator of the other gods and of men.

This theme of the clay origin of the first men is widespread. In the Bible one reads (Genesis, 2.7) that 'the Lord God formed man of the dust of the ground, and breathed into his nostrils the breath of life'. Likewise in the Koran the first men are made of clay, while the parallel series of the *djinn* come from the fire. In Greek mythology, Zeus entrusts Hephaïstos with the moulding of a woman, Pandora, from clay. In Maya mythology, men in the original world are made of earth. In Akkadian mythology, humanity is born from clay commixed with the flesh and blood of god We. In Sumerian mythology, Enky is generated from clay coming from the Apsu, the deep-water abyss. He inseminates his mother with that clay, and she gives birth to the first man.
(The illustration is taken from E. A. Wallis Budge's *The Gods of the Egyptians*, Dover, New York, w.d., vol. II, ch. III.)

regularly.) He had a nose for the rewarding work, i.e. maximizing the number of publications per unit time spent in the lab, a tendency which I fought more or less successfully. I cannot resist the pun: he was a very astute schemer and skimmer. Very smooth, a fast talker, and adept at adapting existing work from the literature to our procedures, he was a born operator. Just to observe him function was a joy.

The group included soft-voiced, quiet and delightful individuals, very witty, such as the aforementioned Maria Balogh and Janusz Baran, such as a young German girl, Suzanne Hoyer, and a French post-doc from Brittany, Pierre Métra (Brittany reclaimed him: once a Breton ...). It had also some forceful characters, none more forceful than Pat Costanzo, an American lady from Buffalo, NY — a force of nature, as we say in French. A go-getter, authoritative, determined, nothing would stand in her way, and in a few weeks she had — notwithstanding the numerous frustrations suffered by any American trying to function as effectively as she would at home in the European anarchy — many parts of the University of Liège collaborating and awaiting her instructions. I have seldom seen anyone launching so many simultaneous experiments, turning out reams of handwritten reports and suggestions daily. She is pure energy — which shows in every sector. She is the youngest grandmother I have ever supervised (or met!), and she had raised a large family, as a single parent, while doing her Ph.D. at the same time!

Productivity-wise, she had two Belgian rivals. Dr Jean Lucchetti, like her and like Eugenio of Italian descent, was one of the very best synthetic organic chemists I have ever come across (he is now a group leader in a pharmaceutical company). He was amazingly fast: a week would suffice him, from having an idea to trying it out on the bench and to having optimized it to perfection! He physically loves experimentation in the laboratory. Dr Arthur Mathy, also an extraordinary person (he commuted on a large motorbike from near Brussels where, a confirmed bachelor, he lived with his mother), would advance on a broad front. He too would have several reactions running simultaneously. Where Dr Lucchetti had a quicksilver temperament, Arthur was slower but very deliberate, methodical and thorough. They both accomplished a lot.

With a Sherman tank such as Pat Costanzo, with a daring hit-and-run commando fighter such as Jean Lucchetti, with an armoured bulldozer such as Arthur Mathy, keeping them supplied at all times was vital. Two technicians cooperated in this daily miracle, Stéphane Chalais and André Gerstmans. While they both had the makings and the fibre of senior scientists, they selflessly gave themselves totally to service, not only juggling compressed gas tanks, chromatography columns and syringes, but also registering foreign co-workers at City Hall, getting their work permits extended or renewed, taking them out to sample Belgian beers, negotiating short-term leases with apartment owners, besides carrying out astronomical numbers of experiments.

One person went calmly through this turmoil to obtain a Ph.D. in record time, Pascal Pennetreau, whom I have cited already. His adaptability is spectacular, which no doubt he turns to advantage in his present function as a group leader in a major Belgian chemical company. Nearly unfazeable (as the episode which I have narrated shows), his forte is also a very sound training in physico-chemical concepts coupled with an admirable gift for the laboratory.

There were others too. Only one or two are best forgotten, because their mentality was exclusively nine-to-fiveish or mercenary.

REFERENCES

[1] See P. Laszlo, *Molecular Correlates of Biological Concepts*, Elsevier, Amsterdam, 1986.

[2] J. D. Bernal, *Proc. Phys. Soc. London*, 1949, **62**, 537; *The Physical Basis of Life*, Routledge and Kegan Paul, London, 1951.

[3] T. H. Jukes, *Nature*, 1973, **246**, 22.

[4] See, for instance, I. Lakatos, in *Criticism and the Growth of Knowledge*, I. Lakatos and A. Musgrave (Eds), Cambridge University Press, 1970, p. 91 *et seq.*

[5] C. C. Addison, N. Logan, S. C. Wallwork and C. Garner, *Chem. Soc. Quart. Rev.*, 1971, **25**, 289.

[6] A. Naumann, *Ber. Dtsch. Chem. Ges.*, 1904, **37**, 4328.

[7] L. Naldini, *Gazz. Chim. Ital.*, 1960, **90**, 1231.

[8] G. Polya, *Math. Ann.*, 1921, **84**, 149.

[9] E. W. Montroll, *Proc. Symp. Appl. Math. Ann. Math. Soc.*, *1964*, **16**, 193; E. W. Montroll and G. H. Weiss, *J. Math. Phys.*, 1966, **6**, 167.

[10] G. Adam and M. Delbrück, in *Structural Chemistry and Molecular Biology*, A. Rich and N. Davidson (Eds), Freeman, San Francisco, 1968, p. 198.

[11] Short for *clay*-supported *f*erric *n*itrate; A. Cornélis and P. Laszlo, *Synthesis*, 1985, 909.

[12] J. Fripiat, J. Cases, M. François and M. Letellier, *J. Coll. Interface Sci.*, 1982, **89**, 378.

[13] J. Thorez, in *Preparative Chemistry Using Supported Reagents*, P. Laszlo (Ed.), Academic Press, San Diego, CA., 1987, p. 177 *et seq.*

[14] H. Bohn, B. McNeal and G. O'Connor, *Soil Chemistry*, 2nd edn, Wiley-Interscience, New York, 1985.

[15] G. S. Manning, *J. Chem. Phys.*, 1969, **51**, 924.

[16] J. Grandjean and P. Laszlo, *J. Magn. Resonance*, 1989, **83**, 128.

[17] D. Y. Curtin and R. J. Crawford, *J. Am. Chem. Soc.*, 1957, **79**, 3156.

[18] S. Chalais, P. Laszlo and A. Mathy, *Tetrahedron Lett.*, 1986, **27**, 2627.

[19] C. Friedel and J. M. Crafts, *Compt. Rend. Acad. Sci. Paris*, 1877, **84**, 1292, 1450.

[20] P. Laszlo and A. Mathy, *Helv. Chim. Acta*, 1987, **70**, 557.

[21] F. Freud and F. Batllo, in *Structure and Active Sites of Minerals*, ACS Symposium Series, 145, Washington DC, 1989.

[22] T. J. Pinnavaia, *Science*, 1983, **220**, 365.

[23] P. Laszlo, *Accounts Chem. Res.*, *1986*, **19**, 121; *Science*, 1987, **235**, 1473.

[24] S. A. Boyd and M. M. Mortland, *Nature*, 1985, **316**, 532.

12

Microwave activation of reactions on inorganic solid supports

G. Bram, A. Loupy and D. Villemin

12.1 MICROWAVES

Microwaves are electromagnetic waves of frequency between 300 MHz and 300 GHz. Microwaves are used in three principal applications: measurement (e.g. radar, measures of dimensions, temperature, etc.); telecommunications (e.g. TV, Hertzien transmissions); and energy applications [1].

International conventions strictly limit the energy applications to special frequencies owing to interference with measurements and telecommunications. Most industrial, scientific and medical (ISM) applications have been made at 2450 MHz [1].

In preparative organic chemistry, microwaves are primarily used either for generation of reactive species such as singlet oxygen [2] or ground-state (triplet) atomic oxygen [3,4] or for acceleration of organic reactions previously studied under conventional heating.

The energy carried by microwaves at 2450 MHz is very small (ca. 1 J mol^{-1} of quanta). In contrast to classic photochemistry, many quanta are necessary to activate a chemical bond.

A simple model of the interaction between microwaves and matter is that in the presence of an electric field polar molecules stretch to align their dipolar moment parallel to the electric field. If the molecules are bound by intermolecular forces like hydrogen bonds or van der Waals interactions, these bindings prevent the periodic adjustments and the energy is dissipated as heat.

12.2 APPLICATION OF COMMERCIAL MICROWAVE OVENS TO ORGANIC SYNTHESIS IN *SOLUTION*

12.2.1 Microwave heating and solvent behaviour

The first reports on the use of domestic microwave ovens as thermal sources for conducting organic reactions in solutions appeared only in 1986 [5,6]. Gedye and Giguere thus demonstrated that organic compounds can be synthesized in sealed

Teflon vessels in a microwave oven with considerable reductions in reaction time of up to three orders of magnitude over conventional heating techniques. Microwave heating was consequently claimed to be the main effect, including the absence of significant temperature gradients within the sample and very short response times [7].

The mechanism of microwave heating is essentially that of dielectric heating [1,8–10]. When polar molecules are submitted to microwave irradiation, i.e. to an electromagnetic field, their dipole moments align themselves with the external field. Part of the energy absorbed in this process is dissipated as thermal energy leading thus to increases in temperature. Gedye *et al.* [11] and Giguere *et al.* [12,13] have measured the rises in temperature for different organic solvents irradiated for one minute. It was found that all polar compounds absorb microwave energy strongly, many reaching their boiling point within one minute. As expected, little or no microwave energy was absorbed by the non-polar carbon tetrachloride and hydrocarbon solvents (Table 12.1).

Systematic solvent studies have disclosed the existence of a linear relationship between the polarity of the molecules (evaluated by their dielectric constants) and the temperature increase induced by microwave irradiation. The effect of microwave activation is therefore strongly dependent on the solvent nature, rather weak when non-polar solvents are employed, considerable with polar solvents, which can be at the origin of very rapid heating. However, as very high temperatures and dangerous pressures can develop in the vessels under microwave oven conditions, reactions have to be carried out in special sealed Teflon vessels (i.e. Berghof or Savillex bottles). Furthermore, if the volume of the reaction mixture is kept to around 10% of the quantity of the container, the pressure will not exceed safe limits [13].

12.2.2 Organic reactions

Microwave heating is now being exploited in an increased number of organic syntheses. Substantial rate increases are observed, attributed essentially to the effect of

Table 12.1 — Temperature of 50 ml of several solvents after irradiation for 1 min at 560 W in an open vessel in a microwave oven

Solvent	Temperature (°C)	Boiling point (°C)
H_2O	81	100
CH_3OH	65	65
C_2H_5OH	78	78
$C_6H_{13}OH$	76	156
CH_3COOH	110	119
$CH_3CO_2C_2H_5$	73	77
DMF	131	153
Hexane	25	68
CCl_4	28	77

higher pressure and superheating of the solvent [5,6,11–13]. Significant examples of such works are described below.

12.2.2.1   Diels–Alder, Claisen and ene reactions

Microwave heating noticeably reduces the time to complete Diels–Alder (e.g. equation (12.1)), Claisen (e.g. equation (12.2)) and ene reactions (e.g. equation 12.3)) [6,12,15].

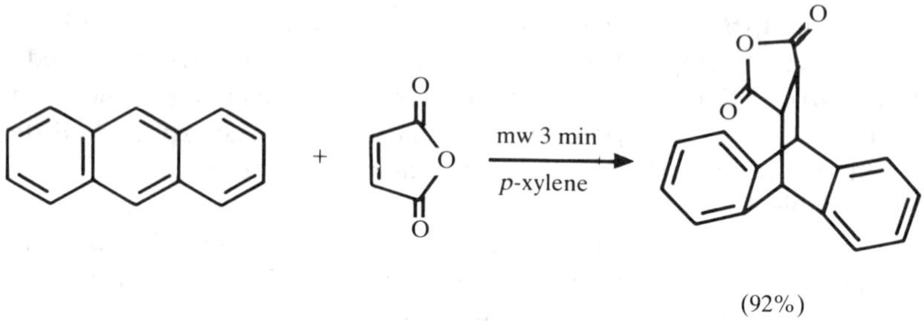

(12.1)

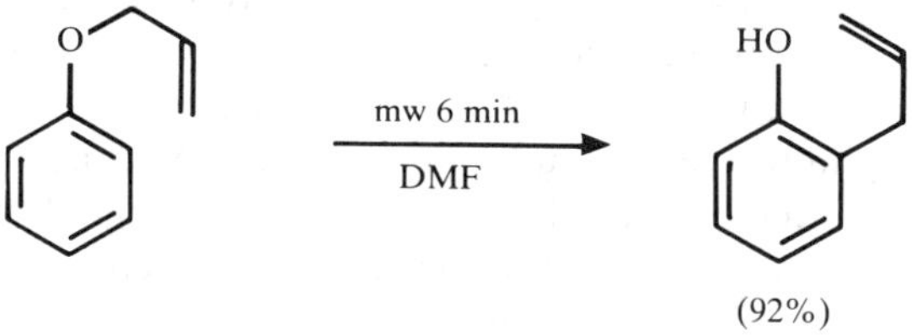

(92%)

(12.2)

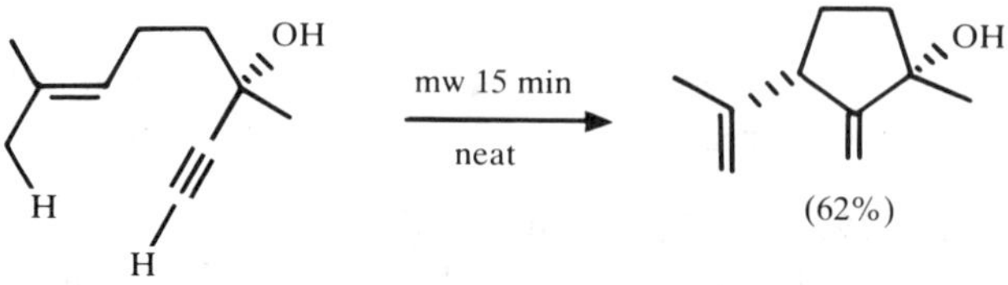

(92%)

(12.3)

(62%)

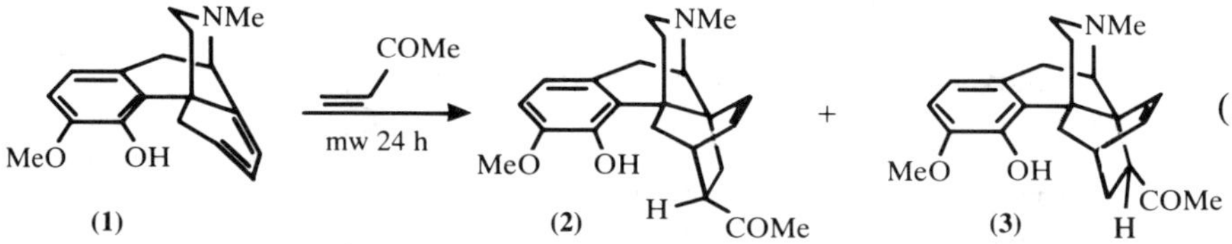

(12.4)

Linders has treated 6-demethoxy-2β-dihydrothebaine **(1)** with an excess of methyl vinyl ketone under atmospheric conditions to give a good yield of the two adducts **(2)** and **(3)** in a ratio of 3:2 (equation (12.4)) [5]. Under conventional heating (60 h at reflux), extensive polymerization of the dienophile occurs. Therefore, performing the reaction in a microwave oven results in a faster and cleaner reaction.

Racemization of (−) vincadifformine via concurrent Diels–Alder cycloreversion and cycloaddition is quantitative in DMF within 20 min using a microwave oven

[16]. It was previously performed for a longer time in a high-boiling solvent, leading, unfortunately, to a considerable amount of uncharacterizable byproducts.

12.2.2.2 *Nucleophilic substitutions*

Several kinds of nucleophilic substitution reactions have been carried out under microwave irradiation in commercial ovens. These include the reactions described in equations (12.5) [5,11], (12.6) [17], (12.7) [13], (12.8) [15], (12.9) [18] and (12.10) [20]. In all cases, considerable accelerations are observed when compared to conventional heating.

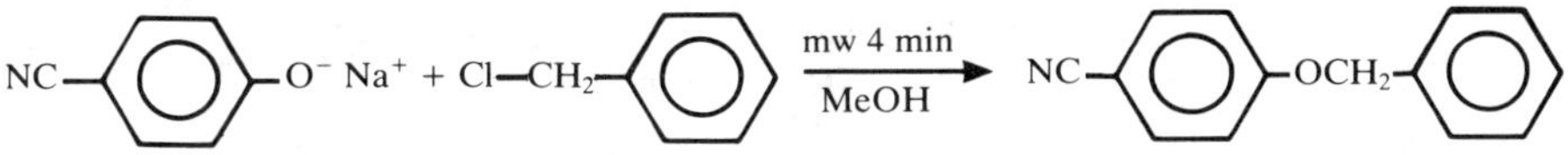

$$\text{NC}-\!\!\!\bigcirc\!\!\!-O^-\,Na^+ + Cl-CH_2-\!\!\!\bigcirc \xrightarrow[\text{MeOH}]{\text{mw 4 min}} NC-\!\!\!\bigcirc\!\!\!-OCH_2-\!\!\!\bigcirc \quad (12.5)$$

(93%)

$$\text{NaN}_3 + R-\overset{\overset{\displaystyle OTs}{|}}{CH}-CH_2-CH\!\!=\!\!CH-R \xrightarrow[\text{DMA}]{\text{mw 15 min}} R-\overset{\overset{\displaystyle N_3}{|}}{CH}-CH_2-CH\!\!=\!\!CH-R \quad (12.6)$$

(86%)

$$\text{NaI} + H_{25}C_{12}-Cl \xrightarrow[\text{Acetone}]{\text{mw 15 min}} H_{25}C_{12}-I \quad (12.7)$$

(93%)

$$\text{ArOCH}_3 + \text{KOH} \xrightarrow[\text{glycol}]{\text{mw 5h}} \text{ArOH} \quad (12.8)$$

(63%)

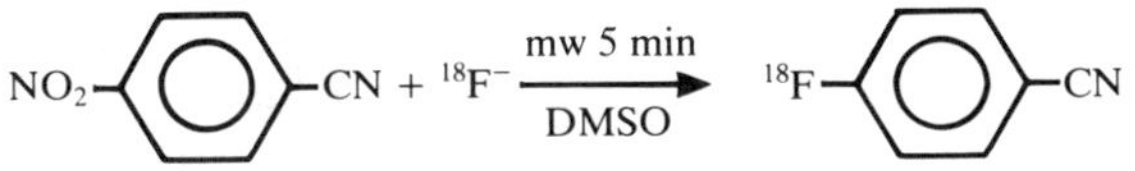

$$\text{NO}_2-\!\!\!\bigcirc\!\!\!-CN + {}^{18}F^- \xrightarrow[\text{DMSO}]{\text{mw 5 min}} {}^{18}F-\!\!\!\bigcirc\!\!\!-CN \quad (12.9)$$

(68%)

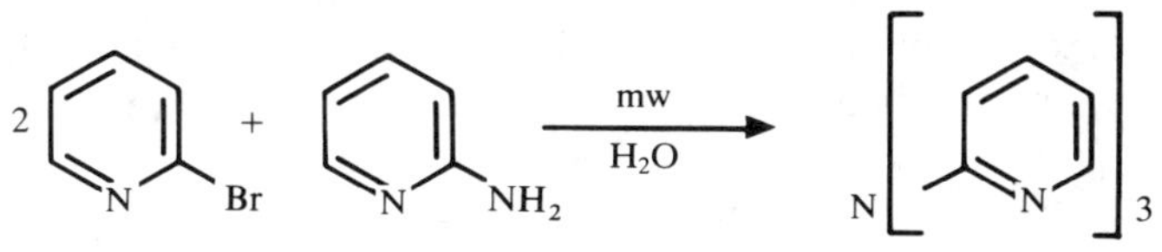

$$2\;\text{(pyridine)}-Br + \text{(pyridine)}-NH_2 \xrightarrow[\text{H}_2\text{O}]{\text{mw}} \left[\text{(pyridine)}\right]_3 \quad (12.10)$$

In the case of the S_NAr reaction (equation (12.9)) [18], fluorination yields after 5 min of microwave heating significantly exceed those from conventional heating for the same interval, and in most cases are approximately equal to those obtained after 30 min at 135°C. Furthermore, degradation of reagents is less under microwave conditions. The technique can be conveniently applied in the synthesis of a wide variety of radiopharmaceuticals or other radiolabelled compounds [18,19].

12.2.2.3   Esterifications

Some benzoic acid esterifications, catalysed by sulphuric acid, have been conducted using microwave heating (Scheme 1) [5,11]. Yields are comparable to those obtained by conventional heating but there are major reductions in time ($\geqslant 100$ fold). Acid-catalysed esterification of free fatty acids (100%, mw = 5 min, in MeOH) [17] and acetylation of cellulose (mw = 10 min, in CH_3COOH) [21] under microwave conditions have also been reported.

$$PhCOOH + ROH \xrightarrow[H_2SO_4]{mw} PhCOOR$$

(R = Me, 5 min, 76% yield)

(R = nBu, 7.5 min, 79% yield)

Scheme 1 — Esterification of benzoic acid under microwave conditions.

12.2.2.4   Oxidations

Oxidation of toluene to benzoic acid (equation (12.11)) [5,11] and epoxidation of unsaturated fatty acid esters by m-chloroperbenzoic acid (equation (12.12)) [17] have been carried out under microwave conditions.

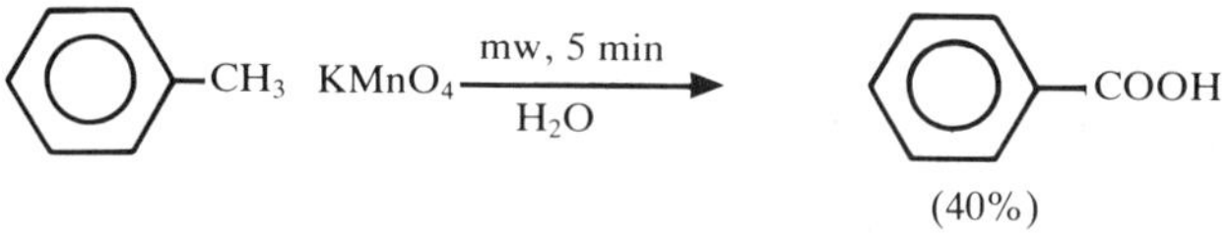

$$\text{Ph}-CH_3 \xrightarrow[H_2O]{KMnO_4 \quad mw, 5 min} \text{Ph}-COOH \qquad (12.11)$$

(40%)

$$CH_3(CH_2)_7CH=CH(CH_2)_7CO_2Me \xrightarrow[CH_2Cl_2]{MCPBA, mw, 3 min}$$

$$CH_3(CH_2)_7\overset{\displaystyle O}{\overset{\diagup\diagdown}{CH-CH}}(CH_2)_7CO_2Me \qquad (12.12)$$

($\geqslant 95\%$)

12.2.2.5   Hydrolysis

Under microwave conditions, benzamide is hydrolysed to benzoic acid in 10 min in the presence of sulphuric acid as catalyst [5,10]. Hydrolysis of castor oil in the presence of KOH in ethanol also requires only 5 min to complete. The hydrolysis of benzodiazepines to benzophenones, normally carried out with HCl at 120°C for more than 1 h, is complete within 5 min under microwave conditions (e.g. equation (12.13)) [22].

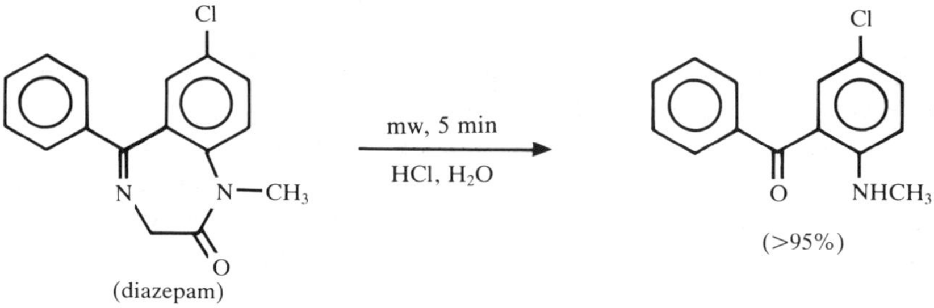

$$\xrightarrow[HCl, H_2O]{mw, 5 min} \qquad (12.13)$$

(diazepam)

(>95%)

In a microwave irradiation period of 4 min, 91% of ATP is hydrolysed (equation (12.14)) while controls heated in a dry block demonstrate only 7% hydrolysis over the same period [23].

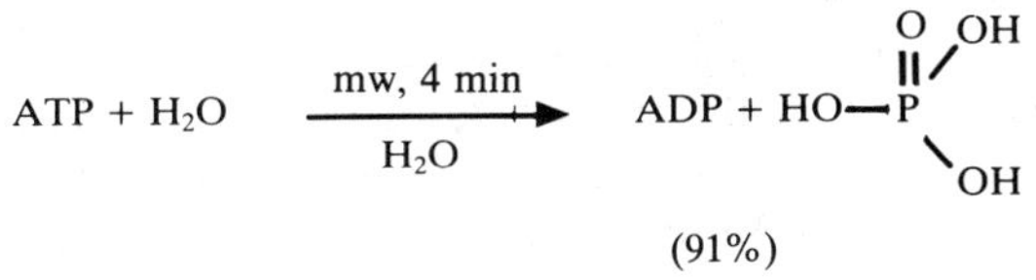

$$ATP + H_2O \xrightarrow[H_2O]{mw,\ 4\ min} ADP + HO-P\ \substack{O\ OH\\ \\ OH} \qquad (12.14)$$

(91%)

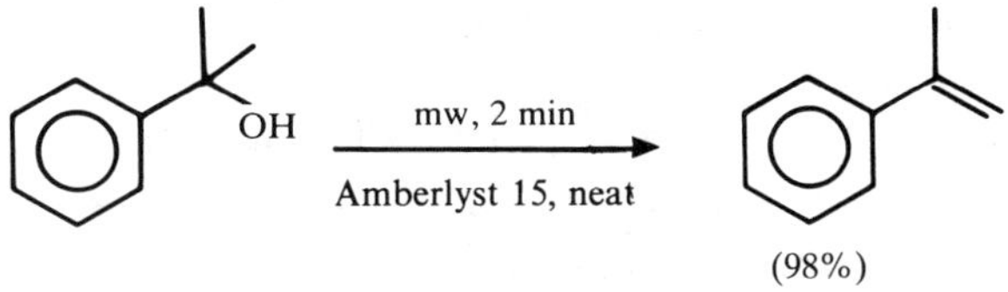

$$\qquad (12.15)$$

(98%)

12.2.2.6 Miscellaneous reactions

A simple method for dehydration of alcohols involves microwave heating of an alcohol in the presence of an easily removable acid source such as Amberlyst 15 (equation (12.15)) [13]. Different products may be obtained after longer irradiation times.

The acid-catalysed hydrolysis (ring opening) of 2,5-dihexylfuran to the corresponding 1,4-dioxo derivative requires three 10 min periods of heating (equation (12.6)) [7].

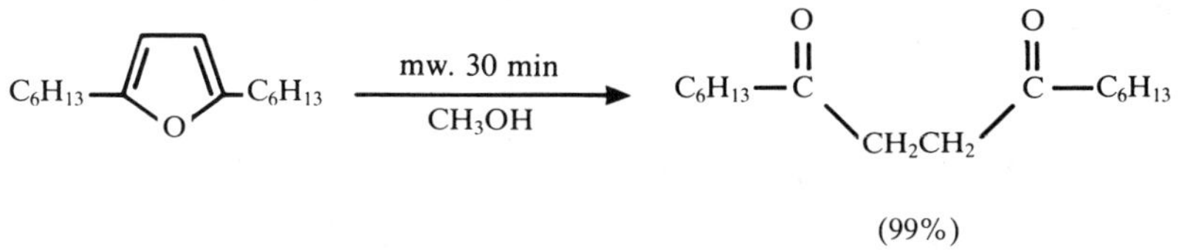

$$C_6H_{13}\text{-furan-}C_6H_{13} \xrightarrow[CH_3OH]{mw.\ 30\ min} C_6H_{13}-C(O)\ CH_2CH_2\ C(O)-C_6H_{13} \qquad (12.16)$$

(99%)

The pharmacologically interesting heterocycle dihydrodibenzazepine is prepared from 2,2'-diammoniodibenzyl diphosphate under microwave irradiation (equation (12.17)) [10].

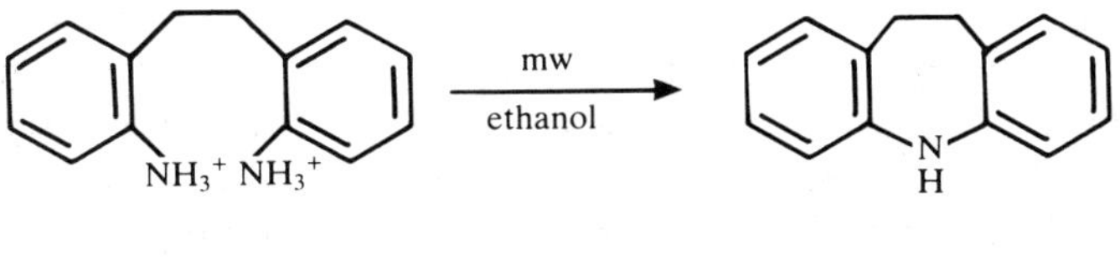

$$\xrightarrow[ethanol]{mw} \qquad (12.17)$$

12.2.2.7 Applications in amino acid chemistry

Peptides and proteins can be hydrolysed by the microwave technique [24–27].

The method has been applied to complete amino acid analysis using a single non-volatile solvent, methanesulphonic acid. It provides a radical expedition within 5 min of protein and peptide hydrolysis via commercial microwave ovens and specially designed Teflon–Pyrex vials, circumventing the tedious conventional procedures using vacuum-sealed Pyrex tubes and heating at 110°C for more than 24 h [28]. This rapid peptide–bond cleavage is of great potential in the automation of the complete process of amino acid analysis starting from the preparation of protein hydrolysates.

Rapid racemization of optically active amino acids has been developed in a Teflon vial reactor by using a commercial microwave oven as heating source. In a typical example, L-α-aminobutyric acid was quantitatively racemized within 2 min in acetic acid [29]. Under conventional conditions, the reaction needs longer incubation and heating in strong base or acid solution, thus inducing some decomposition of labile compounds during the process of racemization.

12.2.3   Applications in inorganic and organometallic chemistry

12.2.3.1   Synthesis and modification of organometallic compounds

McWhinnie and collaborators have described the use of a domestic microwave oven for mercuration of aromatics, ligand redistribution reactions and the reaction of aluminate with propane-1,2-diol [20]. Yields obtained in PTFE containers are generally of the same order as those obtained via conventional methods, but times of reaction are shorter.

Mingos and collaborators have prepared, in excellent yields, diene-rhodium(I) and -iridium(I) complexes by reactions of dienes (norbornadiene, cyclooctadiene) in aqueous alcohol using a sealed Teflon container and a commercial microwave oven (e.g. equation (12.18)) [30]. The sandwich cation $[Rh(C_5H_5)_2^+]$ is obtained under microwave conditions by the reaction of $RhCl_3$ and cyclopentadiene, whereas the conventional method of synthesis has been achieved previously only by using a Grignard reagent, C_5H_5MgBr. Bis(benzeneruthenium dichloride) is obtained from hydrated $RuCl_3$ and 1,3-cyclohexadiene (equation (12.19)) [30].

$$Rh Cl_3 . xH_2O + C_8H_{12} \xrightarrow[\text{50 s, 500 W}]{\text{Et OH,H}_2\text{O}} Rh(C_8H_{12})_2Cl$$
$$(91\%) \hspace{4cm} (12.18)$$

$$RuCl_3 . xH_2O + C_6H_8 \xrightarrow[\text{35 s, 500 W}]{\text{Et OH,H}_2\text{O}} [Ru(\eta - C_6H_6)Cl_2]_2$$
$$(89\%) \hspace{4cm} (12.19)$$

12.2.3.2   Intercalation

The intercalation compound $VO(PO_4)pyr_{0.85}$ has been synthesized two orders of magnitude more quickly using microwaves (equation (12.20)) than with conventional thermal methods [31]. $VO(PO_4).2H_2O$ does not absorb microwaves strongly, but the intercalated product and pyridine solutions show rapid heating effects when exposed to microwaves.

$$VO(PO_4). 2 H_2O + 2 \hspace{0.3cm} \bigcirc\hspace{-0.4cm}N \xrightarrow[\text{200°C, 80 atm}]{\text{5 min, 650 W}} VO(PO_4)pyr_{0.85} \hspace{1cm} (12.20)$$
$$(100\%)$$

12.2.4 Advantages and limitations

In all the examples described above, the use of commercial microwave ovens leads to *considerable rate increases* of the reactions. It has been well demonstrated that organic compounds can be synthesized up to 1000 times faster in sealed Teflon vessels in a domestic microwave oven than by conventional (reflux) techniques. Such microwave methods often avoid tedious procedures and provide at least comparable yields. Furthermore, the use of microwaves not only decreases considerably the reaction time but also leads to cleaner products. As heating time is largely reduced, it causes less degradation of reagents and generation of side-products, which generally make isolation of the desired compound difficult. However, severe limitations appear and are often evidenced. They involve dealing with *safe precautions*.

Owing to the very high pressures that can be developed in the reaction vessels, violent explosions may occur in ordinary glassware. Thus, a change in the design of reaction vessel is required. Sealed Teflon vessels are strongly recommended. Furthermore, it is useful to keep the volume of the reaction mixture to about 10% of the volume of the container, so that the pressure will not exceed safe limits. Consequently, the method is limited by the *small quantities* that can be utilized. These limitations arise because of superheating and pressure buildup of the solvent. Thus, it would be of great interest and potential to apply this method to 'dry reactions', i.e. performed in the absence of organic solvents.

Microwaves also increase the mobility of polar molecules and cause heating when intermolecular bindings exist between polar molecules (e.g. water). In the case of a polar liquid in a closed vessel (autoclave, pore of solid absorbent), fast overheating gives local pressures, and reactions can be accelerated. Increase of molecular mobility of polar molecules is an important aspect of *activation of molecules absorbed on a solid catalyst*. Microwaves allow activation of organic polar molecules absorbed on thermal inorganic insulating materials (such as silica, alumina) [62,65]. This *activation is instantaneous* and thermal diffusion plays no role in it in contrast to the classic thermic activation. In the case of reactions in polar liquids with an open vessel, the most numerous molecules (solvent) are activated and shield the molecules of reagents from the microwaves. Thus, no specific microwave effects take place and the reactions are similar to those obtained by a classic heating. *Specific activation by microwaves can be found only in reactions without polar solvents.*

12.3 SOLID MATERIALS UNDER MICROWAVES

12.3.1 Synthesis of organometallics and intercalation

Internal metallation by a transition metal (Pd, Pt) has been achieved by microwave irradiation of solid bipyridyl complexes of Pd and Pt [32]. This reaction (equation (12.21)) is the first example of activation of $C - H$ bonds by transition metals under microwave irradiation.

Metallic porphyrines and phthalocyanines intercalated in clays have been obtained by direct synthesis using microwaves [33].

$$(M = Pt, Pd; X = Cl, Br)$$

12.3.2 Application to regeneration of catalysts and adsorbents

Microwaves have been used for the regeneration of zeolites [36], other catalysts [31] and absorbents such as silica, alumina, and charcoal [34].

Microwaves can be used in the laboratory [37] for regeneration of drying solid such as molecular sieves, silica gel or calcium chloride, and chromatographic absorbents (silica, alumina). Drying of glassware and thin layer chromatography plates have also been described.

The microwave heating of zeolites gives rise to efficient dehydration. It has been shown [36] that water molecules are directly desorbed by the electromagnetic field, this process being independent of the temperature of the solid. This microwave-induced dehydration process has been shown to proceed through a two-step pathway, with a discontinuity in the kinetic parameters. A mathematical model describing the energy balance and the heat transfer between the microwaves, the granular material and the gas has been proposed [36].

Microwaves also have the potential to preserve the structure and reactivity of clay materials used as supports and catalysts in organic transformations.

12.3.3. Solid inorganics

Microwaves have been used for selectively heating minerals, many of which become heated whereas most common host rock does not. Thus, selective oxidation or reduction is possible by passing reactive gas through the ore during microwave irradiation [38].

CuCl reaches 425°C and begins melting within 3 min at 500 W input power. $ZnCl_2$ is melted (723°C) after 5 min at 500 W. CuO and Fe_3O_4 are extremely good microwave receptors, reaching temperatures higher than 1000°C in less than a minute at 1000 W. High temperatures (> 1000°C) are also obtained with Cr_2O_3, FeS_2 and PbS_2. By contrast, SiO_2 and $CaCO_3$, which are common gangue constitutents, are not microwave receptive. The highest temperatures are obtained with the metal oxides: NiO, MnO_2, Fe_2O_3, Co_2O_3, CuO and WO_3. Most sulphides heat well, and although metals generally reflect microwaves, metal powders also heat well. Metal powder heating is ascribed to inductive effects. Rapid microwave heating has been used to stress fracture ore samples. In this manner, iron ores containing haematite, magnetite and goethite have been subjected to microwave energy in bath operations at 3 kW and heated to average temperature between 840 and 940°C. Grindability tests showed that microwave heating had reduced the work index of iron ore so treated by 9.9 to 23.7% [39].

Curing, heating, tempering, sintering, calcining and drying are among operations with microwaves that have been successfully demonstrated on a wide range of products. Microwave heating is often suggested as a means of improving the processing of ceramic products, particularly the drying operation. A large number of processes in ceramic plants have been shown to be technically feasible with micro-waves. Slip casting of clay and alumina, of clay, alumina and silicon carbide composites, and of ferrites; and drying and sintering of ceramics such as alumina, zirconium oxide (ZrO_2), silica (SiO_2) and ferrites (barium ferrite and calcium vanadium garnet) have all been demonstrated [40].

12.3.4   Organic materials

12.3.4.1   *Polymerization and depolymerization*
A large amount of work has been devoted to the study of microwave-induced polymerization reactions [1]. Polyesters, polyurethanes, epoxyresins and amino-plastes have been thus successfully prepared. It has been observed that cross-linking, hardening and, in general, the mechanical properties of the polymers are considerably affected by the microwave process, in comparison with the classical thermal one [41–43].

Some interesting depolymerization reactions have been obtained under micro-wave irradiation. Thus, 1,6-anhydro-β-D-glucopyranose (**4**), a useful chiral synthon in organic chemistry, has been thus obtained from cellulose starch or various (1 → 4)-D-glucans (amylopectin, amylose maltodextrin, β-cyclodextrin) [44,45].

(**4**)

The starting materials are inexpensive, and the ease of the method outweights the low yields ($\leqslant$ 1%) when the production of a small amount ($\sim$ 1 g) of the synthon (**1**) is required. A recent Japanese patent claims the saccharification of starch using microwave heating [46].

12.3.4.2   *Solid–liquid solvent-free reactions*
Solid–liquid phase transfer catalysis (PTC) without solvent has been advocated as an effective method in organic synthesis when anionic activation is concerned [47–49]. Such reactions, carried out in the absence of any solvent, are very effective and can be accomplished under milder conditions with a very easy workup. In some cases, in competitive reactions, a peculiar selectivity results under these conditions. It has been shown that it is possible to perform solid–liquid PTC without solvent in a domestic microwave oven, and thus to take advantage of the cumulative effects of the two

methodologies. It is notworthy that the safety problems due to the presence of an organic solvent during the microwave irradiation (see section 12.2.4) are thus ruled out.

The first selected model reaction consisted in alkylation of acetate anion by alkyl halides, leading to esters (Scheme 2) [50].

$$CH_3CO_2{}^-K^+ + RX \xrightarrow[\text{mw, 600 W, 1 min}]{\substack{\text{Aliquat 336 [51],}\\ \text{no solvent}}} CH_3CO_2R + KX$$

RX = n-C_8H_{17}Br   yield = 98%
RX = n-C_8H_{17}Cl   yield = 98%
RX = n-$C_{16}H_{33}$Br   yield = 98%

Scheme 2 — Alkylation of acetate anion under solvent-free, microwave PTC conditions.

Yields are excellent, and times of reaction are very short, even with rather long chain halides, which are known to be poor electrophiles. Considerable accelerations (relative rates of 120–180) are observed when compared to classical heating. Alkylations of long chain (equation (12.22)) or hindered (equation (12.23)) carboxylate anions in similar conditions also lead to excellent yields of esters [52].

$$C_{17}H_{35}CO_2^-K^+ + n\text{-}C_8H_{17}Br \xrightarrow[\text{mw, 600 W, 1 min}]{\text{Aliquat 336, no solvent}} C_{17}H_{35}CO_2C_8H_{17} \quad (12.22)$$
$$(96\%)$$

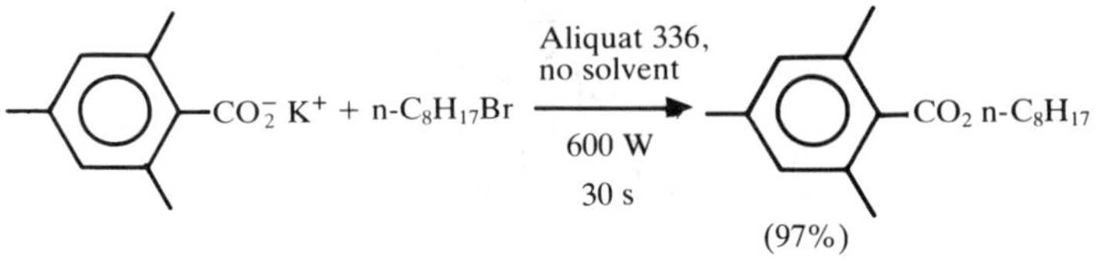

$$(12.23)$$
$$(97\%)$$

12.4 MICROWAVE-ACTIVATED ORGANIC SYNTHESIS ON SOLID SUPPORTS

12.4.1 General considerations

Reactions involving reagents supported on inorganic supports such as alumina, silica and clays were initially usually run in the presence of an organic solvent. However, it has been shown that it is possible to react very efficiently supported reagents in the absence of any organic solvent, in so-called 'dry media' conditions [53,54]. Many advantages result from performing syntheses in 'dry media': faster reactions than in the presence of a solvent, different selectivities observed and more economical conditions due to the absence of solvent during the reaction. In our two groups, we have undertaken to couple reactions on inorganic supports and microwave activation in the absence of any organic solvent to take advantage of the cumulative effects of these two methodologies. We reasoned that performing reactions in 'dry media' conditions would overcome the difficulties and limitations due to the presence of an organic solvent during microwave irradiation (see section 12.2.4). It would thus become possible to work with standard open vessels in commercial domestic microwave ovens.

One of the first model reactions tested involved the alkylation of acetate anion by long chain alkyl halides performed on alumina selected as the inorganic solid support (equation (12.24)) [55,56].

$$CH_3CO_{22}{}^- K^+ + R - X \xrightarrow[\text{mw}]{\text{alumina, no solvent}} CH_3CO_2R + KX \qquad (12.24)$$

The results obtained from this model reaction reveal some general features.
- Microwave efficiency is highly dependent on the shape and dimensions of the irradiated samples: simple-shaped (cylindrical, spherical or cubic) vessels with dimensions (height and diameter) between λ and $\lambda/10$ ($\lambda = 12.2$ cm for the 2450 MHz domestic microwave oven) appear optimal.
- Studies of the thermal behaviour of alumina (Fig. 12.1) indicate that the temperatures reached depend on the amount of irradiated solid. In a commercial domestic oven, a maximum temperature is obtained for about 200 g of alumina, 4 g appearing as about the minimum amount for an appreciable thermal effect to be observed [a 1 g sample of alumina is unable to reach 100°C even after a long irradiation time (20 min)].
- Solid potassium acetate, after loss of water by dehydration which induces a temperature rise, only slightly absorbs microwave energy. It behaves as a weakly

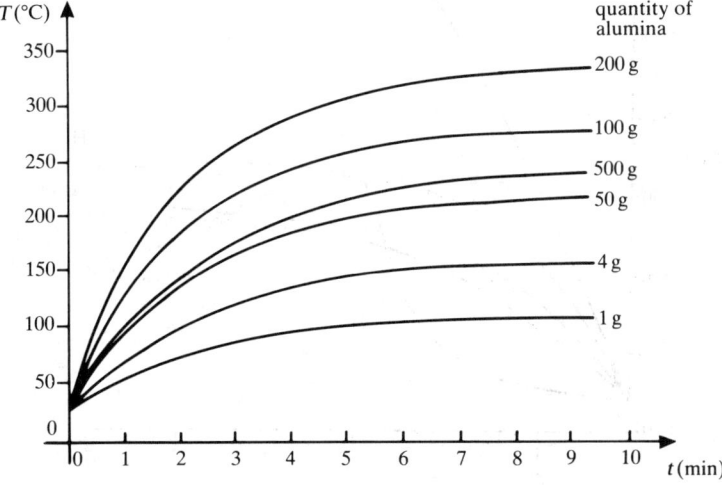

Fig. 12.1 — Thermal behaviour of neutral alumina (T90 Merck) as a function of irradiation time and quantity ($P = 600$ W).

polar compound, probably in connection with its compact tridimensional aggregated strcuture in the solid state. In contrast, the same reactant impregnated, from an aqueous solution, on alumina adsorbs microwaves strongly and the temperature exceeds 400°C after a few minutes of irradiation (Fig. 12.2). This is probably due to variations in dielectric properties of the medium [1,57,58] related to specific interactions between alumina and potassium acetate, giving rise to some disaggregation and even a partial ionic dissociation of potassium acetate. This could be at the origin of the activation of this salt and other anionic species) at the contact with the surface of alumina. However, this conclusion needs a precise dielectric analysis of all materials using a microwave cavity technique [59].

- Neat *n*-octyl bromide alone absorbs microwave energy a little more effectively than solid potassium acetate, but, when impregnated on alumina, *n*-octyl bromide is almost without effect on the medium temperature (Fig. 12.2).

12.4.2 Nucleophilic substitution reactions

12.4.2.1   *Alkylation of carboxylate and sulphinate salts*
In a preliminary communication [55] it was shown that microwave irradiation on alumina gave yields for alkylation of potassium acetate by *n*-octyl bromide which were comparable to those obtained by classical heating, but with a significant acceleration (× 30). However, owing to the limited amount of reactants utilized, the microwave technique was rather limited and the reproducibility appeared poor. With

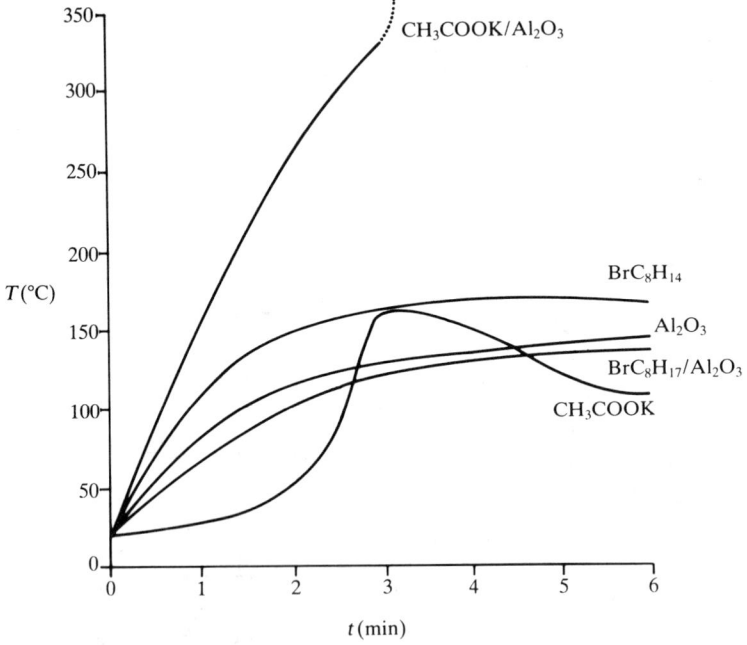

Fig. 12.2 — Behaviour of reactants, alone and impregnateed on alumina (4 g) ($P = 600$ W).

Table 12.2 — Microwave activation (600 W) of the reaction of n-OctBr (7.5 mmol) and CH_3COOK/alumina (3 g; 5 mmol CH_3COOK)

Time (min)	1	1.5	2	3
Temperature (°C)	182	192	196	184
Yield (GC) (%)	63	91	99	98

larger amounts of reactants (at least 4 g of a 1/5 CH_3COOK–Al_2O_3 mixture), in a well shaped open pyrex vessel, however, good yields of esters have been obtained with good reproducibility. Thus, when CH_3COOK is used in excess (50%), a 99% yield is obtained within 2 min at 600 W (Table 12.2).

A comparison between yields obtained under microwave irradiation or classical heating indicates that the microwave-induced reaction is 150 times faster than an oil-bath thermally induced one. In order to try to determine the origin of the microwave effect, the reaction was performed in an oil bath at 190°C (i.e. very close to microwave conditions). Comparison between yields after 2 min of reaction under microwave (99%) or classical heating ($< 2\%$) demonstrates that the special reactivity observed under microwave activation is not due only to a simple heating effect. As previously suggested [60], this observation can be related to the poor thermal diffusion of alumina, which is an obstacle for thermal activation of the reaction, whereas transmission of microwaves is operative.

When small quantities ($\leqslant 1$ g) of supported reagents are involved, the use of additional alumina in an external bath allows the reaction to occur effectively (92%, 4 min, 600 W) and with high reproducibility [56]. When compared to classical heating [47], rate enhancement is 100 times at least.

Longer chain hexadecyl halides are less electrophilic than n-octyl bromide. Nevertheless, it has been shown that use of an excess of CH_3COOK permits the production of good yields with hexadecyl bromide (95%, 75 s, 600 W) and chloride (90%, 2.5 min, 600 W) (Fig. 12.3) [56].

It is noteworthy that after 20 h at 85°C (classical oil bath method) the yield is 0% with hexadecyl chloride. It appears that when microwave activation on alumina in dry media is involved, alkyl chloride reactivity is only slightly inferior to that of alkyl bromide. Thus, it becomes possible to use alkyl chlorides (cheaper and more available) instead of alkyl bromides. This is often difficult classically owing to the intrinsic inferior electrophilicity of alkyl chlorides compared to alkyl bromides.

Sodium benzenesulphinate on alumina is alkylated under microwave conditions (160 W, 5 min) to give sulphones (equation (12.25). Other supports are less efficient than alumina. No reaction occurs in the absence of support and no reaction is observed without activation. Ultrasound gives a similar activation effect.

$$R - CH_2X \xrightarrow[\text{mw or ultrasound}]{C_6H_5-SO_2^- \, Na^+/Al_2O_3} R - CH_2 - SO_2 - C_6H_5 \qquad (12.5)$$

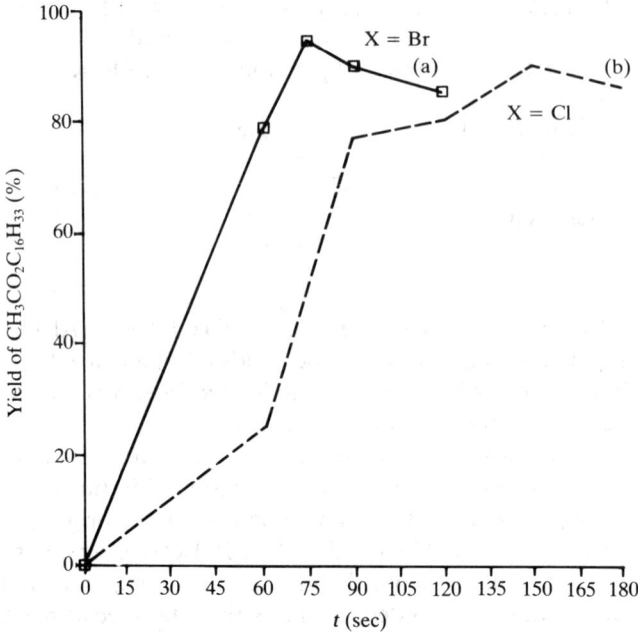

Fig. 12.3 — Microwave activation (600 W) of the reaction of $C_{16}H_{33}X$ (5 mmol) and CH_3COOK/alumina = 4.5 g (7.5 mmol CH_3COOK).

12.4.2.2 Anionic condensations

Microwave activation allows many dry condensations which do not occur (or occur with poor yield) at room temperature [60].

For example, condensation of methylenesulphones with aldehydes on KF-alumina is achieved in 5 min under microwave irradiation (550 W) (equation (12.26)) [60,75].

$$ Y-CH_2-S(O)_2-C_6H_5 + R^1,R^2-C_6H_3-CHO \xrightarrow[\text{5 min, 550 W}]{Al_2O_3-KF} R^1,R^2-C_6H_3-C(Y)=CH-S(O)_2-C_6H_5 $$

$$ (Y=COOC_2H_5, CN, COC_6H_5) \qquad (56-94\,\%) \qquad (12.26) $$

Many cyclic compounds with an acidic methylene group can be condensed with aldehydes by absorption on basic KF-alumina and microwave irradiation. For example, rhodamine gives 5-alkylidene products of biological interest, which are also precursors of β-substituted α-mercaptoacrylic acids (equation (12.27)) [60,73].

$$ ArCHO + \text{(rhodamine)} \xrightarrow[\text{350 W, 4 min}]{Al_2O_3-KF} \text{(5-alkylidene)} \xrightarrow[\text{ii, } H_3O^+]{i, Al_2O_3-NaOH, 210\,W, 5\,min} Ar-CH=C(SH)-COOH $$

$$ (Ca.\ 100\%) \qquad (12.27) $$

Piperazinediones can also be condensed with aldehydes (equation (12.28)) [73]. Pure (Z)-isomer can be obtained by using the N,N'-diacetyl derivative. Mono- or dicondensation products are produced, depending on the stoichiometry. Many naturral products contain this type of structure and albonursin and natural analogues of *Actinomyces* can be synthesized by this method [73].

$$\text{Ar—C—H} + \text{CH}_3\text{CO N} \quad \text{N—COCH}_3 \xrightarrow[\text{280 W, 5 min}]{\text{Al}_2\text{O}_3\text{–KF}} \text{H—N} \quad \text{N—COCH}_3 \qquad (12.28)$$

12.4.3   Cationic reactions

12.4.3.1   Friedel–Crafts synthesis of anthraquinone
Anthraquinone is an important starting material for the dyestuffs industry. Its manufacture involves the acid-catalysed cyclodehydration of o-benzoylbenzoic acid in hot concentrated sulphuric acid (equation (12.29)).

$$\xrightarrow{\text{H}^+} \qquad (12.29)$$

This method is accompanied by the formation of much waste sulphuric acid, not to be desired from the point of view of environmental pollution. It has been shown that it is possible to perform the reaction by heating in dry media o-benzoylbenzoic acid deposited on a catalytic clay support [63]. Anthraquinone is obtained, with a bentonite clay, in 89% yield (30 min, 350°C), but the turnover is low: after a few cycles, yields decrease sharply and it is not possible to regenerate the catalytic activity of the clay either by washing or by roasting. On the other hand, in a commercial microwave oven, with the same bentonite clay, a rapid and efficient reaction takes place (95%, 5 min, 600 W) [64–66]. After 15 cycles with the same batch, the yield is still 85% and washing with an organic solvent brings about the recovery of the whole catalytic efficiency of the clay. With classical heating, the clay obviously suffers a partial degradation of its microstructure, with a loss of the catalytic properties of the support. Microwave activation seems to induce the reaction without such degradation of the structure.

Large rate enhancements have also been obtained for pinacol rearrangements using 'dry media' microwave irradiation using montmorillonite clays as supports.

12.4.3.2   Other electrophilic substitutions
Phthalodinitrile impregnated on montmorillonite exchanged with a metallic cation (Cu^{2+}, Co^{2+}, Mn^{2+}, Fe^{2+}) gives metal complexes of phthalocyanines (equation

(12.30)) [33] intercalated in the sheets of clay. The phthalocyanines can be extracted by dissolving the material in concentrated sulphuric acid, but the intercalated species may be promising as redox catalysts [33].

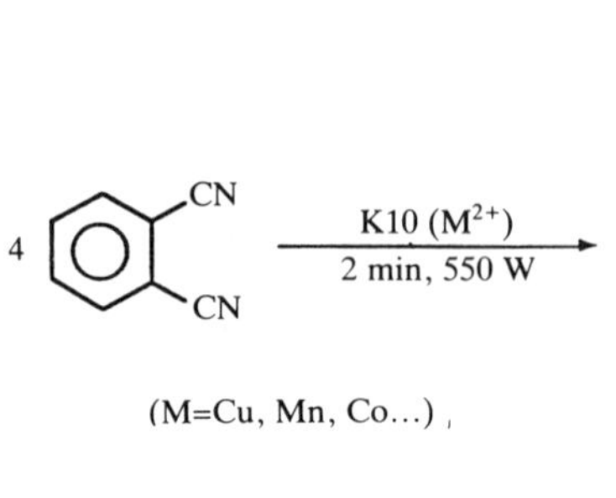

(12.30)

$(M=Cu, Mn, Co...)$,

A mixture of an aromatic aldehyde and pyrrole absorbed on metal-exchanged clay also gives strongly coloured porphyrins under microwave irradiation (equation (12.31)) [33]. Most of the product is again absorbed or intercalated in the clay. The experiment is interesting as a model of possible prebiotic synthesis of metalloporphyrins.

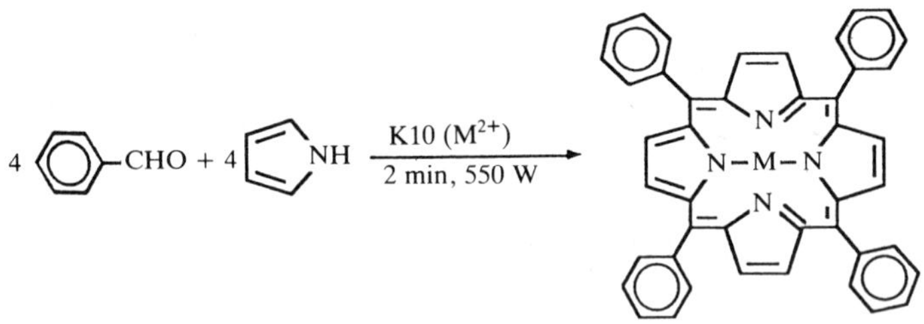

(12.31)

12.4.3.3   Cationic dehydration and hydration

Tertiary alcohols such as cedrol absorbed on acidic clay are easily dehydrated under microwave irradiation (equation (12.32)) [33]. A secondary alcohol such as cholestanol is more difficult to dehydrate in these conditions [33].

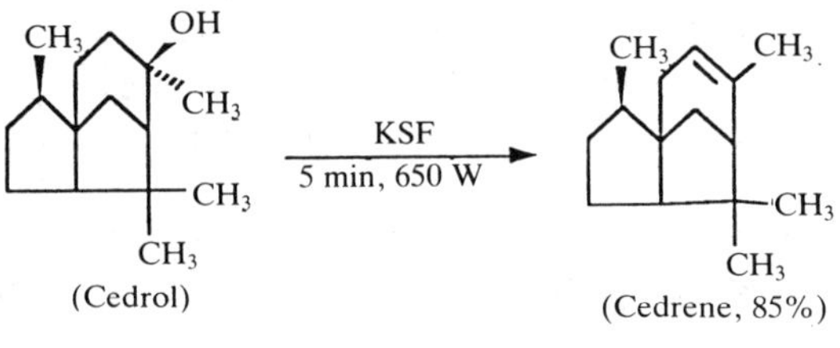

(12.32)

Surprisingly, in the case of 1-ethynylcyclohexanol absorbed on KSF clay, hydration of the acetylenic bond (by water of clay?) occurs more rapidly than dehydration of the tertiary alcohol (equation (12.33)) [60].

$$ (12.33) $$

41%   51%

12.4.3.4 Cationic rearrangements

Pinacol absorbed on acidic clay is easily rearranged to pinacolone (equation (12.34)) [55]. Table 12.3 shows the results of conversion of pinacol to pinacolone with the starting reagent adsorbed on different homoionic montmorillonites. The percentage of pinacol transformed by microwave irradiation (450 W) is compared with that achieved by classical thermal activation (100°C). The durations used were chosen so as to achieve the optimal yields, at least with the most active montmorillonites. They were 15 h for conventional heating and 15 min for microwave activation. The dramatic saving in time evidences the high enhancement of the reaction rate.

$$ (12.34) $$

1,2-Diphenylethane-1,2-diol absorbed on montmorillonite KSF gives 2,2-diphenylethanal by a similar pinacolic rearrangement [60].

Acetylenic alcohols absorbed on acidic clay (K10 or KSF) are rearranged to enones under microwave irradiation (equation (12.35)) [60]. A temperature of 170°C was observed for the solid after 5 min of microwave irradiation, but thermic heating of a similar reaction mixture of 170°C for 5 min gives only traces of rearranged product.

Table 12.3 — Molecular rearrangement of pinacol on M^{n+} — montmorillonites. Conversion (%) of pinacol to pinacolone (GC yields with internal standard)

M^{n+}	Conventional heating (100°C, 15 h)	Microwave activation (450 W, 15 min)
Na^+	5	38
Ca^{2+}	2	23
Cu^{2+}	30	94
La^{3+}	80	94
Cr^{3+}	99	98
Al^{3+}	98	99

Prolonged heating (1 h), however, allows thermal diffusion, giving similar yields to microwave-activated reactions.

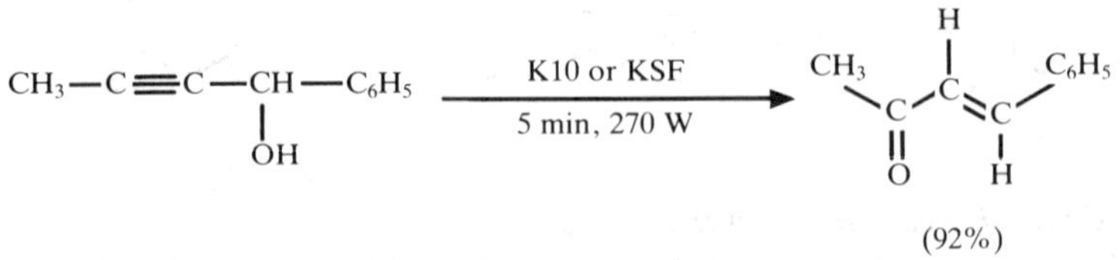

$$CH_3-C\equiv C-CH-C_6H_5 \xrightarrow[\text{5 min, 270 W}]{\text{K10 or KSF}} \quad (12.35)$$
$$\underset{OH}{|}$$

(92%)

During the cationic dehydration of lanosterol on K10 or alumina, a cationic rearrangement to isolanostatriene occurs [33].

12.4.3.5 Esterification
The acid-catalysed esterificaion of carboxylic acids with alcohols under microwave activation has been investigated. Theoretically, it is possible to shift the equilibrium by evaporation of water. On acidic solids such as KSF, K10 and zeolites, the complete displacement of the equilibrium does not occur. Use of a copper-exchanged K10 clay does not improve this result. Water chemisorbed on solid is certainly activated by microwaves and esterification is presumably limited by hydrolysis of the ester before evaporation of water. Substitution of inorganic solid by a catalytic trace of p-toluenesulphonic acid greatly improves the yield of esterification and the quasi complete shift of equilibrium is observed (Scheme 3) [67].

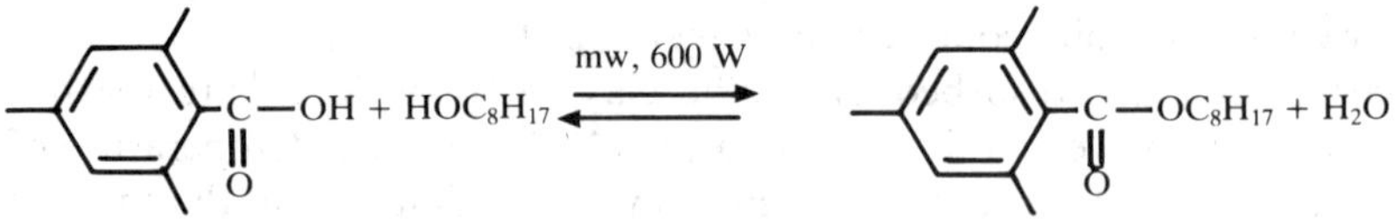

$$\text{(aryl)}C-OH + HOC_8H_{17} \underset{}{\overset{\text{mw, 600 W}}{\rightleftharpoons}} \text{(aryl)}C-OC_8H_{17} + H_2O$$

catalyst	yield of ester (%)
montmorillonite KSF	45
montmorillonite K10(H$^+$)	48
montmorillonite K(10)(Cu^{2+})	50
zeolite 5A	33
p-toluenesulphonic acid	97

Scheme 3 — Esterification using microwave activation.

12.4.3.6 Cationic opening of epoxides
Epoxides, absorbed on K10 clay and in the presence of DMSO, ring open and undergo a subsequent Kornblum oxidation to α-hydroxyketones under microwave irradiation (equation (12.36)) [33].

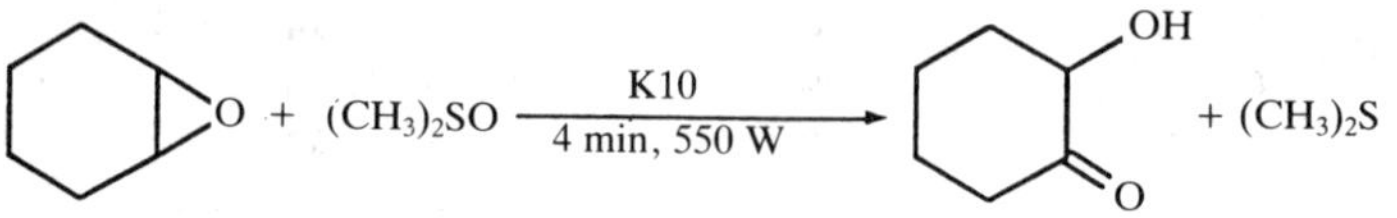

$$\text{(cyclohexene oxide)} + (CH_3)_2SO \xrightarrow[\text{4 min, 550 W}]{\text{K10}} \text{(2-hydroxycyclohexanone)} + (CH_3)_2S \quad (12.36)$$

12.4.3.7 *Cationic condensation*

Tetronic acid and aromatic aldehydes adsorbed on KSF or K10 clays, silica or alumina yield 3-(arylmethylene) derivatives under microwave activation (equation (12.37)) [73]. This type of structure is known in many biologically active compounds such as lignans, some of which are active against several tumours in mice.

$$\text{(equation 12.37)} \tag{12.37}$$

12.4.4   Thermal reactions

12.4.4.1 *Cope rearrangement*

Acidic clay catalysis and microwave activation of a [3,3]-sigmatropic reaction is effective (equation (12.38)) [60]. The reaction appears to be general for allylic alcohols [33].

$$\text{(equation 12.38)} \tag{12.38}$$

Alkylation by benzyl chloride of methyl dithiopropanoate by adsorption on KF-alumina at room temperature gives a good yield of the ketene dithioacetal (equation (12.39)) [68].

$$\text{(equation 12.39)} \tag{12.39}$$

The corresponding reaction of a mixture of methyl dithiopropanoate and allyl bromide gives an adsorbed ketene dithioacetal which is easily rearranged under microwave irradiation into methyl 2-methylpent-4-enedithioate (equation (12.40)) [69].

$$\text{(equation 12.40)} \tag{12.40}$$

Condensation-alkylation of carbon disulphide with ketones by adsorption on KF-alumina (equation (12.41)) has already been described [70].

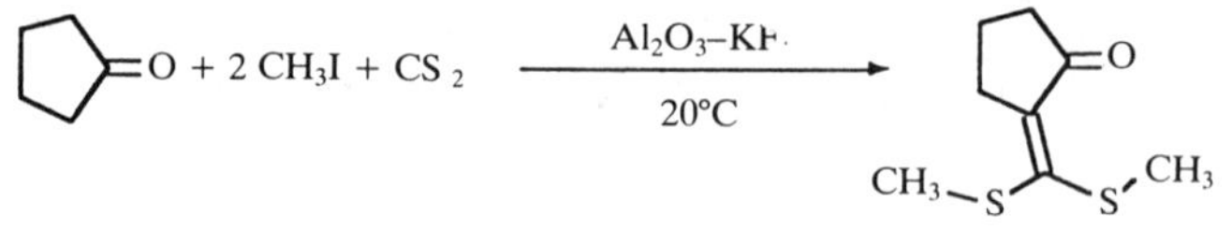

$$\text{(12.41)}$$

A mixture of cyclopentanone, carbon disulphide and allyl bromide (2 equiv.) adsorbed on KF-alumina and then irradiated with microwaves gives rise to compound **(5)** [69]. This result can be rationalized by a thio-Cope rearrangement followed by a [3-3]-sigmatropic rearrangement and proton shift (Scheme 4) [69].

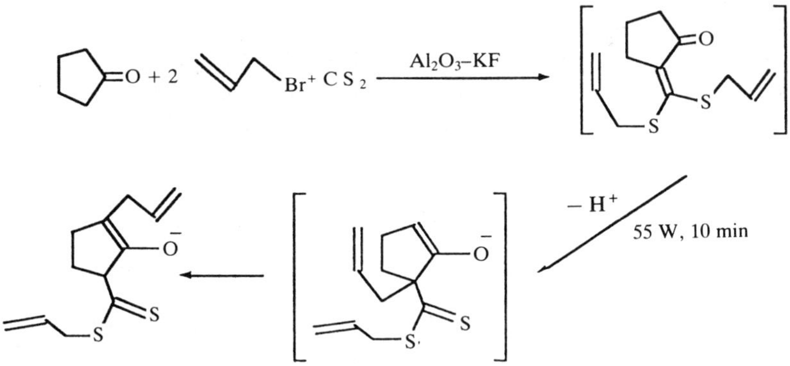

(5)

Scheme 4 — A complex microwave-induced production of a dithioester.

12.4.4.2   *Fischer indole synthesis*

Indoles can be synthesized by the Fischer reaction from phenylhydrazones, or from a mixture of a carbonyl compound and phenylhydrazine, by use of solid acids such as clays. Without solvent, yield of indole is poor with classical heating, and purification of impure product is difficult. Equally, homogeneous reaction under microwave irradiation leads to degradation of indoles. However, the use of microwave activation without solvent has largely improved the yield of the reaction. In a typical example, the ketone and phenylhydrazine are adsorbed on KSF and irradiated to give indoles in 5 min (equation (12.42)) [62].

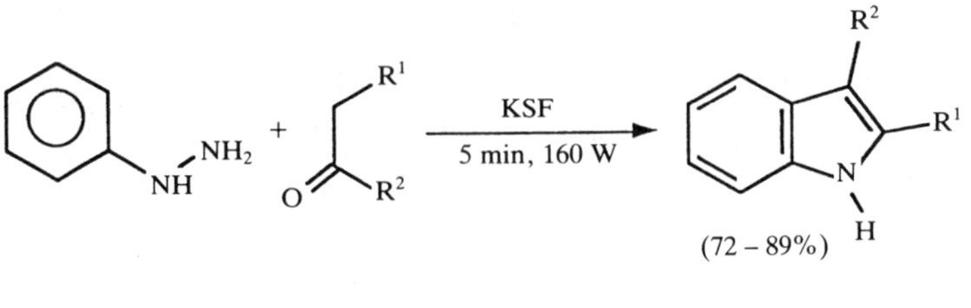

$$\text{(12.42)}$$

(72 – 89%)

12.4.4.3   Saytseff elimination

Scheme 5 shows the results of microwave irradiation of a lanosteryl xanthate supported on different solids. Lanostriene (6) was obtained on KF-alumina and isolanostriene (7) on alumina itself [69].

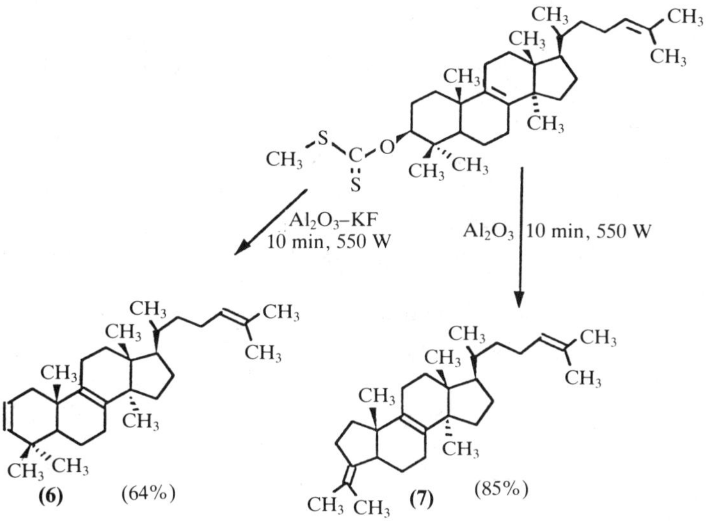

Scheme 5 — Saytseff elimination of a lanosteryl xanthate by use of a solid support and 'dry media' microwave induction.

12.4.5   Technical implications of reactions on solid supports using microwaves

Ferric oxide or other microwave-receptive minerals can be used as heat baths when organic compounds do not absorb microwaves directly; in these cases, activation is essentially thermic.

Commercial microwave ovens can be used in preparative chemistry, but for industrial applications or kinetic studies a more sophisticated, and more expensive, apparatus is necessary [71].

With a commercial oven, the uniform distribution of microwaves is difficult to obtain; ovens with a rotary tray are preferred. Corrosion of the walls of the oven is also a problem when reactions are conducted in an open flask; ovens with an 'inox' wall are preferred. Reactions can be conducted in material transparent for microwaves: Teflon or Pyrex glass can be used [72]. Teflon vessels with security valves for conducting reactions under pressure are commercially available. Hazards are in part suppressed by the use of dry reactions, though inflammable products must be eliminated *in vacuo* before microwave irradiation. Modification of commercial microwave ovens is difficult and very sensitive [37]. Microwave leaks are the principal problem when the microwave oven wall is pierced. Linders has described a modified apparatus [15]. In all cases, emission of microwaves must be controlled with an appropriate detector.

REFERENCES

[1]   J. Thuery, *Les Micro-ondes et leurs Effects sur la Matière, Technique et Documentation*, Lavoisier, 1989, Chapter 3, 2nd edn.

[2]   K. Gollnick and G. Schade, *Tetrahedron Lett*, 1973, 857.

[3]   E. Zadok, D., Amar and Y. Mazur, *J. Am. Chem. Soc.*, 1980, **102**, 6369.

[4]   E. Zadok, S. Rubinraut, F. Frolow and Y. Mazur, *J. Org. Chem.*, 1985, **50**, 2647.

[5]   R. Gedye, F. Smith, K. Westaway, H. Ali, L. Baldisera, L. Laberge and J. Roussel, *Tetrahedron Lett.*, 1986, **27** 1279.

[6]   R. J. Giguere, T. L. Bray, S. M. Duncan and G. Majetich, *Tetrahedron Lett.* 1986, **27**, 4945.

[7]   M. Bacci, M. Bini, A. Checcuci, A. Ignesti, L. Millanta, N. Rubino and R. Vanni, *J. Chem. Soc., Faraday Trans. 1*, 1981, **77**, 1503.

[8]   D. A. Copson *Microwave Heating*, The AVi Publishing Company, Westport, CT, 1975, pp. 8–18.

[9]   S. Lefeuvre, *Rev. Gen. Elec.*, Nov. 1981, 793.

[10] S. Agod, I. Gyoken, G. Bene, A. Puskas, J. Vari, A. Pinter, S. Takas and T. Hung, HU 47,613; *Chem. Abstr.*, 1989, **111**, 193766.

[11] R. Gedye, F. Smith and K. Westaway, *Can. J. Chem.*, 1988, **66**, 17.

[12] R. J. Giguere, A. M. Namen, B. O. Lopez, A. Arepally, D. E. Ramos, G. Majetich and J. Defauw, *Tetrahedron Lett.*, 1987, **28**, 6553.

[13] R. J. Giguere, in *Organic Synthesis: Theory and Application*, JAI Press Inc, London, 1989, vol. 1, pp. 153–172.

[14] F. Smith, B. Cousins, F. Bozic and W. Flora, *Anal. Chim. Acta*, 1985, **177**, 243.

[15] J. T. M. Linders, J. P. Kokje, M. Overhand, T. S. Lie and L. Maat, *Rec. Trav. Chim. Pays Bas*, 1988, **107**, 449.

[16] S. Takano, A. Kijima, T Sugihara, S. Satoh and K. Ogasawara, *Chem. Letters*, 1989, **87**.

[17] M. S. F. Lie Ken Jie and C. Yan-Kit, *Lipids*, 1988, **23**, 367.

[18] D. R. Hwang, S. M. Moerlein, L. Lang and M. J. Welch, *J. Chem. Soc., Chem. Comm.*, 1987, 1799.

[19] C. Prenant, J. Sastre, C. Crouzel and A. Syrota, *Labelled Comp. Radiopharm.*, 1987, **24**, 227.

[20] M. Ali, S. P. Bond, S. A. Mbogo, W. R. McWhinnie and P. M. Watts, *J. Organometal. Chem.*, 1989, **371**, 11.

[21] Anonymous, *Res. Disel*, 1988, **294**, 741, *Chem. Abstr.*, 1989, **110**, 9882m.

[22] M. De Le Guardia, A. Salvador, M. J. Gometz and Z. A. De Benzo, *Anal. Chim. Acta*, 1989, **224**, 123.

[23] W. C. Sun, P. M. Guy, J. H. Jahngen, E. S. Rossomando and E. G. E. Jahngen, *J. Org. Chem.*, 1988, **53**, 4414.

[24] S. T. Chen, S. H. Chiou, Y. H. Chu and K. T. Wang, *Int. J. Peptide Protein Res.*, 1987, **30**, 572.

[25] H. M. Yu, S. T. Chen, S. H. Chiou and K. T. Wang, *J. Chromtogr.*, 1988, **456**, 357.

[26] S. H. Chiou, *J. Chin. Chem. Soc.*, 1989, **36**, 435.

[27] S. H. Chiou and K. T. Wang, *J. Chromatogr.*, 1989, **491**, 424.

[28] C. H. W. Hirs and W. H. Moore, *J. Biol. Chem.*, 1954, **211**, 941.

[29] S. T. Chen, S. H. Wu and K. T. Wang, *Int. J. Peptide Protein Res.*, 1989, **33**, 73.

[30] D. R. M. Baghurst, D. M. P. Mingos and M. J. Watson, *J. Organometal. Chem.*, 1989, **368**, C43; D. R. M. Baghurst and D. M. P. Mingos, *J.C.S. Chem. Comm.*, 1988, 829; D. R. M. Baghurst, A. M. Chippindale and D. M. P. Mingos, *Nature*, 1988, **332**, 311.

[31] K. Chatakondu, M. L. H. Green, D. M. P. Mingos and S. M. Reynolds, *J. Chem. Soc. Chem. Commun.*, 1989, 1515.

[32] P. Castan, D. Villemin, F. L. Wimmer, S. Wimmer and B. Labiad, *The Chemistry of Platinum Group Metals*, International Congress, Cambridge, 1990.

[33] D. Villemin and B. Labiad, to be published.

[34] Many patents: see *Chem. Abstr.*, **101**, P9087v; **102**, P64287f; **102**, 150472u; **102**, P27396v; **106**, P12063t; **99**, P197041n; **93**, P241684b.

[35] G. Berrebi, *L'actualité chimique*, October 1983, 29.

[36] C. Roussy, A. Zoulalian, M. Charreyre and J. M. Thiebaut, *J. Phys. Chim.*, 1984, **88**, 5702.

[37] D. Villemin and F. Thibault-Staryk, *J. Chem. Educ.*, in press.

[38] S. L. McGill, J. M. Walkiewicz and G. A. Smyres, *Mater. Res. Soc. Symp. Proc.*, 1988, **124**, 247.

[39 S. L. McGill, J. M. Walkiewicz and L. A. Moyer, *Mater. Res. Soc. Symp. Proc.*, 1988, **124**, 297.

[40] J. M. Walkiewicz, G. Kazonich and S. L. McGill, *Miner. Metall. Proc.*, 1988, **5**, 39.

[41] S. M. Singer, J. Jow, J. D. Nelong and M. C. Hawley, *Polym. Mater. Sc. Eng.*, 1989, **60**, 869.

[42] F. M. Thuillier and H. Jullien, *Macromol. Chem. Macromol. Symp.*, 1989, **25**, 63.

[43] F. M. Thuillier, H., Jullien and M. F. Grenier-Loustalot, *Macromol. Chem. Macromol. Symp.*, 1987, **9**, 57.

[44] G. G. Allan, B. R. Krieger and D. W. Work, *J. Appl. Polym. Sc.*, 1980, **25**, 1839.

[45] A. J. J. Straathof, H. van Bekkum and A. P. G. Kieboom, *Rec. Trav. Chim. Pays Bas*, 1988, **107**, 647.

[46] J. Azuma, T. Tomiya and T. Katada, JP 01,225,601, *Chem. Abstr.*, 1990, **112**, 8941.

[47] J. Barry, G. Bram, G. Decodts, A. Loupy, P. Pigeon and J. Sansoulet, *Tetrahedron Lett.*, 1982, **23**, 5407.

[48] G. Bram, A. Loupy and J. Sansoulet, *Israel J. Chem.*, 1985, **26**, 291 and quoted references.

[49] G. Bram, H. Galons, S. Labidalle, A. Loupy, M. Miocque, A. Petit and J. Sansoulet, *Bull. Soc. Chim. Fr.*, 1989, 247 and quoted references.

[50] G. Bram, A. Loupy and M. Majdoub, *Synthetic Comm.*, 1990, **20**, 125.

[51] C. M. Starks, *J. Am. Chem. Soc.*, 1971, **93**, 195.

[52] G. Bram, B. Labiad, A. Loupy, M. Majdoub, Y. Ouhilal and D. Villemin, *Symposium Société Française de Chimie*, Palaiseau, 1989.

[53] A. Foucaud, G. Bram and A. Loupy, in P. Lazlo (Ed.) *Preparative Chemistry using Supported Reagents*, Academic Press, 1987, p. 317 and quoted references.

[54] G. Bram and A. Loupy, in P. Lazlo (Ed.) *Preparative Chemistry using Supported Reagents*, Academic Press, 1987, p. 387 and quoted references.

[55] E. Gutierrez, A. Loupy, G. Bram and E. Ruiz-Hitzky, *Tetrahedron Lett.*, 1989, **30**, 945.

[56] G. Bram, A. Loupy, M. Majdoub, E. Gutierrez and E. Ruiz-Hitzky, *Tetrahedron*, 1990, **46**, 5167.

[57] R. J. Giguere, T. L. Bray, S. M. Duncan and G. Majetich, *Tetrahedron Lett.*, 1986, **27**, 4945.

[58] R. J. Giguere, in T. Hudlicky (Ed.) *Organic Synthesis Theory and Applications*, JAI Press, London, 1989, vol. 1, pp. 153–172.

[59] M. Delmotte and H. Jullien, work in progress.

[60] A. Ben-Alloum, B. Labiab and D. Villemin, *J. Chem. Soc., Chem. Comm.*, 1989, 386.

[61] D. Villemin and A. Ben-Alloum, *Synth. Commun.*, 1990, in the press.

[62] D. Villemin B. Labiad and Y. Ouhilal, *Chem. Ind. (London)*, 1989, 607.

[63] J. P. Schirmann, M. Devic, A. Decarreau, G. Bram, A. Loupy and A. Petit: (a) *3rd National Congress of Société Française de Chimie*, September 1988; (b) *New J. Chem.*, in press.

[64] G. Bram, M. Devic, A. Loupy, A. Petit and J. P. Schirmann, *Symposium Société Française de Chimie (Palaiseau)*, September 1989.

[65] *Pour la Science*, 1990, **148**, 30.

[66] G, Bram, A. Loupy, M. Majdoub and A. Petit *Chem. Ind. (London)*, in press.

[67] A. Loupy, M. Majdoub, A. Petit, M. Ramdani, C. Yvanaeff, B. Labiad and D. Villemin, to be published.

[68] D. Villemin, *J. Chem. Soc., Chem. Comm.*, 1985, 870.

[69] D. Villemin and A. Ben-Alloum, to be published.

[70] D. Villemin, *Chem. Ind. (London)*, 1985, 166.

[71] J. M. Thiebaut, H. Ammor and C. Roussy, *J. Chim. Phys.*, 1988, **85**, 799.

[72] Z. Sulcek, J. Novak and J. Vyskocil, *Chemicke listy*, 1989, **83**, 388.

[73] D. Villemin and A. Ben-Alloum, *Synth. Commun.*, 1990, **20**, 3325.

[74] D. Villemin and B. Labiad, *Synth. Commun.*, 1990, **20**, 3207.

[75] D. Villemin and A. Ben-Alloum, *Synth. Commun.*, 1991, **21**, 63.

INDEX